Environmental DNA

Environmental DNA

For Biodiversity Research and Monitoring

Pierre Taberlet
Centre National de la Recherche Scientifique and Université Grenoble Alpes, France

Aurélie Bonin
Centre National de la Recherche Scientifique and Université Grenoble Alpes, France

Lucie Zinger
Ecole Normale Supérieure de Paris, France

Eric Coissac
Université Grenoble Alpes, France

OXFORD
UNIVERSITY PRESS

OXFORD
UNIVERSITY PRESS

Great Clarendon Street, Oxford, OX2 6DP,
United Kingdom

Oxford University Press is a department of the University of Oxford.
It furthers the University's objective of excellence in research, scholarship,
and education by publishing worldwide. Oxford is a registered trade mark of
Oxford University Press in the UK and in certain other countries

Published in the United States of America by Oxford University Press
198 Madison Avenue, New York, NY 10016, United States of America

British Library Cataloguing in Publication Data
Data available

Library of Congress Control Number: 2017959944

ISBN 978–0–19–876722–0 (hbk.)
ISBN 978–0–19–876728–2 (pbk.)

DOI: 10.1093/oso/9780198767220.001.0001

Printed and bound by
CPI Group (UK) Ltd, Croydon, CR0 4YY

Preface

For a few years, the scientific community has acknowledged that environmental DNA contains a huge amount of information about all types of organisms found in ecosystems. It is thus very attractive to try to at least partially read this information encoded in DNA molecules. The development of next-generation DNA sequencing boosted this research area, and opened the possibility of high-throughput data acquisition. However, implementing an eDNA study is not as simple as carrying out a DNA analysis on a single species. It also requires putting together many different skills, including obviously all the classical corpus of knowledge in ecology, but also skills in the field for sampling, skills at the bench for producing the sequence results, and skills in bioinformatics for dealing with massive amounts of sequence data.

As for any emerging scientific area, many preliminary or pilot experiments have been published thus far, as well as many reviews highlighting the potential of eDNA in ecology, paleoecology, archaeology, forensics, and biodiversity management. Unfortunately, despite this literature, extracting pertinent information stored in eDNA is not straightforward, and not exempt from biases. There are many potential traps in eDNA studies. The experiments must be designed very carefully, taking into account all the problems likely to occur at any step of the analysis, and include appropriate controls to ensure accurate final results. In light of these observations, we decided a few years ago to organize DNA metabarcoding schools, with now at least one session per year. Seven editions have occurred since 2012. At the beginning, the number of applicants was reasonable, however it is no longer the case due to the recent increasing interest in eDNA. The number of applicants is now far more than our teaching capacities, and we do not feel comfortable rejecting applications from many motivated young scientists. This book is an attempt to cope with the high demand from ecologists for developing research involving eDNA.

Our aim was to write a book for ecologists, who do not necessarily have a strong background in molecular genetics. As a result, some parts might look naive to experienced molecular ecologists. Moreover, it was our wish not to give very precise protocols, but to provide the background information that will ultimately enable the design of sound experiments. This book is technically oriented, as the difficulties in eDNA studies mainly derive from the technical aspects, with which ecologists are usually not familiar. When working with eDNA in ecology, the questions and concepts remain the same as in any ecological study, and we did not emphasize these aspects, as many excellent ecological textbooks are already available. Incidentally, it is interesting to note that eDNA now allows a few questions that could not be previously addressed to be tackled. Another objective of this book was to deliver a relatively comprehensive bibliographic orientation, allowing the readers to refer to the primary literature. However, due to the recent burst of eDNA studies, it becomes very difficult to be complete, and clearly, we did not cite all pertinent papers.

Finally, we must also recognize that this book reflects our view of eDNA analysis, and does not necessarily represent all the different opinions expressed in the scientific community. Globally, our objective was to favor simple and robust solutions that might not be optimal in terms of accuracy, but which can be implemented at large scales for analyzing hundreds or thousands of environmental samples.

Contents

Acknowledgments

It would not have been possible to write this book without the fruitful discussions about environmental DNA and metabarcoding we have had over the years with colleagues and participants in the different metabarcoding schools. Among them, we are particularly thankful to Miklós Bálint, Guillaume Besnard, Julien Bessière, Kristine Bohmann, Frédéric Boyer, Christian Brochmann, Anthony Chariton, Jérôme Chave, Corinne Cruaud, Bruce Deagle, Marta De Barba, Tony Dejean, Mary Edwards, Francesco Ficetola, Roberto Geremia, Simon Jarman, Stefaniya Kamenova, Håvard Kauserud, Carla Martin Lopes, Christelle Melo de Lima, Céline Mercier, Ludovic Orlando, Johan Pansu, Jan Pawlowski, François Pompanon, Dorota Porazinska, Gilles Rayé, Tiayyba Riaz, Maurizio Rossetto, Heidy Schimann, Wasim Shehzad, Min Tang, Philippe Thomsen, Wilfried Thuiller, Philippe Usseglio-Polatera, Alice Valentini, Alain Viari, Eske Willerslev, Patrick Wincker, Meng Yao, Nigel Yoccoz, Douglas Yu, and Xin Zhou. We gratefully acknowledge the help of Amaia Iribar, Philippe Gaucher, Ludovic Gielly, Christian Miquel, Delphine Rioux, and Marie-Odile Taberlet in the field or at the bench. We are also very grateful to Bryony Taberlet for her help with language proofreading. Finally, we must acknowledge that we greatly appreciated the help and encouragement of Ian Sherman, Bethany Kershaw, and Lucy Nash from Oxford University Press throughout the process of completing this book.

July 2017
Aurélie Bonin
Eric Coissac
Pierre Taberlet
Lucie Zinger

CHAPTER 1

Introduction to environmental DNA (eDNA)

Environmental DNA (eDNA) is becoming a key component of the ecologists' and environmental managers' toolbox. Such a strong enthusiasm was recently sparked by the development of next-generation sequencers. In 10 years, our sequencing capabilities have indeed been multiplied by at least four orders of magnitude.

Environmental microbiology aside, the emergence of eDNA studies has been surprisingly slow when considering the small number of articles published during the few years following the commercialization of the first next-generation sequencers in 2005. This is probably due to the relatively high costs associated with this new type of sequencing and to the characteristics of eDNA. As a complex mixture of DNA from different organisms, possibly degraded and in low concentrations, eDNA can indeed be more difficult to analyze than DNA originating from the fresh tissues of a single organism. Another impediment is that eDNA analysis requires the combination of many different skills, from classical ecology to bioinformatics, including molecular biology techniques.

The aim of this introductory chapter is to give definitions of what eDNA is, to present its history, to highlight the different steps of an eDNA study, and to give an overview of the different types of eDNA methods implemented in research or biodiversity management. It also points the reader to the chapter(s) where the important aspects of eDNA analyses are addressed in detail.

1.1 Definitions

Environmental DNA is a complex mixture of genomic DNA from many different organisms found in an environmental sample (Taberlet et al. 2012a). Soil, sediment, water, or even feces are considered as environmental samples, which can also include the material resulting from filtering air or water, from sifting sediments, or from bulk samples (e.g., the whole insect content of a Malaise trap). Alternatively, environmental DNA can be defined from another perspective (i.e., the objective of the study). In this case, eDNA corresponds to DNA extracted from an environmental sample with the aim of obtaining the most comprehensive DNA-based taxonomic or functional information as possible for the ecosystem under consideration. Total eDNA contains both intracellular and extracellular DNA (Levy-Booth et al. 2007; Pietramellara et al. 2009). Intracellular DNA originates from living cells or living multicellular organisms that are present in the environmental sample. Extracellular DNA results from cell death and subsequent destruction of cell structures, and can be degraded through physical, chemical, or biological processes. For example, DNA molecules can be cut into smaller fragments by nucleases. After its release, extracellular eDNA may be adsorbed by inorganic or organic surface-reactive particles such as clay, sand, silt, and humic substances.

If we are to identify the taxa present in an eDNA sample, two approaches can be considered, mainly based on PCR (polymerase chain reaction; Mullis & Faloona 1987; Saiki et al. 1985, 1988). When the aim is to determine the presence or absence of a single species, the best solution is often to favor a species-specific approach, generally based on quantitative PCR (e.g., Logan et al. 2009). Alternatively, a more general approach based on targeted PCR (Saiki

Environmental DNA for Biodiversity Research and Monitoring. Pierre Taberlet, Aurélie Bonin, Lucie Zinger, & Eric Coissac, Oxford University Press (2018). © Pierre Taberlet, Aurélie Bonin, Lucie Zinger, & Eric Coissac 2018. DOI: 10.1093/oso/9780198767220.001.0001

et al. 1988; White *et al.* 1989) or on shotgun sequencing (Deininger 1983) has the potential of revealing the presence of all species within a clade. This last approach is called "DNA metabarcoding."

The expression "DNA metabarcoding" was first used in 2011 (Pompanon *et al.* 2011; Riaz *et al.* 2011). It corresponds to the simultaneous DNA-based identification of many taxa found in the same environmental sample. Generally, DNA metabarcoding involves examining metabarcode sequences amplified from eDNA. A metabarcode consists of a short and taxonomically informative DNA region flanked by two conserved regions serving as primer anchors for the PCR (see Chapter 2). DNA metabarcoding can also be performed by shotgun sequencing of eDNA, without any metabarcode amplification (Taberlet *et al.* 2012b). Shotgun sequencing involves the sequencing of random DNA fragments from a DNA extract, generally using next-generation sequencers (Glenn 2011). However, taxa identification based on shotgun sequencing is difficult to achieve as it requires high sequencing depths and extensive reference databases for taxonomic assignment.

In the early days of DNA metabarcoding, many different terminologies were coined to designate PCR-based identification of multiple taxa at the same time: ecometagenetics (Porazinska *et al.* 2010), ecogenomics (Chariton *et al.* 2010), environmental barcoding (Hajibabaei *et al.* 2011), metataxogenomics (Terrat *et al.* 2012), and metasystematics (Hajibabaei 2012). Microbiologists, who first routinely analyzed eDNA to have access to uncultivable microorganisms, often improperly used metagenomics to mean DNA metabarcoding. Indeed, when working with eDNA, microbiologists have three main objectives: (i) identifying the microbial taxa present in environmental samples; (ii) identifying their potential biochemical functions via the analysis of coding genes; and (iii) assembling whole genomes of uncultivable microorganisms. The expression "DNA metabarcoding" is more appropriate when referring to taxa identification, while the two last objectives (i.e., the functional aspects and the assembly of genomes) are more clearly called to mind by the word "metagenomics." The confusion between these two terminologies has arisen from a seminal article with "metagenomics" in the title, which combined shotgun sequencing and 16S rDNA-based taxonomic identification (Tringe *et al.* 2005).

Aside from these concepts, metatranscriptomics is the analysis of a complete set of ribonucleic acid (RNA) molecules extracted from environmental samples to examine gene expression and regulation at the sampling time. This methodology is hence usually employed to assess both the taxonomic and functional components of the examined sample. Metatranscriptomics remains challenging, primarily because the half-life of messenger RNA (mRNA) is short (Selinger *et al.* 2003) and because total RNA is mainly composed of ribosomal RNA (rRNA) that does not provide direct information about functional aspects.

1.2 A brief history of eDNA analysis

Figure 1.1 is a chronology illustrating the brief history of eDNA analysis and its milestones. The history of eDNA started in 1987, with the report of an extraction protocol for eDNA found in sediments (Ogram *et al.* 1987). Only three years later, and surprisingly about 10 years earlier than the subsequent papers, the first DNA metabarcoding study was published (Giovannoni *et al.* 1990). This pioneering work analyzed the diversity of the 16S rRNA gene in bacterioplankton sampled in the Sargasso Sea, using PCR followed by cloning (but see also Ward *et al.* 1990 using a similar idea, but starting from rRNA). Metagenomics was initiated in 1998, by cloning and sequencing fragments of soil eDNA for identifying new pathways for the synthesis of bioactive molecules in uncultivated microorganisms (Handelsman *et al.* 1998). At the beginning of the 2000s, metagenomics and DNA metabarcoding based on cloning became commonplace in microbiology. In 2003, the first DNA metabarcoding article focusing on macroorganisms showed that it was possible to retrieve megafaunal (mammoth, bison, horse) and ancient plant DNA from permafrost, and DNA of extinct ratite moa from cave sediments (Willerslev *et al.* 2003).

The first metagenomics and metabarcoding studies relied on cloning to isolate single DNA fragments from a complex mixture prior to Sanger sequencing. DNA fragments were inserted into cloning vectors (e.g., plasmids or bacteriophages), which allowed for their isolation and multiplication

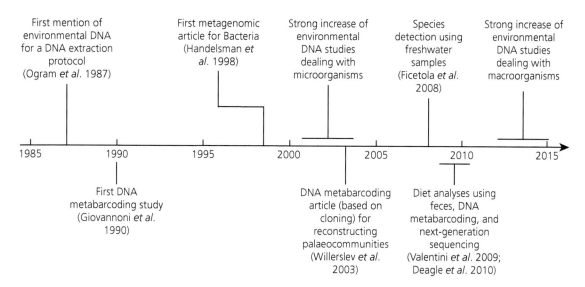

Fig. 1.1 Overview of the emergence of eDNA studies.

in the cells of a suitable host, such as *Escherichia coli*. Fortunately, this expensive and time-consuming cloning step was made unnecessary by the outbreak of next-generation sequencing after 2005 (Shendure & Ji 2008), which further stimulated metabarcoding and metagenomics. By the end of the 2000s, application of DNA metabarcoding was extended to macroorganisms, first for diet analyses using feces as a source of DNA (Deagle *et al.* 2010; Pompanon *et al.* 2012; Valentini *et al.* 2009), and then for water and soil eDNA studies (Ficetola *et al.* 2008; Sønstebø *et al.* 2010). More recently, eDNA analysis from freshwater macroorganisms led to many publications, dealing mainly with single-species detection (Dejean *et al.* 2012; Goldberg *et al.* 2011; Jerde *et al.* 2011), but also aiming at identifying multiple taxa through metabarcoding (Thomsen *et al.* 2012b; Valentini *et al.* 2016). In parallel, marine macrofauna and meiofauna were also analyzed using either water samples (Thomsen *et al.* 2012a), multilayered settlement surfaces (autonomous reef monitoring structures; Leray & Knowlton 2015), or sifted sediments (Chariton *et al.* 2010, 2015). If studies of soil microorganisms using eDNA can now be considered as routine, this is different for soil meio and macrofauna, for which only a handful of studies have been completed (Baldwin *et al.* 2013; Bienert *et al.* 2012; Pansu *et al.* 2015a; Wu *et al.* 2011). How-

ever, there is no doubt that soil macroorganisms will also be extensively investigated via eDNA in the near future.

After Willerslev *et al.*'s seminal study in 2003, analysis of ancient eDNA has also begun to be a very attractive approach for gaining insight into past communities, using either DNA metabarcoding (Epp *et al.* 2015; Giguet-Covex *et al.* 2014; Pansu *et al.* 2015b) or metagenomics (Smith *et al.* 2015). However, despite this recent enthusiasm of the scientific community for eDNA analysis, some of the technical aspects remain challenging, both at the bench and bioinformatics levels.

1.3 Constraints when working with eDNA

One of the main characteristics of eDNA is the heterogeneity of the extracts obtained from environmental samples. Thus, working with eDNA is usually not as straightforward as when working with DNA extracted from a tissue sample of a known plant or animal species. A wide range of situations can be encountered: from concentrated high-quality DNA without enzyme inhibitors (comparable to DNA extracted from tissues), to highly diluted and degraded DNA (similar to the extracts obtained in ancient DNA studies). Furthermore, it has been shown that different DNA extraction protocols are not as equally efficient at removing PCR inhibitors depending on the sample type (e.g., varying amounts of humic substances in soil samples) and that this can lead to different results (Frostegård *et al.* 1999; Martin-Laurent *et al.* 2001). As a consequence, there is no simple and standard protocol suitable for the analysis of all types of eDNA. In this context, the objective of the following chapters is to help adjust the experimental protocols according to the question and experimental constraints, in order to carry out a sound eDNA study.

1.4 Workflow in eDNA studies and main methods used

Figure 1.2 describes the main steps of an eDNA study and the alternative strategies that can be adopted for eDNA analysis. The first one consists in targeting a single species using standard or quantitative PCR and is already popular for many taxa (e.g., Ficetola *et al.* 2008; Goldberg *et al.* 2011; Jerde *et al.* 2011; Thomsen *et al.* 2012b). The second strategy relies on PCR-based assays aiming at detecting all taxa from a given taxonomic group such as Bacteria (e.g., Tringe *et al.* 2005; Sogin *et al.* 2006), Fungi (e.g., Blaalid *et al.* 2012; Tedersoo *et al.* 2014), plants (e.g., Kartzinel *et al.* 2015; Sønstebø *et al.* 2010; Yoccoz *et al.* 2012), eukaryotes (e.g., Baldwin *et al.* 2013; Chariton *et al.* 2010, 2015; De Vargas *et al.* 2015), earthworms (Bienert *et al.* 2012; Pansu *et al.* 2015a), fish (e.g., Kelly *et al.* 2014a; Thomsen *et al.* 2012a; Valentini *et al.* 2016), and so on. Metagenomics, the third strategy which is based on shotgun sequencing of eDNA without any targeted PCR, is extensively employed for studying the functional characteristics of genomes, mainly of microorganisms (see review in Simon & Daniel 2011).

At the beginning of an eDNA study, when uncertainties remain about the experimental protocols to be implemented, it is highly advisable to carry out pilot experiments. Such exploratory trials allow for the adjustment of parameters in order to design a reliable full experiment. At this stage, it is important to foresee all possible errors and artifacts that can happen during the course of the full experiment, including problems during sampling, eDNA

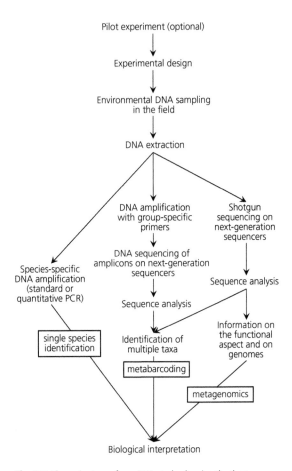

Fig. 1.2 The main steps of an eDNA study, showing the three possible approaches: single-species identification, metabarcoding, and metagenomics. The same molecular method, shotgun sequencing, can lead to metabarcoding or to metagenomics (if the objective is the taxonomic or functional aspects, respectively).

extraction, PCR, sequencing, and data analysis. With these potential problems in mind, the full experiment must be designed in a way that will prove that the obtained results do not originate from errors or artifacts. Ideally, it should include negative extraction controls, negative PCR controls, and positive PCR controls. For example, an inappropriate bioinformatic treatment of eDNA sequences might induce erroneous results. Therefore, it is extremely important to be able to detect these during the data analysis step using a few positive controls, so as to modify the bioinformatic pipeline accordingly if this is necessary.

Planning the full eDNA experiment is obviously crucial and any mistake at this stage can strongly compromise the entire study (see Appendix 3 for a checklist concerning the experimental design). Key parameters to be decided are:

(i)   the different controls to be included at various steps of the process (extraction negative controls, PCR negative controls, PCR positive controls of known composition, replicated samples, blind samples, and so on);
(ii) the sampling strategy (how many samples, how many sample replicates, how to spatially distribute the samples, at which time of the year, and so on);
(iii) the sample preservation method and the DNA extraction protocol (should the samples be preserved before DNA extraction, or do they have to be extracted immediately in the field to avoid degradation and/or microorganism development; which extraction protocol to choose according to the scientific question, logistic constraints, financial aspects, and so on);
(iv) the protocol for DNA amplification in DNA metabarcoding (which metabarcode(s) to analyze, which multiplexing strategy to adopt according to the number of samples and the sequencing platform, and so on);
(v)   the sequencing strategy (should the sequencing be done in-house, or should it be outsourced; with which sequencing platform, and so on);
(vi) the strategy for data analysis.

The following chapters provide key information for each step of an eDNA study, including design of new metabarcodes (Chapter 2), choice of reference databases for taxonomic assignment (Chapter 3), sampling design (Chapter 4), DNA extraction (Chapter 5), DNA amplification (Chapter 6), DNA sequencing (Chapter 7), bioinformatic analysis of metabarcoding data (Chapter 8), methods for single-species identification (Chapter 9), and all the different aspects of metagenomics (Chapter 10).

Outsourcing one or several steps of the eDNA workflow is sometimes a realistic option to consider. Sequencing can now be easily and relatively cheaply outsourced to one of the many companies or common facilities proposing next-generation sequencing services. A few structures already offer full eDNA analyses, either including or excluding the sampling step, but obviously without the biological interpretation. In any case, any outsourced eDNA study should include blind samples, for which the provenance or any other characteristics are not known by the person in charge of the experiments. These are very helpful for assessing the reliability and reproducibility of the process.

1.5 Environmental DNA as a monitoring tool

Beyond research in ecology, eDNA has also proven to be a useful material for biodiversity monitoring purposes. For example, single-species detection is often applied to track rare species, or invasive species in the early stages of invasions, mainly in aquatic ecosystems (e.g., Deagle *et al.* 2003; Dejean *et al.* 2012; Jerde *et al.* 2011; Mächler *et al.* 2014; Nathan *et al.* 2015; Tréguier *et al.* 2014).

DNA metabarcoding also has a huge potential for the biomonitoring of different types of ecosystems. Several studies have already tried to adjust and standardize experimental protocols. In marine environments, for instance, it is now possible to assess the impact of pollution on eukaryotes in sediments (Chariton *et al.* 2015), or of fish farming on benthic Foraminifera communities (Pawlowski *et al.* 2014). In freshwater ecosystems, standardized metabarcoding protocols are already available for surveying fish and amphibians (Valentini *et al.* 2016), as well as diatoms (Apothéloz-Perret-Gentil *et al.* 2017; Visco *et al.* 2015; Zimmermann *et al.* 2015). Monitoring macroinvertebrates for assessing water quality is also a field of intense research (Hajibabaei *et al.* 2011, 2012; Thomsen *et al.* 2012b) that should soon lead to normalized approaches. In terrestrial ecosystems, metabarcoding can identify thousands of insects collected in Malaise traps in a single experiment, with the aim of taking decisions in restoration ecology and systematic conservation planning (Ji *et al.* 2013). More surprising is the application of metabarcoding to the limitation of birdstrike hazards at airports via the analysis of bird gut content (Coghlan *et al.* 2013). However, the development of environmental management via DNA metabarcoding is highly dependent upon the availability of extensive reference databases for taxonomic assignment for the examined metabarcodes (Chapter 3), and upon the robustness of the experimental protocols (Chapter 19).

Decisions guiding environmental management can be based not only on taxonomic information, but also on functional capability. Metagenomics thus has a role to play in assessing the effects of anthropogenic pressures on the potential functions of microorganisms in ecosystems. Integration of metagenomics into environmental monitoring campaigns should therefore allow a better management of human impact on ecosystems (Kisand *et al.* 2012).

CHAPTER 2

DNA metabarcode choice and design

DNA barcodes are short, standardized genetic markers used for the taxonomic identification of isolated specimens (e.g., CBoL Plant Working Group 2009; Hebert *et al.* 2003a). Their characteristics are optimized for this purpose, and do not necessarily meet the requirements for a DNA metabarcoding experiment where many species must be identified simultaneously, often by analyzing low quality DNA. For this application, markers with other properties must be selected. To avoid confusion, hereafter we distinguish between DNA barcodes for classical taxonomic identification, and DNA metabarcodes for eDNA-based biodiversity surveys.

2.1 Which DNA metabarcode?

In any DNA metabarcoding study, the choice of the metabarcode is crucial and can greatly impact the end results. For example, when testing a metabarcoding protocol to assess freshwater invertebrate biodiversity, Elbrecht and Leese (2015) found that their metabarcode, derived from the standard cytochrome c oxidase subunit I (COI) barcode for animals, was unsuccessful in retrieving species abundance and biomass. This was due to differential polymerase chain reaction (PCR) efficiency among species. It is thus necessary to overcome the idea that a marker published for a given clade and application is always the best marker in a different context. Most importantly, such reflection should come early in the experimental process, as the decisions made at this stage can influence other aspects of the study (sampling, lab experiments, ecological interpretation, and so on).

Several elements should be carefully considered to make an informed choice. First, the taxonomic group of interest should be clearly defined, as well

as the level of taxonomic resolution required to answer the question at hand. For example, in the diet analysis of the omnivorous brown bear, De Barba *et al.* (2014) characterized the overall plant component of the diet using a universal plant metabarcode. As it is not highly resolutive within a few plant families (Asteraceae, Cyperaceae, Poaceae, Rosaceae), they also resorted to family-specific metabarcodes to increase resolution within these families. The aim of the experiment is another point to consider. Preferential amplifications are more easily tolerated when it comes to detect a few indicator species, however, they are detrimental for exhaustive or quantitative biodiversity surveys. Another criterion to consider is the potential presence of DNA from non-target organisms in the collected samples, and whether its amplification is prejudicial. For example, several primers designed for amplifying fungal markers are well known to co-amplify plants. If one is conducting a study on anaerobic Fungi found in the rumen of ruminants, it might be better to select a specific primer pair that does not amplify plant DNA coming from the ruminant diet. Alternatively, when the target and non-target groups are too closely related, like in the analysis of a vertebrate predator's diet, it is possible to use blocking oligonucleotides that hinder amplification of the predator's DNA (see Chapter 6 for the design and implementation of blocking oligonucleotides, and Chapter 17 for examples). The expected level of DNA degradation will also constrain the size of the selected metabarcode: shorter metabarcodes (<100–150 bp) should be favored in case of highly degraded eDNA, such as eDNA originating from feces, while microbiome studies can usually accommodate longer metabarcodes (<250–300 bp), especially when targeting intracellular DNA. Finally, the better solution is to sometimes

Environmental DNA for Biodiversity Research and Monitoring. Pierre Taberlet, Aurélie Bonin, Lucie Zinger, & Eric Coissac, Oxford University Press (2018). © Pierre Taberlet, Aurélie Bonin, Lucie Zinger, & Eric Coissac 2018.
DOI: 10.1093/oso/9780198767220.001.0001

resort to a custom-made metabarcode especially designed for the examined taxon. However, this solution can only be contemplated if one has access to a sufficient number of reference sequences spanning the entire studied group, with no taxonomic bias, and where it is possible to identify a short variable fragment flanked by two conserved regions (i.e., a fragment displaying the characteristics of the ideal metabarcode; see Section 2.2). When it comes to metabarcode selection, all the aforementioned points will enter the equation to varying extents, and the final choice will ultimately be a matter of compromises, like many other experimental decisions.

2.2 Properties of the ideal DNA metabarcode

In an ideal world, the perfect DNA metabarcode is a DNA fragment as short as possible, displaying a highly variable sequence, and flanked by two conserved regions (Fig. 2.1). The central variable region is discriminative for all species of the target group, that is, its sequence is uniquely associated to a given species and not shared with others (Fig. 2.2). Note that this definition includes intraspecific polymorphism. On the contrary, the two flanking conserved regions are identical across the target group, but different in non-target taxa. These conserved regions correspond to the sites where the DNA metabarcoding primers will anneal perfectly, ensuring un-

biased amplification of the targeted species, while preventing that of undesirable taxa. For this reason, it is usually a bad idea to design primers in protein-coding regions, where variation typically occurs every three nucleotides due to the redundancy of the genetic code. Indeed, DNA barcode markers, like COI for animals, have been shown to perform poorly and favor amplification of some target taxa over others (Clarke *et al.* 2014; Deagle *et al.* 2014). Furthermore, our experience shows that degenerating too many bases in primers is not a very satisfactory way to deal with variation in primer-annealing regions. Finally, an ideal metabarcode should be located in a genomic region that is exhaustively and accurately documented across all targeted taxa in reference databases (i.e., no missing species, no sequence or assignment errors) in order to obtain unambiguous taxonomic identifications.

The size of the ideal metabarcode is highly dependent on the number of target taxa to distinguish. In theory, a metabarcode of n nucleotides allows discriminating 4^n species. This means that 10 nucleotides should be sufficient to discriminate 1,048,576 different species (for the sake of comparison, this is close to the estimated number of Diptera on Earth). In practice, even if there is no metabarcode approaching this figure, a 16S rRNA fragment as short as 30 nucleotides is enough to differentiate all earthworm species across the world, and it even shows some intraspecific variability (Bienert *et al.* 2012).

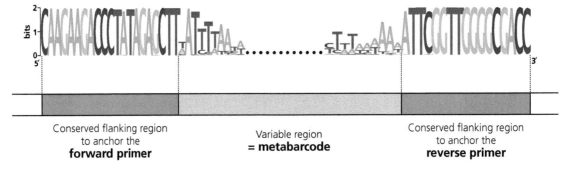

Conserved flanking region
to anchor the
forward primer

Variable region
= metabarcode

Conserved flanking region
to anchor the
reverse primer

Fig. 2.1 Example of a variable metabarcode with its conserved flanking sequences. This example is based on the Lumb01 primer pair (see Bienert *et al.* 2012 and Appendix 1) targeting the suborder Lumbricina (earthworms). All Lumbricina sequences were extracted from the release 126 of EMBL using ecoPCR (1,973 sequences). Each logo consists of stacks of symbols (A, C, G, T), with one stack for each position in the nucleotide sequence. The overall height of the stack corresponds to the nucleotide conservation at that position across the Lumbricina lineage and is expressed in bits (a value of 2 indicates a perfect conservation, while 0 means the same probability for the four nucleotides). The height of each symbol within the stack indicates the relative frequency of each nucleotide at that position.

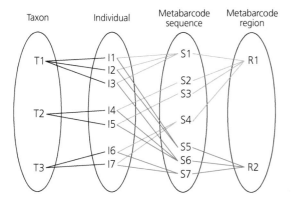

Fig. 2.2 Relationships between taxa, individuals, metabarcode sequences and metabarcode regions (adapted from Figure 1 in Ficetola *et al.* 2010). In a perfect metabarcode system, each metabarcode sequence is uniquely associated with one taxon, and not shared with others. In this figure, this is the case of the barcode system targeting region 1 (in green), and taxon 2 even shows some intragroup polymorphism. Conversely, the system targeting region 2 (in red) is not ideal, as metabarcode sequence S6 cannot discriminate between taxa 1 and 2.

In the real world, one always should compose with either lack of strict conservation of the flanking regions, or lack of taxonomic resolution for the chosen metabarcode, or more generally both. For a given metabarcoding system, two indexes are useful to evaluate the extent of departure from the perfect metabarcode: the coverage index (Bc), corresponding to the ratio between the number of amplified target taxa and the total number of target taxa, and the specificity index (Bs), defined as the ratio between the number of taxonomically discriminated taxa and the number of amplified taxa (Fig. 2.3; Ficetola *et al.* 2010). It must be noted that these two ratios are highly dependent upon the reference database they are estimated from. If this reference database contains sequencing errors, misassigned sequences, over or underrepresented taxa, this will influence the Bc and Bs values. As a result, some metabarcode systems are bound to evolve, as sequence databases are completed and/or curated constantly. For example, our initial primer pair for Fungi, originally published in Epp *et al.* (2012), was recently modified to take into account new Glomeromycota sequences published in the European Molecular Biology Laboratory (EMBL) database (see Section 2.4, and metabarcodes Fung01 and Fung02 in Appendix 1).

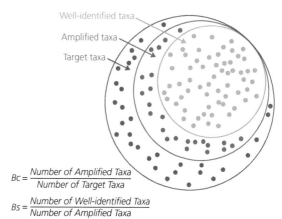

$$Bc = \frac{Number\ of\ Amplified\ Taxa}{Number\ of\ Target\ Taxa}$$

$$Bs = \frac{Number\ of\ Well\text{-}identified\ Taxa}{Number\ of\ Amplified\ Taxa}$$

Fig. 2.3 Calculation of the Bc (coverage) and Bs (specificity) indexes.

2.3 *In silico* primer design and testing

Sometimes, metabarcodes that are already available and tested are not entirely satisfying for the purpose of an analysis. *In silico* primer design is then an alternative increasingly worth considering, especially nowadays where an exponential number of sequences is added every day to public sequence databases like the EMBL database, for a wider range of organisms. Indeed, one prerequisite for designing a robust and efficient metabarcoding system *in silico* is to have access to a set of reference sequences truly representative of the taxonomic group of interest, with reliable taxonomic annotation and low levels of sequencing errors. Additionally, it might be interesting to widen the set of examined sequences to non-target sequences, if the aim is to design primer pairs that preferentially amplify the target clade. It is unfortunate, however, that many of the sequences submitted to public databases do not include the conserved priming sites used for their amplification, which limits their value for primer design.

Obviously, the selected sequences need to correspond to the same locus or genomic regions. One can then decide to build on *a priori* knowledge of the studied organism (i.e., to focus on loci known to display interspecific variation while harboring conserved regions that can serve as primer anchors). It is the case, for instance, of the internal transcribed spacer (ITS) region in Fungi (Schoch *et al.* 2012), or the 16S rRNA gene in Bacteria and Archaea (Fox *et al.* 1977; Pace 1997). Alternatively, one can choose an approach without *a priori*, i.e., without selecting a particular genomic region first. This entails working on whole mitochondrial, chloroplastic, or prokaryote genomes, and possibly even complete eukaryote genomes in the near future.

Once the working set of reference sequences has been obtained, two approaches can be adopted for the primer design itself. The first one relies on sequence alignments, like the method described by Walters *et al.* (2011), and is thus particularly appropriate for small sets of references with few indels. Primer anchoring regions are identified by screening the alignment for conserved regions. The second approach does not require the sequences to be aligned and it is based instead on a pattern search, like in the program ecoPrimers (Riaz *et al.* 2011). As there is no alignment step in this approach, it can work with any set of sequences, even whole or ganelle or prokaryote genomes. In the following paragraphs, we illustrate this approach by showing how to design new metabarcoding primers for Bacteria on whole bacterial genomes using ecoPrimers. We then test these new primers by running an *in silico* PCR with the ecoPCR program. This example is intended as a basic tutorial, so we also provide all the input and output files (available at http://www.oup.co.uk/companion/taberlet), and the associated Unix commands.

2.3.1 Prerequisites

The ecoPrimers (http://metabarcoding.org/ecoprimers) and ecoPCR (http://metabarcoding.org/ecopcr) programs are required to run this tutorial, more exactly to design and test primers *in silico*, respectively. In addition, the OBITools program suite (http://metabarcoding.org/obitools) is helpful for sequence handling, formatting, and filtering.

2.3.2 Reference sequences: description, filtering, and formatting for ecoPrimers

ecoPrimers and ecoPCR require a sequence database and a taxonomy database. The sequence database is constituted by the set of complete bacterial genomes downloaded from the European Bioinformatics Institute (EBI) Ensembl genome ftp site (ftp://ftp.ensemblgenomes.org/pub/bacteria/) in June 2012. The taxonomy database was downloaded from the National Center for Biotechnology Information (NCBI) ftp site (ftp://ftp.ncbi.nlm.nih.gov/pub/taxonomy/taxdump.tar.gz) at the same date.

In a taxonomy database, each taxon is designated by a unique integer value, commonly known as a taxid. A taxid identifies unambiguously a taxon, but is neither universal nor permanent. It is valid for a given taxonomy database (e.g., NCBI or SILVA) and a given version of this database. From a release of the database to another, some taxids can be added, changed, or removed. Working with different databases or different versions of the database can lead to inconsistencies. It is therefore important to keep track of the taxonomy database used for an analysis and of its version. When formatted for the OBITools, the taxonomy database consists of at least three files describing the tree structure of the taxonomy: Taxonomy.ndx, Taxonomy.rdx, and Taxonomy.tdx.

To avoid overrepresentation of well-studied bacterial genera (e.g., *Streptococcus* or *Clostridium*), the downloaded sequences were further subsampled to produce a database containing only one randomly selected sequence per bacterial genus. Ultimately, our set of reference sequences contained 517 whole-genome sequences in a FASTA format (file name: ReferencesSequences.fasta). Each sequence is annotated with a compulsory taxid.

First, the ReferenceSequences.fasta file was converted to a format combining sequence and taxonomic information, the ecoPCR database format, which is required to run ecoPrimers (Command 2.1).

Command 2.1

```
obiconvert -d Taxonomy --fasta --ecopcrdboutput=FILE1_ReferenceDatabase \
ReferenceSequences.fasta
```

This command produces five files whose names start with the "FILE1_ReferenceDatabase" prefix.

2.3.3 *In silico* primer design with `ecoPrimers`

In the second step, the `ecoPrimers` program was run on the `ecoPCR` database just created (Command 2.1). The name of the `ecoPCR` database was specified with the *-d* option (Command 2.2).

Command 2.2

```
ecoPrimers -d FILE1_ReferenceDatabase -e 3 -l 30 -L 280 -3 3 > \
FILE2_BacteriaPrimers.ecoprimer
```

The only two other mandatory parameters are the minimum and maximum metabarcode lengths (excluding primers) specified by the *-l* and *-L* options. All the other parameters have default values (see http://metabarcoding.org/obitools/doc/scripts/ecoPrimers.html for details), but it might be important to adjust them to fine-tune the primer design process. In particular, the default values can be too stringent to search for primers in a large taxonomic group like Bacteria. Here, we allowed each primer to exhibit at most three mismatches with the priming site on the amplified sequence (*-e* option). To ensure good amplification, we also disallowed mismatches within the three last nucleotides at the primer 3'-end of each primer (*-3* option) as this would strongly impede PCR efficiency (Wu *et al.* 2009). Another important parameter is the *-r* option that specifies the taxid of the clade for which the primer pair is optimized. The taxid of the clade whose amplification should be avoided can be defined via the *-i* option. This last option is interesting when there is a risk of amplifying a substantial proportion of non-target organisms.

2.3.3.1 The `ecoPrimers` output

The first part of the `ecoPrimers` output file summarizes the different parameter settings used for the primer design. The second part is a table listing the primer pairs identified with these settings, as well as the characteristics associated with the corresponding metabarcode system (Box 2.1).

A complete description of the `ecoPrimers` output can be found in http://metabarcoding.org/obitools/doc/scripts/ecoPrimers.html, so hereafter, we will focus only on the characteristics echoing the properties of the ideal metabarcode (see Section 2.2). The *Bc* and *Bs* indexes (columns 16 and 18, respectively, in red in Box 2.1) should of course be maximized, while the maximum and average metabarcode lengths (columns 20 and 21, respectively, in blue in Box 2.1) should be minimized.

Besides, it should be noted that `ecoPrimers` will identify all potential primers in a given conserved region, and combine them with all suitable primers from other conserved regions, as long as the resulting metabarcode fulfills the requirements. As a result, the different metabarcodes proposed by the program are often variants of the same metabarcode systems (Fig. 2.4). For example, there are only minor differences among primer pairs 0, 1, 2, 3, and 6 listed in Box 2.1 as regards their *Bc* and *Bs* indexes, or their maximum and average metabarcode lengths, because they all target the same variable region (Fig. 2.4).

This feature of the `ecoPrimers` output can actually be exploited to refine the selected primer pair. Indeed, it is sometimes advantageous to extend or shift the primer(s) from one or a few nucleotides in 5' or 3' to equilibrate the two primer melting temperatures (*Tm*). This can also be done to avoid primer dimers, 3' complementarity, or secondary structures like hairpins (Fig. 2.5). Moreover, a good rule is, if possible, to avoid thymines at the 3'-end, to ensure good primer specificity (Kwok *et al.* 1990). For Bacteria, we ultimately selected two primers representing a good compromise in terms of *Bc*, *Bs*, metabarcode length, having similar *Tm*s, and being not susceptible to secondary structures (see the primer sequences in Fig. 2.4). These primers happen to target the V5–V6 regions of the 16S rRNA gene, a locus where conserved regions are known to be interspersed with highly variable ones (Schloss 2010).

2.3.4 *In silico* primer testing with `ecoPCR`

Now that bacterial primers have been designed, it can be useful to have a better indication of their

Box 2.1 Top of the ecoPrimers output file FILE2_PrimersForBacteria.ecoprimer. This corresponds to the first seven primer pairs suggested by ecoPrimers, out of a total of 683.

```
# ecoPrimer version 0.3
# Rank level optimisation: species
# max error count by oligonucleotide: 3
#
# strict primer quorum: 0.70
# example quorum: 0.90
#
# database: FILE1_ReferenceDatabase
# Database is constituted of 517 examples corresponding to 517 species
# and 0 counterexamples corresponding to 0 species
#
# amplifiat length between [30,280] bp
# DB sequences are considered as linear
# Pairs having specificity less than 0.60 will be ignored
#
```

0	ACGACACGAGCTGACGAC	GGATTAGATACCCTGGTA	60.5	44.8	49.5	25.3	11	8	GG	514	0	0.994	514	0	0.994	508	0.988	246	275	258.68
1	CACGACACGAGCTGACGA	GGATTAGATACCCTGGTA	60.5	44.8	49.5	25.3	11	8	GG	514	0	0.994	514	0	0.994	508	0.988	247	276	259.68
2	CACGACACGAGCTGACGA	GATTAGATACCTGGTAG	60.5	44.8	48.2	30.4	11	8	GG	513	0	0.992	513	0	0.992	507	0.988	246	275	258.68
3	ACGACACGAGCTGACGAC	ATTAGATACCCTGGTAGT	60.5	44.8	48.5	35.7	11	7	GG	513	0	0.992	513	0	0.992	507	0.988	244	273	256.68
4	CGTGCCAGCAGCCGCGGT	GACTACCAGGGTATCTAA	69.6	53.6	49.5	25.6	14	8	GG	516	0	0.998	516	0	0.998	507	0.983	254	259	255.94
5	CGTGCCAGCAGCCGCGGT	GGACTACCAGGGTATCTA	69.6	53.6	51.6	28.3	14	9	GG	516	0	0.998	516	0	0.998	507	0.983	255	260	256.94
6	ACGACACGAGCTGACGAC	GATTAGATACCCTGGTAG	60.5	44.8	48.2	30.4	11	8	GG	513	0	0.992	513	0	0.992	507	0.988	245	274	257.68

7 . . .

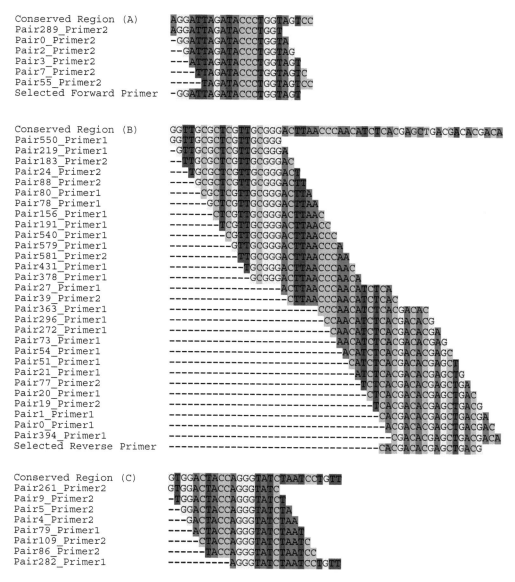

Fig. 2.4 The three conserved regions (A, B, C) where the first seven primer pairs mentioned in Box 2.1 were identified by `ecoPrimers` (together with others). Some primers are missing because they are strictly identical to other primers already indicated in this figure (e.g., Primer 2 of pair 1 is identical to primer 2 of pair 0, see Box 2.1). The two primers ultimately selected for Bacteria are also listed for conserved regions A and B.

conservation across bacterial taxa, or an estimation of the resolutive power of the corresponding metabarcoding system at different taxonomic levels. This can be done by performing an *in silico* PCR with the `ecoPCR` program (http://metabarcoding. org/ecopcr), either on the set of reference sequenc-es used for primer design, or on a larger set of sequences, like those available in EMBL. Here, we run `ecoPCR` on an `ecoPCR` database built from release 126 of EMBL (specified with the *-d* option), using the two primer sequences as compulsory parameters (Command 2.3).

```
5'-GGTATGGCGAAATCGCTAGC-3'
                ||||||
         3'-CGATCGCTAAAGCGGTATGG-5'
```
3'-end complementarity

```
  5'-GGTATGGCGACCATACTAGC-3'
            ||||  |||||
      3'-ATCCTGGGATGATGAGGTCAAT-5'
```
complementarity

```
    ⌜CGGTATGG-5'
    G ||||||
    ⌞ACCATACTAGC-3'
```
hairpin

Fig. 2.5 Motifs that must be avoided when designing PCR primers.

Command 2.3

```
ecoPCR -d embl_126 -e 6 -r 2 -L 1000 GGATTAGATACCCTGGTAG#T# \
CACGACACGAGCTGAC#G# > FILE3_PrimerTest.ecopcr
```

In this command, the *-e* option sets the maximum number of mismatches allowed per primer. We deliberately chose a high number of mismatches (6) in order to later evaluate how conserved the primers are across bacterial taxa. The "#" symbol means that there must be a perfect match for the preceding nucleotide, in this example for the two nucleotides in 3′ of the primers. The *-r* option restricts the amplification to bacterial sequences (Bacteria taxid: 2). The *-L* option limits the maximum metabarcode size to 1,000 bp.

2.3.4.1 The ecoPCR output

The ecoPCR output file contains two parts (Box 2.2). The first part is a reminder of the options specified in the ecoPCR command. The second part is a "|"-delimited table listing the characteristics of the *in silico* amplified sequences, like the taxid of the amplified sequence, the metabarcode sequence, the number of mismatches on each primer, and the sequences of the priming sites. More details on the table contents can be found at http://metabarcoding.org//obitools/doc/scripts/ecoPCR.html.

Box 2.2. Top of the ecoPCR output file FILE3_PrimerTest.ecopcr. This corresponds to the first seven *in silico* amplicons obtained by ecoPCR on EMBL release 126, out of a total of 1,809,393.

```
#@ecopcr-v2
#
# ecoPCR version 0.2
# direct strand oligo1 : GGATTAGATACCCTGGTAG#T#; oligo2c: C#G#TCAGCTCGTGTCGTG
# reverse strand oligo2 : CACGACACGAGCTGAC#G#; oligo1c: A#C#TACCAGGGTATCTAATCC
# max error count by oligonucleotide: 6
# optimal Tm for primers 1 : nan
# optimal Tm for primers 2 : nan
# database: /Volumes/R0/Barcode-Leca/R126/embl_r126
 # amplifiat length smaller than 1000 bp
```

Continued

Box 2.2 *Continued*

\# output in superkingdom mode

\# DB sequences are considered as linear

\#

AB003391; | 1453 | 77,133 | species | 77,133 | uncultured bacterium | -1 | ### | -1 | ### | 2 | Bacteria | D | GGATTAGATACCCTGGTAGT | 0 | nan | CACGACACGAGCTGACG | 0 | nan | 258 | CCACGCCGTAAACGATGAGT GCTAAGTGTTAGAGGGTTTCCGCCCTTTAGTGCTGAAGTTAACGCATTAAGCACTCCGCCTGGGGAGTACGGCCGCAAGGCT GAAACTCAAAGGAATTGACGGGGGCCCGCACAAGCGGTGGAGCATGTGGTTTAATTCGAAGCAACGCGAAGAACCTTACCAG GTCTTGACATCCTCTGAAAACCCTAGAGATAGGGCTTCTCTTCGGGAGCAGAGTGACAGGTGGTGCATGGTTGT | Uncultured bacterium gene for 16S rRNA, partial sequence, clone:sp6

AB004575; | 1342 | 77,133 | species | 77,133 | uncultured bacterium | -1 | ### | -1 | ### | 2 | Bacteria | D | GGATTAGATACCCTGGTAGT | 0 | nan | CACGACACGAGCTGACG | 0 | nan | 260 | CCTAGCTGTAAACGATGTTCACTA GATGTGGGAGGTATCGACCCCTTCCGCGTCGACGCTAACGCATTAAGTGAACCGCCTGGGGAGTACGACCGCAAGGTTGAAACT CAAAGGAATTGACGGGGGCCCGCACAAGCGGTGGAGTATGTGGTTTAATTCGACGCAACGCGAAGAACCTTACCTGGGCTT GAACTGCTAATGGTAAAAGCCGGAAACGGTGATGACCCGCAAGGGAGTTAGCACAGGTGCTGCATGGCTGT | Uncultured bacterium gene for 16S ribosomal RNA, partial sequence, clone: TG13

AB004576; | 1343 | 77,133 | species | 77,133 | uncultured bacterium | -1 | ### | -1 | ### | 2 | Bacteria | D | GGATTAGATACCCTGGTAGT | 0 | nan | CACGACACGAGCTGACG | 0 | nan | 260 | CCTAGCTGTAAACTATGTTCACTA GATGTGGGAGGTATCGACCCCTTCCGCGTCGACGTTAACGCATTAAGTGAACCGCCTGGGGAGTACGGCCGCAAGGTTGAAACT CAAAGGAATTGACGGGGGCCCGCACAAGCGGTGGAGTATGTGGTTTAATTCGACGCAACGCGAAGAACCTTACCTGGGCTT GAACTGCTGATGGTAAAAACCGGAAACGGTGATGACCCGCAAGGGAGTCAGCAGAGGTGTTGCATGGCTGT | Uncultured bacterium gene for 16S ribosomal RNA, partial sequence, clone: TG14

AB004577; | 1343 | 77,133 | species | 77,133 | uncultured bacterium | -1 | ### | -1 | ### | 2 | Bacteria | D | GGATTAGATACCCTGGTAGT | 0 | nan | CACGACACGAGCTGACG | 0 | nan | 260 | CCTAGCTGTAAACTATGTTCACTA GATGTGGGAGGTATCGACCCCTTCCGCGTCGACGTTAACGCATTAAGTGAACCGCCTGGGGAGTACGGCCGCAAGGTTGAAACT CAAAGGAATTGACGGGGGCCCGCACAAGCGGTGGAGTATGTGGTTTAATTCGACGCAACGCGAAGAACCTTACCTGGGCTT GAACTGCTGATGGTAAAAACCGGAAACGGTGATGACCCGCAAGGGAGTCAGCAGAGGTGTTGCATGGCTGT | Uncultured bacterium gene for 16S ribosomal RNA, partial sequence, clone: TG25

AB004578; | 1343 | 77,133 | species | 77,133 | uncultured bacterium | -1 | ### | -1 | ### | 2 | Bacteria | D | GGATTAGATACCCTGGTAGT | 0 | nan | CACGACACGAGCTGACG | 0 | nan | 260 | CCTAGCTGTAAACTATGTTCACTA GATGTGGGAGGTATCGACCCCTTCCGCGTCGACGTTAACGCATTAAGTGAACCGCCTGGGGAGTACGGCCGCAAGGTTGAAACT CAAAGGAATTGACGGGGGCCCGCACAAGCGGTGGAGTATGTGGTTTAATTCGACGCAACGCGAAGAACCTTACCTGGGCTT GAACTGCTGATGGTAAAAACCGGAAACGGTGATGACCCGCAAGGGAGTCAGCAGAGGTGTTGCATGGCTGT | Uncultured bacterium gene for 16S ribosomal RNA, partial sequence, clone: TG33

AB004759; | 1448 | 77,133 | species | 77,133 | uncultured bacterium | -1 | ### | -1 | ### | 2 | Bacteria | D | GGATTAGA TACCCTGGTAGT | 0 | nan | CACAACACGAGCTGACG | 1 | nan | 257 | CCACGCCGTAAACGATGTCGACTTGGAGGTTGTTC CCTTGAGGAGTGGCTTCCGGAGCTAACGCGTTAAGTCGACCGCCTGGGGAGTACGGCCGCAAGGTTAAAACTCAAATGAATT GACGGGGGCCCGCACAAGCGGTGGAGCATGTGGTTTAATTCGATGCAACGCGAAGAACCTTACCTACTCTTGACATCCACG GAACTTTGCAGAGATGCTTTGGTGCTTCGGGAACCGTGAGACAGGTGCTGCATGGCTGT | Uncultured bacterium gene for 16S ribosomal RNA, partial sequence, Bacterium T

AB004760; | 1435 | 77,133 | species | 77,133 | uncultured bacterium | -1 | ### | -1 | ### | 2 | Bacteria | D | GGATTAGATACCCTGGTAGT | 0 | nan | CACGACACGAGCTGACG | 0 | nan | 256 | CCACGCCGTAAACGATGTCAAC TAGCCGTTGGGAGCCTTGAGCTCTTAGTGGCGCAGCTAACGCATTAAGTTGACCGCCTGGGGAGTACGGCCGCAAGGTTAAAACT CAAATGAATTGACGGGGGCCCGCACAAGCGGTGGAGCATGTGGTTTAATTCGAGCAACGCAGAGAACCTTACCAGGCCTTGA CATCCAATGAACTTTCTAGAGATAGATTGGTGCCTTCGGGACATTGAGACAGGTGCTGCATGGCTGT | Uncultured bacterium gene for 16S ribosomal RNA, partial sequence, Bacterium U. . .

2.3.4.2 Filtering of the `ecoPCR` output

In generalist DNA sequence databases like EMBL, many sequences are not taxonomically annotated at the species level. Including such sequences will jeopardize the evaluation of the primer pair specificity and conservation. Command 2.4 keeps only metabarcodes obtained from sequences with a species level taxonomic assignment. First, it adds taxonomic information at the order, family, and genus levels to each sequence (*—with-taxon-at-rank* options of the `obiannotate` program) based on the associated taxonomy (*-d* option); and second, it selects only the sequences for which these taxonomic ranks are actually defined (*—require-rank* options of the `obigrep` program). Note that Command 2.4 will not exclude sequences with an incorrect taxonomic assignment.

Command 2.4

```
obiannotate -d embl_r126 --with-taxon-at-rank=order \
--with-taxon-at-rank=family --with-taxon-at-rank=genus \
FILE3_PrimerTest.ecopcr | obigrep -d embl_r126 --require-rank=order \
--require-rank=family --require-rank=genus > \
FILE4_PrimerTest_filtered.fasta
```

In the newly created `FILE4_PrimerTest_filtered.fasta` file, we can select one sequence (*-n* option) at random per genus (*-c* option), to avoid overrepresentation of some bacterial genera (Command 2.5).

Command 2.5

```
obiselect -c genus -n 1 FILE4_PrimerTest_filtered.fasta > \
FILE5_PrimerTest_final.fasta
```

2.3.4.3 Evaluation of primer conservation

To evaluate primer conservation within the target taxonomic group, sequence logos can be built using the *forward_match* and *reverse_match* attributes of each sequence of the `FILE5_PrimerTest_final.fasta` file. These logos are composed of stacks of letters symbolizing the nucleotides, one stack for each position in the sequence. The overall height of the stack is an indication of nucleotide conservation at that position, while letter heights within the stack are proportional to nucleotide frequencies at that position (Schneider & Stephens 1990). Sequence logos are thus good graphical representations of primer conservation across the examined taxonomic group. Figure 2.6 shows the sequence logos for both the forward and reverse primers we previously selected for Bacteria, with three mismatches allowed per primer and no fixed nucleotides at the 3′-end. In Figure 2.7, we can see the distribution of the number of mismatches when allowing up to six mismatches

(A) Forward primer

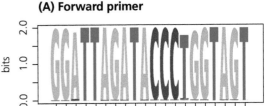

(B) Reverse primer

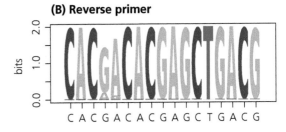

Fig. 2.6 Sequence logos for the forward (A) and reverse (B) primers selected for Bacteria. The primer sequences are indicated under the logos.

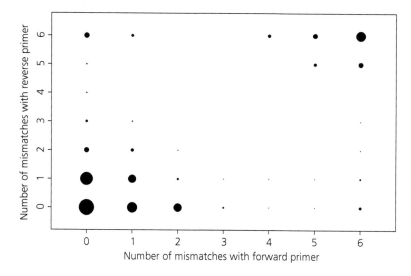

Fig. 2.7 Distribution of the number of mismatches with the forward and reverse bacterial primer sequences. For each combination, the size of the dot is proportional to the number of amplicons. The dots with a high number of mismatches for both primers correspond to non-specific amplifications.

Command 2.6

```
obigrep -p 'forward_error < 4' -p 'reverse_error < 4' \
FILE5_PrimerTest_final.fasta | ecotaxspecificity -d embl_r126
```

---

with the forward and/or reverse Bacteria primer sequences and when the two last nucleotides are fixed. These two figures indicate an elevated level of primer conservation across Bacteria, with at most two mismatches per primer for most amplicons (i.e., genera since we selected one amplicon per genus) and most genera having no mismatch at all for both primers. Two mismatches remain a reasonable value and should ensure limited amplification bias for the affected bacterial genera.

2.3.4.4 Taxonomic resolution and *Bs* index

The `ecotaxspecificity` program of the OBI-Tools program suite can be run to assess the resolution of our metabarcode system at different taxonomic ranks. First, we have to restrict the set of examined EMBL sequences to those for which the number of mismatches is at most 3 for each primer (-*p* "*forward_error* < 4" and -*p* "*reverse_error* < 4" options of the `obigrep` program; Command 2.6). Indeed, in practice, amplification is likely to be poor above three mismatches on any primer.

The `ecotaxspecificity` output is reported in Table 2.1 for different taxonomic levels. Our me-

tabarcoding system appears to perform well, even at lower taxonomic levels since it can distinguish more than 90% of the bacterial genera present in our set of EMBL sequences. It has to be noted, however,

Table 2.1 Output of the `ecotaxspecificity` program at different taxonomic levels. *Taxon_total* is the number of taxa to distinguish, and *Taxon_OK* is the number of taxa that can be distinguished based on the metabarcode sequences

Taxonomic rank	Taxon_ok	Taxon_total	Percentage (*Bs* index * 100)
Superkingdom	1	1	100.00
Superphylum	2	3	66.67
Phylum	27	29	93.10
Subphylum	1	1	100.00
Class	52	58	89.66
Subclass	1	1	100.00
Order	143	154	92.86
Suborder	5	5	100.00
Family	320	342	93.57
Subfamily	1	1	100.00
Genus	2,131	2,281	93.42

Table 2.2 List of primer codes and corresponding target taxonomic groups presented in Appendix 1

Code	Scientific name	Common name	NCBI taxid
Bact01	Bacteria	Eubacteria	2
Bact02	Bacteria	Eubacteria	2
Bact02	Archaea	Archaea	2157
Bact03	Bacteria	Eubacteria	2
Bact03	Archaea	Archaea	2157
Bact04	Bacteria	Eubacteria	2
Bact04	Archaea	Archaea	2157
Cyan01	Cyanobacteria	Blue green algae	1117
Arch01	Archaea	Archaea	2157
Euka01	Eukaryota	Eukaryotes	2759
Euka02	Eukaryota	Eukaryotes	2759
Euka03	Eukaryota	Eukaryotes	2759
Euka04	Eukaryota	Eukaryotes	2759
Fung01	Fungi	Fungi	4751
Fung02	Fungi	Fungi	4751
Baci01	Bacillariophyta	Diatoms	2836
Sper01	Spermatophyta	Seed plants	58024
Bryo01	Bryophyta	Mosses	3208
Trac01	Tracheophyta	Vascular plants	58023
Poac01	Poaceae	Grass family	4479
Aste01	Asteraceae	Daisy family	4210
Cype01	Cyperaceae	Sedge family	4609
Apia01	Apiaceae	Carrot family	4037
Rosa01	Rosaceae	Rose family	3745
Sapo01	Sapotaceae	Sapodilla family	3737
Meta01	Metazoa	Metazoans	33208
Echi01	Echinodermata	Echinoderms	7586
Echi02	Echinodermata	Echinoderms	7586
Moll01	Mollusca	Mollusks	6447
Gast01	Gastropoda	Gastropods	6448
Poly01	Polychaeta	Polychaetes	6341
Olig01	Oligochaeta	Oligochaetes	6382
Lumb01	Lumbricina	Earthworms	6391
Lumb02	Lumbricina	Earthworms	6391

Table 2.2 *Continued*

Code	Scientific name	Common name	NCBI taxid
Ench01	Enchytraeidae	Potworms family	6388
Arth01	Arthropoda	Arthropods	6656
Arth02	Arthropoda	Arthropods	6656
Inse01	Insecta	True insects	50557
Isop01	Isoptera	Termites	7499
Cole01	Coleoptera	Beetles	7041
Culi01	Culicidae	Mosquitoes	7157
Coll01	Collembola	Springtails	30001
Acar01	Acari	Mites and ticks	6933
Aran01	Aranae	Spiders	6893
Opil01	Opiliones	Harvestmen	43271
Cope01	Copepoda	Copepods	6830
Pera01	Peracarida	Isopods/amphi-pods	6820
Pera02	Peracarida	Isopods/amphi-pods	6820
Amph01	Amphipoda	Amphipods	6821
Vert01	Vertebrata	Vertebrates	7742
Chon01	Chondrichthyes	Cartilaginous fishes	7777
Elas01	Elasmobranchii	Rays, sharks, skates	7778
Elas02	Elasmobranchii	Rays, sharks, skates	7778
Tele01	Teleostei	Teleost fishes	32443
Tele02	Teleostei	Teleost fishes	32443
Tele03	Teleostei	Teleost fishes	32443
Cypr01	Cyprinidae	Carp family	7953
Gadi01	Gadidae	Cods	8045
Batr01	Batrachia	Frogs and sala-manders	41666
Test01	Testudines	Turtles	8459
Aves01	Aves	Birds	8782
Aves02	Aves	Birds	8782
Aves03	Aves	Birds	8782
Mamm01	Mammalia	Mammals	40674
Mamm02	Mammalia	Mammals	40674

that public databases suffer from many assignment errors, especially in groups where taxonomic identification remains a challenge (e.g., Bacteria, Fungi, Nematoda, and so on). To obtain a better estimation of the actual resolution power and coverage of the examined metabarcoding system, it is always better to work on well-curated sequence data sets when these are available.

2.4 Examples of primer pairs available for DNA metabarcoding

This section is dedicated to the description of different primer pairs available for DNA metabarcoding analysis of a wide range of taxonomic groups. We also provide statistics about the primers, as well as the characteristics of the associated DNA metabarcode. We mainly focus on short metabar-

codes, allowing the analysis of degraded eDNA, and deliberately overlooked primers designed in protein-coding genes (Deagle *et al.* 2014).

For each primer pair described in Appendix 1, we indicate a six-digit code (composed of the first four letters of the target taxonomic group scientific name, followed by a two-digit serial number) and the NCBI taxid of the target taxonomic group (Table 2.2). For the three following primer pairs, Bact02, Bact03, and Bact04, the analyses were carried out considering either Bacteria or Archaea as the target group. All the statistics were estimated after running ecoPCR (Bellemain *et al.* 2010; Ficetola *et al.* 2010) on the release 126 of the EMBL database, and by keeping only the most common sequence per species. Such a strategy ensures that overrepresented species in the EMBL database do not skew the statistics. For all statistics except the

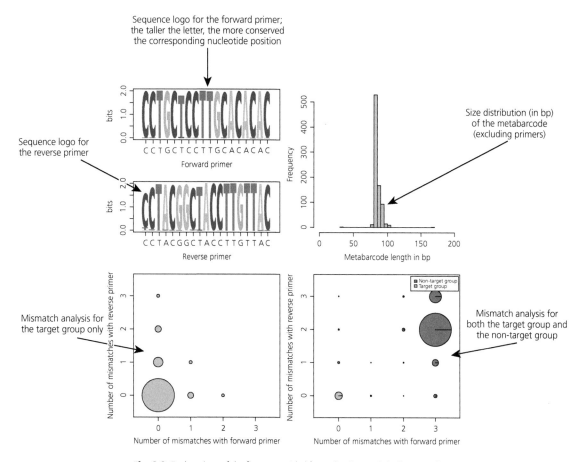

Fig. 2.8 Explanations of the figures provided for each primer pair in the Appendix 1.

conservation logos (Schneider & Stephens 1990), mismatches on the two last nucleotides on the 3'-end of the primers were disallowed. We plotted the size distribution of the metabarcode (excluding primers) in the target taxonomic group, and performed a mismatch analysis for both the target and the non-target groups (Fig. 2.8). This allowed the conservation of the primers within the target group to be estimated and to see if other non-target groups can be amplified. Finally, we present information about the coverage within the target taxonomic group (*Bc* index; Riaz *et al.* 2011), the taxonomic resolution within the target group (*Bs* index; Riaz *et al.* 2011), and the minimum, mean, and maximum lengths of the metabarcode (excluding primers). The coverage was assessed based on the available organelle genomes (on January 5, 2016). We did not evaluate the coverage for markers based on nuclear ribosomal DNA (rDNA) because no databases of full length transcribed rDNA are currently available. The estimate of the metabarcode resolutions at the different taxonomic levels (species, genus, family, order) is highly dependent on the errors of taxonomic assignment in public databases. Such errors are not randomly distributed among the different groups. Without errors, the resolution must increase between each level, from the species level to the order level. However, this is not always the case. For example, for the diatoms (primer pair Baci01), the taxonomic resolution goes from 82.1% at the family level to 71.9% at the order level. This is a strong indication that a few identical metabarcode sequences have been attributed to different species belonging to different orders. The impact of such mistakes is higher at the order level simply because the number of orders is lower. However, the resolution at the species level is also underestimated in most cases. Finally, the estimated resolution was carried out at the worldwide level, and is much better if a local reference database is available.

Obviously, all the above-mentioned statistics strongly depend on the data available in public databases. For example, the mismatch analysis for the primer pair Batr01 shows that it will also amplify a lot of other vertebrates. This phenomenon is exaggerated by the fact that public databases contain far more mammalian sequences than batracian sequences.

Published primers are not always fully optimized. For example, the Elas01 and Tele03 primer pairs published by Miya *et al.* (2015) can be further optimized *in silico* (see Elas02 and Tele02, respectively).

## CHAPTER 3

Reference databases

One of the primary objectives of DNA metabarcoding is to retrieve the taxonomic composition of environmental samples from sequence data. Taxonomic identification of DNA sequences basically consists in comparing an unknown eDNA sequence against a metabarcode reference database containing taxonomic information. In this context, it relies heavily on the completeness and quality of DNA sequence databases. Although we are currently far from having "metabarcoded" most species living on Earth, the considerable existing efforts and initiatives engaged to build such reference libraries, together with the decreasing cost of sequencing, will undoubtedly reduce this gap in the near future.

In this chapter, we will provide an overview of existing DNA reference databases, indicate where to find these useful resources and discuss their pros and cons. We will also provide brief guidelines for building new reference databases. The process of comparing DNA sequences against reference databases will be addressed in Chapter 8. Note also that this chapter is dedicated to reference databases for DNA metabarcoding applications. We will discuss those employed in metagenomics and metatranscriptomics in Chapter 10.

3.1 Extracting reference databases from EMBL/GenBank/DDBJ

Since the beginning of the 1980s, the European Molecular Biology Laboratory (EMBL; established in October 1980; Hamm & Cameron 1986) and GenBank (created in 1982; Bilofsky *et al.* 1986) databases, as well as the DNA Data Bank of Japan (DDBJ) have been accumulating the vast majority of DNA sequences published by scientists. As the three

members of the International Nucleotide Sequence Database Collaboration (INSDC), these three public databases are synchronized daily and to date (February 2017), the EMBL contains about 803 million sequence entries. Most of these sequences are annotated functionally but also taxonomically, and constitute valuable resources to build reference databases. Taxonomic assignment can be carried out in several ways. The easiest is to rely on the Basic Local Alignment Search Tool (BLAST) server provided by the National Center of Biotechnology Information (NCBI; Johnson *et al.* 2008). This eliminates the need to maintain a local copy of the database, and transfers the computational effort to public servers, as long as the data set to be analyzed is not too big. However, this approach has drawbacks. As the database is continuously updated, it becomes difficult to compare results obtained at different dates. Another drawback is the high computation time due to the need to compare the query to the whole public database. To compensate for this extra computation time, approximative (heuristic) alignment methods must be used. As no global heuristic alignment algorithm exists, the only choice is to use local alignment algorithms like BLAST (Altschul *et al.* 1997). This choice requires post-processing the results to avoid misidentification due to partial matches between query sequences and database sequences. To limit most of these drawbacks, the following sections will provide a set of basic principles for setting up a reference database for a given metabarcode on a local server.

3.1.1 Downloading a local copy of EMBL

A local copy of the EMBL database can be downloaded from the European Bioinformatics Insti-

Environmental DNA for Biodiversity Research and Monitoring. Pierre Taberlet, Aurélie Bonin, Lucie Zinger, & Eric Coissac, Oxford University Press (2018). © Pierre Taberlet, Aurélie Bonin, Lucie Zinger, & Eric Coissac 2018.
DOI: 10.1093/oso/9780198767220.001.0001

Table 3.1 The EMBL data classes. Most of the time it is only the STD class that is useful when producing a reference database

Class acronyms	Class names
CON	Constructed
EST	Expressed sequence tag
GSS	Genome sequence scan
HTC	High-throughput cDNA sequencing
HTG	High-throughput genome sequencing
PAT	Patents
STD	Standard
STS	Sequence tagged site
TSA	Transcriptome shotgun assembly
WGS	Whole genome shotgun

Table 3.2 The EMBL taxonomical divisions. Note that some clades are distributed over several divisions, for example, the mammalian sequences are the union of the MAM, HUM, ROD, and MUS divisions

Division acronyms	Division names
ENV	Environmental samples
FUN	Fungi
HUM	Humans
INV	Invertebrates
MAM	Other mammals
MUS	Mus musculus
PHG	Bacteriophages
PLN	Plants
PRO	Prokaryotes
ROD	Rodents
SYN	Synthetic
TGN	Transgenic
UNC	Unclassified
VRL	Viruses
VRT	Other vertebrates

tute (EBI) ftp site (ftp://ftp.ebi.ac.uk). The whole EMBL database consists of many text files, each of them containing several thousand sequences. The EMBL database is first broken down by data class (Table 3.1), then by taxonomical division (Table 3.2). Most of the time, only the standard (STD) data class is required when building reference databases, as it contains sequences with the best functional and taxonomic annotations. The STD class includes about 28 million sequences over the approximate 803 million described in the complete release of the EMBL database (Release 131, February 2017). If the objectives are restricted to a single clade, it is possible to download only the relevant division(s) of the EMBL. A sequence in the EMBL database belongs to a single division. For example, human sequences are only stored in the HUM division but not in the MAM division. Therefore, if the mammalian sequences are of interest, the divisions MAM as well as HUM, ROD, and MUS should be downloaded.

Specific class and division files can be downloaded through a standard web browser but it is more convenient to use a Unix command line approach. The wget Unix command allows downloading a set of files in a single instruction. wget is often part of standard Unix systems. If it is absent, packages for most of the systems are available. wget can also be downloaded directly from the GNU web site (https://www.gnu.org/software/wget/). For example, Command 3.1 downloads the complete standard (STD) data class.

For downloading only the human (HUM) division, the pattern must be modified as in Command 3.2.

Command 3.1

```
wget "ftp://ftp.ebi.ac.uk/pub/databases/embl/release/std/rel_std_*.dat.gz"
```

Command 3.2

```
wget "ftp://ftp.ebi.ac.uk/pub/databases/embl/release/std/rel_std_hum_*.\
dat.gz"
```

3.1.2 Identifying sequences corresponding to the relevant metabarcode

If having a local copy of the EMBL database addresses some of aforementioned problems, building a reference database dedicated to a single metabarcode is even more useful. Sequences corresponding to a specific gene can be selected by using annotations. Unfortunately, sequences in the EMBL database are not consistently annotated. A more reliable solution is to base the selection on sequence similarity. Several methods use this approach to select a subset of sequences corresponding to the metabarcode of interest. An initial set of sequences can be obtained based on the annotations, with this initial set used as queries for matching sequences into generalist databases using programs like BLAST (Altschul *et al.* 1997). A common strategy is to use sequences extracted from the first BLAST, as queries for a second search. This allows catching more diverse sequences, and this can be repeated until no new sequences are retrieved. To efficiently implement these strategies, *ad hoc* scripts must be developed to parse BLAST outputs, to filter matches, to extract sequences, and eventually to loop the process until all the sequences have been extracted.

For extracting only full-length markers, a convenient method is to rely on the ecoPCR program (Ficetola *et al.* 2010). ecoPCR selects sequences that are potentially amplifiable by a given polymerase chain reaction (PCR) primer pair. It can be used to extract sequences from a large database like EMBL based on their similarity with the two PCR primers. Primer matches must be in the correct relative orientation on the target sequence and within the specified minimum and maximum distances. The ecoPCR manual and OBITools tutorials give several examples of how to use ecoPCR to reach this objective. However, if most of the sequences corresponding to a given metabarcode in the EMBL database do not include the priming sites, ecoPCR cannot extract these sequences. This prevents the use of this method to build a reference database for this metabarcode.

3.2 Marker-specific reference databases

Numerous entries in the INSDC databases contain errors in their DNA sequence or associated taxonomic description (Nilsson *et al.* 2006; Pruesse *et al.* 2007;

Vilgalys 2003). Limiting the impact of such errors is essential for both DNA metabarcoding surveys and phylogenetic studies. The establishment of gold-standard genomic regions such as the rRNA gene regions for the whole tree of life (Fox *et al.* 1977; Pace 1997, 2009; Schoch *et al.* 2012), COI for animals (Hebert *et al.* 2003a), or *rbc*L and *mat*K for plants (CBoL Plant Working Group 2009; Clegg *et al.* 1991) allows the scientific community to mutualize curation efforts to establish high-quality public databases for these markers (Table 3.1; also see Santamaria *et al.* 2012 for a review). Here, we will briefly explain their properties, how these databases are constructed, and which criteria ensure their quality and reliability in terms of taxonomic annotations.

3.2.1 Nuclear rRNA gene reference databases

Carl Woese's vision of a classification of all living organisms through their rRNA gene sequence has considerably changed biological thinking. The Ribosomal Database Project directly emerged from this vision and has constituted the first publicly available resource dedicated to a specific genomic region (Olsen *et al.* 1992). The trivialization of the use of rRNA genes for studying the microbial diversity and typing of microbial strains generated a deluge of rRNA data. This led to the emergence of similar, yet independent initiatives: the ARB/SILVA project (Ludwig *et al.* 2004; Pruesse *et al.* 2007), which includes large subunit rRNA sequences; Greengenes, which was initially developed for probe design (DeSantis *et al.* 2006); and EzTaxon, which was initially aimed at refining the taxonomy of Bacteria (Chun *et al.* 2007). The recent growing interest for protist diversity led to initiatives following the same rationale, the PR2 project (Guillou *et al.* 2013) and PhytoREF (Decelle *et al.* 2015).

These databases follow the same building scheme and merely differ in the tools and criteria used in that respect. These features are summarized in Santamaria *et al.* (2012) or provided in the databases' references or website (Table 3.3). Briefly, they consist of rRNA sequences that are collected from the INSDC databases by searching for combinations of keywords (typically 16S, rRNA, and so forth) in the sequence descriptions and with sufficient length (typically >500 nucleotides) to allow robust taxonomic identification. They are also provided by

Table 3.3 High-quality reference databases overview (only the most recent references are shown)

Database name	Organisms covered	DNA region targeted	Website	Reference
SILVA	Bacteria, Archaea, Eukaryota	16S/23S rRNA genes 18S/28S rRNA genes	www.arb-silva.de	Quast *et al.* (2013)
RDP	Bacteria, Archaea, Fungi	16S rRNA gene 28S rRNA gene, ITS region	rdp.cme.msu.edu	Cole *et al.* (2014)
Greengenes	Bacteria, Archaea	16S rRNA gene	greengenes.lbl.gov	DeSantis *et al.* (2006)
EzTaxon	Bacteria, Archaea	16S rRNA gene	www.ezbiocloud.net	Chun *et al.* (2007)
PR2	Protists	18S rRNA gene	ssu-rrna.org/pr2	Guillou *et al.* (2013)
PhytoREF	Photosynthetic eukaryotes	Plastidial 16S rRNA gene	phytoref.org	Decelle *et al.* (2015)
BOLD	Animals, plants, Fungi	Mitochondrial COI Chloroplastic *rbc*L, *mat*K ITS region	www.boldsystems.org	Ratnasingham & Hebert (2007)
UNITE	Fungi	ITS region	unite.ut.ee	Kõljalg *et al.* (2013)
MaarjAM	Arbuscular mycorrhizal Fungi	Multiple	maarjam.botany.ut.ee	Öpik *et al.* (2010)
AFTOL	Fungi	Multiple	www.aftol.org	Celio *et al.* (2006)

users when generated from cultured and identified strains. The resulting sequence set is then screened for potential errors or chimeras by comparing them against high-quality sequences (i.e., coming from type cultured organisms of which taxonomic affiliation has been validated with standard criteria; see Lapage *et al.* 1992; Tindall *et al.* 2010). In some cases, the resulting sequence set is simplified through sequence clustering to reduce database size. Sequences then undergo multiple alignments and phylogenetic reconstruction. These ribosomal resources are available for browsing and querying on the websites indicated in Table 3.3. The user can also download unaligned (FASTA or GenBank format) or aligned versions (FASTA or ARB format) of the RDP, SILVA, or Greengenes sequence databases for local applications. Ribosomal resources also include a variety of software and web services that were specifically developed for the rDNA region, ranging from probe-testing to tree-building. We strongly encourage the reader to explore these facilities.

3.2.2 Eukaryote-specific databases

Although most of the reference databases listed here encompass the whole tree of life, they remain largely biased toward Bacteria. It is only in the late 2000s that marker-specific reference data-

bases dedicated to eukaryotes were created; first for Fungi through the UNITE (Kõljalg *et al.* 2005) and AFTOL (Celio *et al.* 2006) projects. Fungi have long been typed using several molecular markers, in particular the ITS region (Horton & Bruns 2001; Schoch *et al.* 2012). More recently, the BOLD system (Ratnasingham & Hebert 2007) was created for metazoa and plants, following the seminal article of Hebert *et al.* (2003a) on DNA barcodes (Table 3.3). The BOLD system is further divided into sub-databases corresponding to particular metazoan clades (see e.g., www.fishbol.org, www.mammali-abol.org, www.barcodingbirds.org).

UNITE and BOLD basically follow the same building scheme as other rRNA resources with similar data curation tools. Sequence quality standards are, however, higher: most sequences included in UNITE and BOLD come from specimens identified by specialists, and are provided with associated voucher/collection/collector information. For the BOLD database, sequence trace files are provided as well. As for rRNA resources, most of the eukaryotic databases are equipped with a querying system (e.g., taxonomy browser or BLAST). UNITE and BOLD include further web services for phylogenetic placement/reconstruction, and sequence data can be downloaded by the user as FASTA or tabular files.

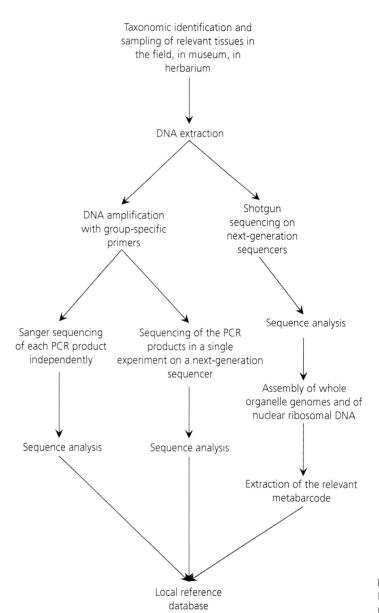

Taxonomic identification and sampling of relevant tissues in the field, in museum, in herbarium

DNA extraction

DNA amplification with group-specific primers

Shotgun sequencing on next-generation sequencers

Sanger sequencing of each PCR product independently

Sequencing of the PCR products in a single experiment on a next-generation sequencer

Sequence analysis

Assembly of whole organelle genomes and of nuclear ribosomal DNA

Sequence analysis

Sequence analysis

Extraction of the relevant metabarcode

Local reference database

Fig. 3.1 The different strategies for building a local reference database suitable for analyzing metabarcoding results.

3.3 Building a local reference database

Figure 3.1 gives an overview of the different steps used when developing a local reference database for metabarcoding studies. The first step is the collection of well-identified samples either from the field, or from existing collections in museums or herbariums. At this stage, it is important to try to obtain homogeneous samples (i.e., same amount, same quality, same type of tissue, and so on) to allow a better standardization during the following steps. The second step involves extracting DNA from all samples using standard protocols. Then, two strategies are possible: either the amplification of the DNA marker followed by subsequent sequencing of the PCR products, or a shotgun sequencing approach called "genome skimming" (Dodsworth 2015; Straub *et al.* 2012).

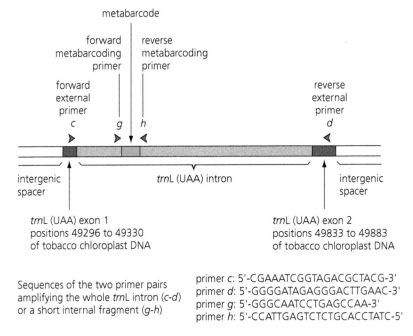

Sequences of the two primer pairs amplifying the whole *trn*L intron (*c-d*) or a short internal fragment (*g-h*)

primer *c*: 5'-CGAAATCGGTAGACGCTACG-3'
primer *d*: 5'-GGGGATAGAGGGACTTGAAC-3'
primer *g*: 5'-GGGCAATCCTGAGCCAA-3'
primer *h*: 5'-CCATTGAGTCTCTGCACCTATC-5'

Fig. 3.2 The use of external primers for building a local reference database: example of the *trn*L intron metabarcode. The metabarcoding study is carried out using the *g-h* primers (Sper01 in Appendix 1; Taberlet *et al.* 2007), while the reference database can be built using the *c-d* primers (Taberlet *et al.* 1991).

3.3.1 PCR-based local reference database

If the DNA extracts are homogeneous, all the samples can be amplified at once using the same PCR conditions. If feasible, the amplification for the local reference database can be performed using external primers and not using the primers that will be used for the metabarcoding study (Fig. 3.2). Thus, information will be obtained not only for the metabarcode itself, but also for the target sequences of the metabarcoding primers. This will allow potential bias during the DNA amplification of the metabarcoding analysis to be assessed. When compared to species exhibiting no mismatches between the primers and the target sequences, a species with a few mismatches will probably produce a lower number of sequence reads.

If the number of samples to be analyzed for building the local reference database is low (a few dozen), it might be cheaper to simply sequence the metabarcode using Sanger sequencing on capillary sequenc-ers. However, as soon as the number of samples becomes high (several hundred to several thousand), it is cheaper and less time-consuming to switch to next-generation sequencers. In this case, external or metabarcode primers must be tagged on their 5'-end to sequence all the samples in a single experiment (using the same tagging system as the one described in Section 6.10). The expected number of sequence reads per sample for building the reference database on next-generation sequencers should be at least 100, and preferably around 1,000, meaning that the sequencing cost per sample corresponds to only a few cents. The sequence analysis can be carried out using the same procedure as for the metabarcoding study (see details in Chapter 8), except that only the most common sequence per sample is kept in the reference database. However, when analyzing a metabarcode localized on the nuclear ribosomal DNA (e.g., on ITS), it might be worth, in some cases, keeping more than a single sequence as different copies of rDNA might persist within the same organism.

3.3.2 Shotgun-based local reference database

For eukaryotes, low coverage shotgun sequencing of genomic DNA (= genome skimming; Dodsworth 2015; Straub *et al.* 2012) represents the "gold" approach. From genome skimming, the whole organelle sequences as well as all the sequences of the repeated nuclear sequences (such as the rDNA cluster) can be assembled using, for example, Organelle Assembler (http://metabarcoding.org/asm). Thus, with a single genome skimming, all the potential metabarcodes and their associated primers' targets can be obtained.

It has been estimated that a sequencing effort of about one gigabase is comfortable for a reliable assembly of nuclear ribosomal DNA in plants and animals, of chloroplastic genomes in plants, and of mitochondrial genomes in animals (Coissac *et al.* 2016). The whole plant mitochondrial DNA is more difficult to obtain through this approach, probably due to its remarkable plasticity in size and structure, with a mosaic of DNA sequences of both endogenous and exogenous origin (Bonen 1998; Bonen & Brown 1993).

Another important advantage of the genome skimming approach for building the reference database is that it allows the use of degraded DNA as often found in herbarium or museum specimens, enhancing the possibility of including data from type specimens. For example, Besnard *et al.* (2014) were able to recover the whole plastid genome using a 100-year-old herbarium specimen of an extinct plant species.

Even though sequencing costs are constantly decreasing, building a reference database based on shotgun approaches is still expensive. The commercial price for sequencing one gigabase is about 80 USD, without taking into account the library preparation (ligation of specific adapters on each side of the fragmented genomic DNA). At present, the cost of the library preparation is also relatively high and roughly equivalent to the cost of the sequencing itself. However, the library cost may significantly decrease with the development of fully automatized systems, either using robots or a microfluidic approach (e.g., Kim *et al.* 2013a). Due to the added scientific value and quality of a shotgun-based reference database (i.e., availability of whole organelle and ribosomal DNA; possibility of using herbarium or museum specimens; less contamination risk than

when using a PCR-based approach), it might be worth considering this option, despite constraints on the costs.

3.4 Current challenges and future directions

Applications of the public or homemade reference databases already listed here go far beyond assigning a taxon name to a metabarcode. They all participate in cataloging the species living on Earth and help to organize them in a coherent way (i.e., within a unified taxonomic framework). Although there are existing collaborations between the INSDC and reference databases listed in Table 3.1., they remain poorly connected from a taxonomic point of view. The problem is particularly important for prokaryotes, for which taxonomy has a much younger history and lower stability when compared to non-microbial eukaryotes. Consequently, the bacterial taxonomy differs among the RDP, Greengenes, and SILVA databases (Yilmaz *et al.* 2014). This overall lack of coherence prohibits the conducting of meta-analyses on metabarcoding-based studies, or with classical taxonomic inventories. Collaborative efforts are therefore still needed to harmonize the taxonomic nomenclature across all the available resources.

More fundamentally, taxonomy is a concept that encompasses not only phylogenetic relationships, but also genetic and phenotypic aspects. Microbiologists, botanists, and zoologists all have independently identified the need for a more integrative taxonomy (Kvist 2013; Rosselló-Móra 2012; Will *et al.* 2005). Progress toward bridging this gap is ongoing notably by the Genomic Standards Consortium (http://gensc.org; Yilmaz *et al.* 2011a). Minimum contextual metadata should be provided with any deposited sequence in the INSDC databases (MIMARKS, geographic location, site environmental characteristics, primers used, and so on; Yilmaz *et al.* 2011b). The BOLD system, UNITE and AFTOL databases also include such information (e.g., pictures, biochemical characteristics, and so on). Although much remains to be done, we foresee that both taxonomists and molecular ecologists will greatly benefit from such integrative databases to improve taxonomic classification and make further ecological inferences from their data.

CHAPTER 4

Sampling

An appropriate sampling design is crucial for the success of any ecological study. Unfortunately, there is no universal approach, and the sampling strategy must be adjusted according to the type of environment (e.g., aquatic vs. terrestrial), the scientific question, as well as statistical, logistical and financial considerations, and eventually the results of a pilot study. This topic has fueled a great body of literature in classical ecology and we will refer the interested reader to it for more details (see e.g., Magurran & McGill 2011).

These considerations also hold true for eDNA studies. However, the problem of sampling is even more difficult to tackle in metabarcoding studies because environmental DNA consists of various forms reflecting different time windows or organism states. The optimal sampling design may also differ among the different organisms that can be retrieved from the same sample. For example, Bacteria, Fungi, and trees can all be the target of the same metabarcoding experiment from soil samples, but the composition of microbial communities will obviously vary on different spatial scales than trees. As a consequence, the overall sampling design will be the result of a compromise between all the aforementioned aspects. Another facet to bear in mind when designing the sampling strategy is the standardization of the approach and the minimization of sample degradation. These considerations are mandatory if we are to obtain sound and realistic comparisons among samples, as well as amounts of DNA and polymerase chain reaction (PCR) inhibitor levels that are as homogeneous as possible among samples to allow the use of the same PCR parameters at the amplification stage. Finally, biological replicates must be included in every study in order to properly estimate the variability introduced during sampling (Prosser 2010).

4.1 The cycle of eDNA in the environment

Having a prior knowledge of the origin, state, fate, and transport of eDNA in the environment is extremely important not only for designing the sampling strategy, but also for interpreting the results (see review in Barnes & Turner 2016).

4.1.1 State and origin

Environmental DNA can be found in multiple forms, but can be divided into two main classes: intracellular and extracellular DNA. Intracellular DNA is present in all living organisms. When present in an environmental sample, it can come from unicellular organisms, such as Bacteria, Archaea, and microeukaryotes that can be either in an active state (cells) or in a dormant stage (spores). It may also come from multicellular individuals such as meiofauna (e.g., nematodes, rotifers, and so on) or dissociated fragments of larger organisms, such as root fragments. Here again, it may correspond to active or dormant organisms (seeds, pupas, pollen, and so on). It is also present within remains of dead cells (e.g., nuclei and organelles surrounded or not by other cell structures). Environmental DNA can be also made of free DNA molecules, solubilized in the aqueous phase, or adsorbed on the surface of different types of organic or mineral particles (Levy-Booth *et al.* 2007; Pietramellara *et al.* 2009). Extracellular DNA is mostly released in the environment from decaying cells or sloughed material (e.g., tissues, feces, and different secretion types). For Bacteria, DNA has also been reported to be released through particular secretion systems (Costa *et al.* 2015), or through spontaneous excretion

Environmental DNA for Biodiversity Research and Monitoring. Pierre Taberlet, Aurélie Bonin, Lucie Zinger, & Eric Coissac,
Oxford University Press (2018). © Pierre Taberlet, Aurélie Bonin, Lucie Zinger, & Eric Coissac 2018.
DOI: 10.1093/oso/9780198767220.001.0001

caused by certain abiotic conditions that affect cell membrane permeability (Thomas & Nielsen 2005).

Due to greater exposure to enzymatic or chemical degradation, the solubilized DNA will most likely contribute less to the total eDNA pool than DNA adsorbed onto particles or cell remains (Pietramellara *et al.* 2009).

4.1.2 Fate

How long can eDNA persist in the environment? This is a key question when interpreting eDNA results. Unfortunately, estimating eDNA persistence in the environment is a complicated task because it depends on the interplay between multiple factors, such as the physical, chemical, and biological properties of the microenvironment (Levy-Booth *et al.* 2007; Pedersen *et al.* 2015; Pietramellara *et al.* 2009). Obviously, increased temperature or microbial activity would inherently shorten eDNA persistence by causing faster degradation and/or recycling. On the contrary, presence of particles with strong affinity to DNA molecules would extend eDNA presence in the environment by preventing leaching or degradation. As such, the habitat in which the study is conducted will be the first indicator of the potential DNA persistence.

In seawater or freshwater, several studies have shown that eDNA detectability ranges between a few days and a couple of weeks (Dejean *et al.* 2011; Green *et al.* 2011; Thomsen *et al.* 2012a). This detectability depends on biotic factors such as targeted population density or enzymatic activities, as well as abiotic conditions such as sunlight intensity or water turbidity (Barnes *et al.* 2014; Green *et al.* 2011; Pilliod *et al.* 2014). Persistence of eDNA is usually found to be longer in sediments when compared to water bodies, and it is likely to be favored by both anoxic conditions and increased adsorption potential on the sediment matrix (Corinaldesi *et al.* 2011; Torti *et al.* 2015; Turner *et al.* 2015). Accordingly, sediment layers can contain DNA that is thousands of years old in both lakes (e.g., Anderson-Carpenter *et al.* 2011; Capo *et al.* 2015; Giguet-Covex *et al.* 2014; Haile *et al.* 2007; Pansu *et al.* 2015b) or marine environments (e.g., Boere *et al.* 2011; Coolen *et al.* 2013; Smith *et al.* 2015; Weiß *et al.* 2015).

For terrestrial ecosystems, Yoccoz *et al.* (2012) showed that even a hundred years after cultivation abandonment, it was possible to find traces of potato, rye, and barley DNA in alpine soils (more details in Section 14.1). Even older DNA traces can be retrieved in colder soil environments, reaching up to 0.5 million years in permafrosts (Bellemain *et al.* 2013; Willerslev *et al.* 2003, 2014). This property has made possible reconstructing the composition and dynamics of past communities (Willerslev *et al.* 2003, 2014). However, such studies may not be possible in warmer environments, where eDNA most likely undergoes a much faster turnover. An experimental study has also shown that mango root DNA would persist only over 2 weeks with tropical edaphoclimatic conditions (Bithell *et al.* 2014). In any case, in soil, the proportion of "old" DNA molecules is much lower than molecules of recent origin. This means that the main part of the biological signal is contemporary. For instance, dramatic shifts of taxonomic composition can be observed between dry and wet seasons in tropical soil samples (Fig. 4.1).

4.1.3 Transport

Most of the eDNA contained in an environmental sample usually has a local origin. However, it might also come from neighboring sites, either through active or passive dispersal of organisms and/or propagules, or through passive transport of extracellular DNA in water (e.g., settling, leaching, or any movement of water). As such, eDNA transport is likely to be very important in aquatic environments. In streams, fish and invertebrate eDNA can be detected in water or sediment samples collected from hundreds of meters to tens of kilometers downstream from the eDNA source, depending on stream flow rate and turbulence (Deiner & Altermatt 2014; Jane *et al.* 2015). A lake contains not only eDNA from organisms living within the water body or sediment layer, but also from organisms living, or that have lived, in the catchment (e.g., Giguet-Covex *et al.* 2014; Pansu *et al.* 2015b; Parducci *et al.* 2013). The latter is transported to the lake through the leaching and settling of eDNA adsorbed to mineral or organic particles. Environmental DNA transported in terrestrial ecosystems has been much less studied, but is expected to contribute poorly to the DNA pool contained in

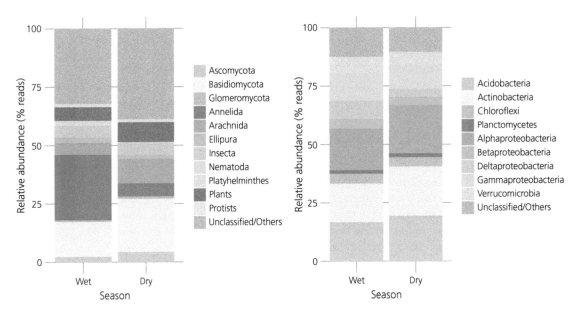

Fig. 4.1 Seasonal shifts in soil communities of Eukaryota (left panel) and Bacteria (right panel) in a one-ha plot in a tropical forest (Nouragues Reserve, French Guiana). The topsoil (0–10 cm) was sampled using a regularly spaced grid of sampling point (10-m mesh size) in April (wet season) and November (dry season) 2012. Eukaryotic and bacterial DNAs were amplified separately using superkingdom-specific primers. Between the wet and the dry seasons, we observed a drastic decrease in annelid sequence frequency. In parallel, the relative abundances of Fungi (in particular Basidiomycota) and Arachnida increased. The seasonal shift is less pronounced for Bacteria, but during the dry season, we still observed a noticeable decline in proportion of sequences affiliated to Deltaproteobacteria, a group predominantly made up of typical anaerobes and most likely associated with waterlogged soils. We also noted an increase in Actinobacteria, whose members have been repeatedly observed to resist drought. Soil community shifts are hence consistent with the climatic conditions, and are further visible at high taxonomic ranks.

a sample. It seems that the contamination of eDNA with material leaching from adjacent sites located upstream is unlikely, or should only account for a minor fraction of the eDNA pool, as shown in arctico-boreal environments (Yoccoz *et al.* 2012). There are several clues suggesting that soils contain eDNA that mainly comes from organisms living below the ground, with little contribution from those living above-ground. For example, plant eDNA in soil does not seem to be mainly released by decaying leaves, but rather from decaying roots. This is suggested by the distribution of plant DNA in soil (Fig. 4.2), as well as by the relatively low ratio of chloroplastic to nuclear ribosomal DNA in soils when compared to leaves.

4.2 Sampling design

From the details given in the previous section, the reader can easily conclude that an eDNA sample consists of various forms of DNA molecules of different

ages, different body sizes (i.e. micro-, meio-, or macroorganisms) and possibly of different geographic origins. Compared to classical field censuses, the information contained in such samples corresponds to a larger time, space, and taxonomic window, whose breadth highly depends on the ecosystem or taxonomic groups studied. This window is hence difficult to precisely identify *a priori*. In this context of partial knowledge, it is highly advisable to carry out pilot experiments to test if the whole experiment can potentially allow the addressed scientific question to be answered. The pilot study should properly identify the organism and eDNA population targeted, and the samples should be collected so as to be as representative as possible of the ecosystem(s) under study. Furthermore, the general physical, chemical, and biological characteristics of the matrix containing eDNA (soil, sediment, water, and so on) should be considered. Finally, field and financial constraints should also be taken into account.

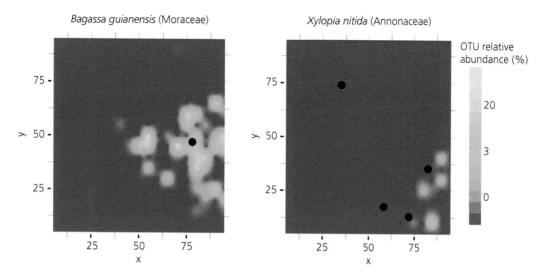

Fig. 4.2 Comparison of soil eDNA footprints corresponding to tree species with a flat root system (left panel, *Bagassa guianensis*) or tap root system (right panel, *Xylipia nitida*) in a one-ha plot in a tropical forest (Nouragues Reserve, French Guiana). Tree stems with diameter at breast height ≥10 cm are indicated with black dots. For *Bagassa guianensis*, the eDNA footprint is consistent with the stem location, the eDNA halo around the stem being likely related to the tree root system. This is not the case for *Xylopia nitida*, which is suspected to not release enough eDNA in the soil surface layer due to its particular rooting system. This species is nevertheless detected in other parts of the plot, which most likely correspond to roots of seedlings or saplings that were not included in the botanical inventory (conducted in Baraloto *et al.* 2012). The maps' color key is in logarithmic scale and represents the relative abundance of the species in the sequence data set.

4.2.1 Focusing on the appropriate DNA population

A single sample can yield very different results in terms of taxonomic composition, depending on how it is processed between its collection and the PCR amplification stage. For example, some differences in taxonomic composition may arise, depending on whether the DNA extraction step targets extracellular or total DNA (Zinger *et al.* 2016).

A first obvious consideration lies in the soil, sediment, or water layer to be sampled, as these might correspond to spatially close, yet functionally highly distinct habitats. For example, the organic layer of soil in a forest consists mainly of decaying material. This habitat differs from the underlying soil layer (next 10 cm) where the activities of plant roots and associated microorganisms dominate. Accordingly, litter and soil samples would display distinct communities. For instance, litter samples seem to harbor greater arthropod diversity and to be enriched in Hexapoda taxa, rather than Arachnida: the Hexapoda to Arachnida operational taxonomic units (OTUs) ratio is 0.9 and 1.7 in soil and litter

samples, respectively (Yang *et al.* 2014). Sediments also display strong vertical structuration. This is the same in large bodies of water, where one can often observe, for instance, high chlorophyll concentration in the euphotic zone (top 10 m; i.e., deep chlorophyll maximum). In a lake or large river, the conditions most likely differ as well along the banks, or in the middle of the body of water, and at different depths. These differences must be considered in the sampling design depending on the initial question. In particular, one should decide whether the sampling should be stratified or not, and pooled or not, according to the different habitats. Indeed, sampling in such heterogeneous environments requires the collection of samples in all the different compartments in order to fully document the complete biodiversity.

Similarly, one can ask whether the sample needs to be sieved or filtered (e.g., Chariton *et al.* 2010) or not (e.g., Guardiola *et al.* 2015). For example, when sifting soil or sediment samples, the passing material will mostly consist of inorganic particles and include a large proportion of microorganisms and propagules (e.g., pupas, pollens, or seeds).

The remaining material will therefore be enriched in meiofauna and root tissues, which can be considered as a bulk sample (Chapter 18). These latter samples provide high quality DNA, as most inorganic and humic substances are excluded, and allow the main focus to be on the living fraction of these macroorganisms, as shown for the benthic meiofauna (Leray & Knowlton 2015) or for plants (Hiiesalu *et al.* 2012). The analysis of passing material would be more appropriate for studying microorganisms. Aquatic microbiologists should use filters with a pore size adapted to their organisms of interest (e.g., viruses vs. Bacteria vs. protists vs. phyto - or zooplankton). On the other hand, if these samples are not sifted, the taxonomic composition will represent a more diverse collection of organisms.

4.2.2 Defining the sampling strategy

Overall, the sampling design should be adjusted so as to maximize the biological signal obtained while minimizing the sampling effort, but defining an appropriate sampling strategy is difficult, as many parameters must be taken into account. Conducting a pilot experiment can help for this purpose, by assessing the representativeness of each sample as well as the richness and the spatial autocorrelation of the biodiversity in the studied area. In this section, we will explain and illustrate sampling considerations for soil diversity analyses. We also encourage the reader to refer to the literature

dedicated to sampling issues in ecology (see e.g., Magurran & McGill 2011 and references within). Note that these reflections are valid for any type of environment and that the spatial considerations can be transposed to a time axis for time-series studies.

When defining a sampling strategy, one key problem is determining the appropriate size of the studied area (extent) and that of the sampling unit (resolution). The studied area encompasses the ecosystem(s) under study, and can refer to a plot, a landscape, a region, or the entire biosphere. It is divided into sampling units, corresponding to a subset of the studied communities. The size of the sampling unit determines the spatial resolution at which the biological pattern will be described. Several samples (i.e., biological replicates), can be collected from a single sampling unit (Fig. 4.3). Biological replicates are important to estimate the variance of results due to local heterogeneity, among others (see e.g., Searle *et al.* 2016). Obviously, all the sampling units belonging to a single study should be collected in a standardized way so as to be comparable. In each sampling unit, different ways of collecting biological material are possible (Fig. 4.3) and several important questions must be answered to determine the best choice. Which soil volume should be collected per sample? Must this volume be collected from a single point, or from several points distributed over the whole sampling unit? Overall, the spatial arrangement of sampling units and the way they are sampled should adequately capture the environmental

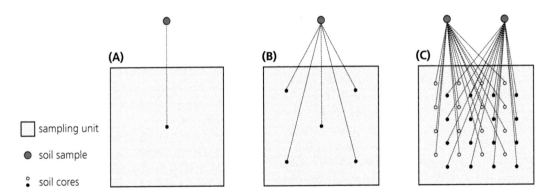

Fig. 4.3 Examples of three different approaches to obtain soil samples from the same sampling units: A, the soil sample is composed of a single soil core within the sampling unit; B, the soil sample is composed of the pooling and mixing of five soil cores; C, each sampling unit leads to two soil samples (i.e., biological replicates), each of them composed of the pooling and mixing of 16 soil cores.

heterogeneity of the studied area, and depends also on whether the study aims to describe patterns, estimate diversity parameters, or detect species. Figure 4.4 shows examples of the three main designs that can be implemented for soil. Systematic sampling is particularly appropriate when the studied area displays linear environmental gradients (see e.g., Barberán *et al.* 2015; Philippot *et al.* 2009). A systematic sampling strategy with a logarithmic spacing of sampling units is particularly appropriate if the goal of the study is to analyze the distance decay (i.e., the decrease of similarity between two samples as the distance between them increases). A systematic sampling can also be done on a transect (linear or belt). Random sampling, as implemented in Ramirez *et al.* (2014), represents another commonly used sampling strategy in ecology. Finally, when the environment is heterogeneous and can be divided into discrete entities, stratified sampling allows improved representa-

tiveness of the collected samples by reducing sampling error. Accordingly, recent works have shown that defining the sampling strategy on the basis of climatic or biological space modeling is more efficient in capturing the biological pattern of interest (Albert *et al.* 2010; Manel *et al.* 2012). In this last case, an equal number of samples is usually taken from each discrete entity, either using a random, a systematic, or a model-based approach. However, equal number of samples per sampling stratum might not be appropriate when considering habitats with highly contrasted diversity or heterogeneity. In this case, it might be worth adapting the sampling effort to maximize the sampling representativeness in each habitat (Cao *et al.* 2002).

For water samples, many studies now use a filtration approach. In this case, the pore size should be adjusted to the turbidity of the water and the expected filtered volume. Small pore sizes of 0.22 µM are particularly appropriate for clean water or small volumes, and larger pore sizes of up to 10 µM for turbid water or large volumes (Robson *et al.* 2016; Turner *et al.* 2014; Valentini *et al.* 2016). Three different filtering strategies can be implemented. The simplest one involves using classical filtering systems in the laboratory. Another strategy consists of filtering the water in the field, using either a small filtering device (e.g., 0.22 µm Sterivex-GP filters, Merck Millipore) for small volumes, or a capsule for tangential filtration (e.g., Envirochek HV 1 µm, Pall Corporation, Ann Arbor, MI, USA) for large volumes, up to 50 L. A precise description of this sampling strategy is described in Civade *et al.* (2016).

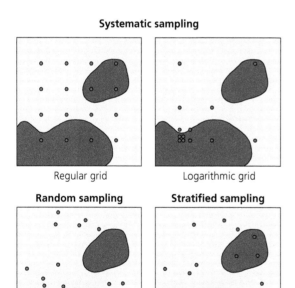

Systematic sampling

Regular grid · Logarithmic grid

Random sampling · **Stratified sampling**

☐ ■ Two different ecosystems

Fig. 4.4 Illustration of the different sampling methods in an area composed of two contrasting ecosystems. Contrary to the random sampling method, the stratified sampling takes the same number of samples in each of the two ecosystems, even if the areas covered by the two ecosystems are very different.

4.3 Sample preservation

To avoid DNA degradation or microbial community changes over time, the most straightforward approach consists in extracting DNA as quickly as possible after sample collection. However, this cannot always be achieved in the field, and in this situation, all biological activities should be blocked as soon as possible after the collection. A first approach consists in freezing the sample (any type of environmental matrix), either in liquid nitrogen or in a freezer. Apart from the problem of access to facilities to freeze and store the samples after collection, freezing and thawing might have consequences on the subsequent DNA extraction, in particular if extracellular DNA is the target. Cells can break during this process, leading to changes in proportions among taxa. This jeopardizes ecological inferences that can be made from the retrieved community and precludes comparisons with samples extracted without any freezing/unfreezing cycle. On the other hand, if total DNA is targeted, this process can facilitate cell lysis and subsequent DNA extraction.

An alternative is to buffer the sample in a solution that stabilizes the DNA. Several solutions can be used for this purpose. For water samples, the addition of 1/10 volume of sodium acetate (3 M, pH 8.0) and two volumes of alcohol has been shown to correctly preserve DNA (Ficetola et al. 2008), but this mix must be stored at -20°C until DNA extraction. Other more efficient buffers are available, such as the cetrimonium bromide (CTAB) or the "Longmire" buffers, which allow the samples to be stored at room temperature. One liter of the "Longmire" buffer can be made by adding (in numerical order): (i) 50 mL of 2 M Tris-HCL, pH 8.0; (ii) 200 mL of 0.5 M EDTA, pH 8.0; (iii) 2 mL of 5 M NaCl; (iv) up to 975 mL double-distilled water; and (v) 25 ml of 20% SDS (w/v) (Longmire et al. 1997; Wegleitner et al. 2015). In this buffer, SDS can be replaced by 1% N-lauryl sarcosine. Such a buffer can be used on any kind of environmental matrix or on filters, provided that the sample is stored in an adequate volume of buffer.

For samples made of whole organisms, such as insect bulk samples, storage in alcohol is a popular preservation method (e.g., Yu et al. 2012). In this case, it is important that the sample volume does not exceed 20% of the alcohol volume (using 96–100% alcohol). Preservation in alcohol is also useful for blocking biological activities in fecal samples when one wants to characterize its microbiome. For instance, the fecal material can be put in 96% or 100% alcohol (Nsubuga et al. 2004), before transfer to silica gel after a few hours.

The last and most convenient option is to dry the sample. This can be done through lyophilization, but this requires having the corresponding facility in the field. An easier option is to use silica gel to dry the sample. This approach can be used on soil, sediment, fecal, or filter samples. In this case, we recommend using silica gel in airtight containers, and at least three times more silica gel than the sample volume to speed up and secure the desiccation process.

## CHAPTER 5

DNA extraction

Many commercial kits or ISO protocols are available for extracting DNA from environmental samples, particularly for soil and feces, as these types of samples are prone to co-extraction of polymerase chain reaction (PCR) inhibitors. For soil, these protocols have been mainly developed for microbiologists to assess the diversity of unculturable Bacteria and Fungi. Therefore, we will not discuss standard DNA extraction procedures based on these approaches (for reviews, see e.g., de Bruijn 2011b; Philippot *et al.* 2012), and will rather emphasize alternative solutions.

The impact of different DNA extraction protocols on the bacterial diversity picture obtained subsequently has already been documented for almost two decades. These protocols are known to influence the observed bacterial community composition, in terms of both presence or absence and relative abundance of taxa (e.g., Frostegård *et al.* 1999; Martin-Laurent *et al.* 2001). Consequently, a reliable comparison of samples can only be made when samples were processed with the same DNA extraction protocol and subsequent procedures (i.e., PCR conditions and primers).

One of the characteristics of DNA metabarcoding is that it allows molecular taxonomic inventories from hundreds, or even thousands of environmental DNA samples to be assembled. Properly preserving all these samples for a DNA extraction at a later stage in a well-equipped laboratory presents a great challenge. Poor preservation of samples can lead to a complete change in the target community through microbial growth and DNA degradation, even at low temperatures. Current methods for preserving DNA in environmental samples involve either sample freezing, buffering/fixing (mixing the sample with e.g., CTAB, alcohol), or drying (e.g.,

silica gel, lyophilization, see Chapter 4 for more details). All these approaches require facilities or amounts of consumables that are incompatible with high-throughput sample processing and field campaigns. In this context, a DNA extraction carried out directly after sampling would be the only reasonable solution from logistic, financial, and even legislative points of view (e.g., samples too heavy to be transported, shipping too expensive, exportation authorization difficult to obtain). In this case, the use of a mobile DNA extraction laboratory, which can be set up with very basic facilities in the field, is preferable. In this chapter, we provide practical solutions for setting up such a laboratory, and for carrying out the DNA extractions immediately, or at least as rapidly as possible, after sample collection. As for the sampling step, it is important to standardize the DNA extraction, in order to allow the use of the same PCR parameters at the amplification stage, and to secure the analysis by comparing data produced in a standardized way. If the DNA extraction step is problematic or highly variable among samples, it might be advisable to add an internal extraction control to the sample before the DNA extraction to test the efficiency of this step (Smets *et al.* 2016). If the samples are heterogeneous, it might produce different results according to the extraction. In this circumstance, it is possible to perform several parallel extractions per sample (technical replicates) to take this heterogeneity into account at the analysis stage.

5.1 From soil samples

Microbiologists often seek to extract the total genomic DNA from soil samples, including both intracellular and extracellular DNA. The

Environmental DNA for Biodiversity Research and Monitoring. Pierre Taberlet, Aurélie Bonin, Lucie Zinger, & Eric Coissac, Oxford University Press (2018). © Pierre Taberlet, Aurélie Bonin, Lucie Zinger, & Eric Coissac 2018.
DOI: 10.1093/oso/9780198767220.001.0001

PowerSoil® DNA Isolation Kit (MoBio Laboratories, Inc., Carlsbad, CA, USA) or the NucleoSpin® Soil Kit (Macherey-Nagel, GmbH, Düren, Germany) are fairly popular in environmental microbiology. Indeed, these kits target high quality and pure DNA through optimized procedures that (i) break down soil particles and microbial cells that would be recalcitrant to chemical lysis, and (ii) remove humic acids, the typical PCR inhibitors occurring in soil. But these commercial kits differ in their efficiency to properly lyse cells and to remove inhibitors (Dineen et al. 2010; Wagner et al. 2015). Additionally, these kits usually utilize a relatively low amount of starting material (typically ~0.25 g of soil). If such a low amount might be appropriate for studying bacterial diversity (but see Ranjard et al. 2003), it might not be

the case when targeting macroorganisms (Andersen et al. 2012). MoBio Laboratories developed another extraction kit (PowerMax® Soil DNA Isolation Kit) for larger amounts of starting material (typically 5 g, and up to 10 g with sandy soils). However, the price per extraction using this kit (about €25) and the time spent for a batch of samples is prohibitive, especially for large-scale implementations.

An alternative approach to increase the amount of starting material in a high-throughput way is to target only extracellular DNA. Extracellular DNA adsorbs to negatively charged soil particles through their phosphate and cation bridging. It can hence easily be desorbed from soil particles by a saturated phosphate buffer. Extracellular DNA desorption can be seen as the first step in the DNA

Box 5.1 Material and consumables for extracting extracellular DNA from 15 g of soil with a mobile laboratory

Material for the mobile laboratory

1. Scale for weighing 15 g of soil in a 50 mL Falcon tube
2. Rack for a single Falcon tube
3. Large bottle equipped with a bottletop dispenser for distributing 15 mL of phosphate buffer
4. Racks for 24 Falcon tubes
5. Rotators for 24 Falcon tubes (e.g., Intelli-mixer)
6. Several racks for Eppendorf tubes
7. Centrifuge for 24 Eppendorf tubes (or two centrifuges for 12 tubes)
8. Vacuum pump
9. QIAvac 24 plus (Qiagen ref. 19,413)
10. Tubing for connecting the vacuum pump and the QIAVac (with a trap)
11. 1,000 µL pipette (for loading the extraction columns)
12. Three Distriman® Gilson (Gilson, ref. F164120C) for distributing the different washing buffers (see item 21)
13. Optional: mobile power supply

Consumables for extracting DNA in the field (250 extractions including controls)

14. Laboratory gloves
15. 250 disposable plastic spoons
16. 250 Falcon tubes
17. DNA-free water (5 L)
18. Five powder mixes for preparing 5 L of saturated phosphate buffer pH 8 (each mix containing 1.97 g of NaH$_2$PO$_4$ and 14.7 g of Na$_2$HPO$_4$)

19. Disposable plastic Pasteur pipettes with wide mouth for transferring the liquid mud from the 50 mL Falcon tube to a 2 mL tube
20. 250 two mL tubes with caps (for centrifuging the soil/phosphate buffer mix)
21. From the NucleoSpin® soil kit (Macherey-Nagel, ref. 740,780.250)
    a. SB buffer
    b. SW1 buffer
    c. SW2 buffer with ethanol
    d. 250 NucleoSpin® Soil Columns
    e. 250 collection tubes without caps
22. 250 connectors (for connecting the NucleoSpin® Soil Columns to the QIAvac; Qiagen ref. 19,407)
23. 250 Eppendorf tubes (for dispatching the SB buffer)

Consumables for completing the extraction in the laboratory

24. From the NucleoSpin® soil kit (Macherey-Nagel, ref. 740,780.250)
    a. SW2 buffer with ethanol
    b. Elution buffer SE
    c. 250 collection tubes with caps
25. 500 collection tubes without caps
26. A few 12.5 mL and 1250 µL distritips (Gilson ref. F164150 and F164140)

extraction procedure (see Boxes 5.1 and 5.2 for a detailed protocol). It basically involves homogenizing and desorbing a given volume of soil in the same amount of saturated phosphate buffer for 15 minutes. Then 2 mL of the resulting sludge are transferred into 2 mL tubes with caps using a disposable plastic Pasteur pipette. Finally, the remaining part of a classical DNA extraction procedure can be carried out as described in Box 5.2 from step 8.

This procedure makes it possible to extract extracellular DNA from larger volumes of starting material in a cost- and time-effective way: the price of reagents composing the phosphate buffer is

---

Box 5.2 Protocol for extracting extracellular DNA from 15 g of soil with a mobile laboratory

Phase 1 (as soon as possible after soil sampling)

1. Prepare 1 L of saturated phosphate buffer by adding the phosphate mix (1.97 g of NaH_2PO_4 and 14.7 g of Na_2HPO_4) to 1 L of DNA-free water (the phosphate buffer must be prepared just before use and cannot be kept for more than one day)
2. Distribute 250 μL of SB buffer into Eppendorf tubes (one tube per extraction)
3. Put connectors on the QIAvac, and put NucleoSpin® Soil Columns on connectors
4. Homogenize the collected soil in zip bags, and transfer 15 g of each soil sample into Falcon tubes using disposable plastic spoons (if the soil is very dry, it might be advisable to transfer less soil, in order to obtain the correct proportions between soil and buffer in the next step)
5. Add 15 mL of saturated phosphate buffer with the dispenser in each Falcon tube (including a negative extraction control containing only the phosphate buffer)
6. Thoroughly mix Falcon tubes and rotate them for 15 minutes using the Falcon rotator (this step can be reduced to 5 minutes if the columns clog at steps 10 to 16 or if the final DNA extract is heavily colored)
7. Transfer 2 mL of the soil/phosphate buffer mix into 2 mL tubes with caps using a disposable plastic Pasteur pipette
8. Centrifuge the 2 mL tubes at maximum speed during 5–10 minutes
9. For each sample, transfer 400 μL of the clear supernatant into the Eppendorf tube containing 250 μL of SB buffer using a 1,000 μL filter tip; mix the supernatant with the SB buffer using the same filter tip, and load the 650 μL to the relevant NucleoSpin® Soil Columns
10. Put the vacuum on and the liquid will pass through the columns; break the vacuum when all liquid has passed through;
11. Load 500 μL of SB buffer to each column with a Distriman

12. Put the vacuum on until all the liquid has passed through
13. Load 550 μL of SW1 buffer to each column with a Distriman
14. Put the vacuum on until all the liquid has passed through
15. Load 750 μL of SW2 buffer to each column with a Distriman; it is important to break the vacuum before loading the SW2 buffer in order to allow this buffer to rinse the upper part of the columns
16. Put the vacuum on until all the liquid has passed through
17. Close the columns, put each of them into a 2 mL collection tube without cap, and centrifuge for 2 minutes at 11,000 × g to dry the silica membrane
18. Discard the collection tubes, and store the columns on silica gel at room temperature until completing the DNA extraction in the laboratory. To avoid cross contamination among the different samples, it might be worth storing the columns individually, or to protect the bottom in an appropriate way (e.g., by using a small piece of aquarium pipe)

Phase 2 (later on in a properly equipped laboratory)

19. Put each column on a 2 mL collection tube without cap
20. Add 700 μL of SW2 buffer, close the columns, and vortex for 2 seconds
21. Centrifuge the columns for 30 seconds at 11,000 × g
22. Discard the collection tube and put each column on a new 2 mL collection tube without cap
23. Centrifuge for 2 minutes at 11,000 × g for drying the silica membrane
24. Put each column on a collection tube with cap
25. Add 100 μL of elution buffer SE (heated at 80°C) to each column
26. Wait for 1 minute at room temperature
27. Centrifuge for 30 seconds at 11,000 × g
28. Remove the columns from the collection tubes, and store the DNA extract for subsequent experiments (note that such extracts might have to be diluted—usually 10 times—before use to limit the influence of PCR inhibitors that are co-extracted together with the DNA)

negligible (ca. 1.5 €/L), the process does not require soil grinding or chemical/enzymatic cell lysis, and the classical extraction kits used downstream for the extracellular DNA desorption are the same as those used for small samples. This approach is particularly efficient when using 15 g of soil, as the desorption can be carried out in 50 mL Falcon tubes, hence allowing a large number of samples (hundreds to thousands) to be processed. It can easily be implemented outside of a well-equipped laboratory, using a mobile laboratory for extracting DNA directly after soil collection (see Fig. 5.1, and Boxes 5.1 and 5.2). Furthermore, the DNA extraction can be stopped before the elution step, hence allowing the soil DNA adsorbed on the silica membrane of the extraction column (Box 5.2), to be brought back to the laboratory at room temperature, provided that it is stored in dry conditions (e.g., in small envelopes with silica gel). In these conditions, it is possible to store the dry columns several months

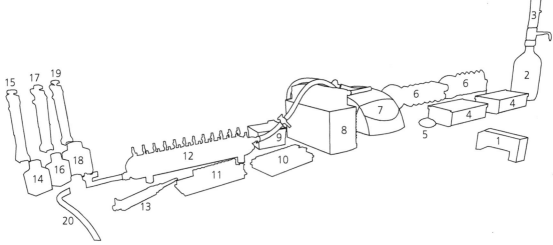

Figure. 5.1 Example of mobile DNA extraction laboratory used for soil samples during a field trip in Australia. 1. Gloves; 2. Bottle with saturated phosphate buffer; 3. Dispenser adjusted on 15 mL; 4. Two Falcon racks for a total of 24 Falcon tubes; 5. Timer; 6. Two rotators for a total of 24 Falcon tubes; 7. Centrifuge for 24 Eppendorf tubes; 8. Vacuum pump; 9. 1 mL filter tips; 10. Rack for Eppendorf tubes containing 250 μL of SB buffer; 11. Rack for Eppendorf tubes (extraction columns) on 2 mL collection tube; 12. QIAVac; 13. Pipet adjusted on 400 μL; 14. SB buffer; 15. Distriman Gilson adjusted on 500 μL; 16. SW1 buffer; 17. Distriman Gilson adjusted on 550 μL; 18. SW2 buffer; 19. Distriman Gilson adjusted on 750 μL; 20. Evacuation pipe.

(or even several years) before completing the extraction. It is also much more convenient to come back to the laboratory with dry columns than with eluted DNA. Using this procedure, we found that it gave comparable results for surveys of plants, soil microflora and meiofauna diversity when compared to the PowerMax® Soil DNA Isolation Kit from Mo Bio Laboratories, despite yielding lower amounts of DNA (Zinger *et al.* 2016).

In addition, the phosphate buffer protocol represents the only convenient alternative for extracting DNA from kilograms of soil (Taberlet *et al.* 2012c). It is merely the size of the container in which the desorption is carried out that limits the amount of soil. For about 5 kg of soil, we use either disposable autoclave plastic bags (e.g., Greiner Bio One GmbH, Frickenhausen, Germany, ref. 644,401) or wide-neck barrels (e.g., Roth Sochiel EURL, Lauterbourg, France, 15.4 L; ref. 0789.1). Using such a procedure, we were able to successfully retrieve the plant species occurring in alpine tundra ecosystems from composite samples of soil (see Chapter 4) of up to several kilograms in Taberlet *et al.* (2012c). Handling more than 10 kg of soil (i.e., 20 kg of soil and buffer mix) begins to pose handling problems during the desorption.

5.2 From sediment

Environmental DNA can also be extracted from sediments, including permafrost samples and lake or marine cores. Basically, extracting DNA from sediment can be carried out using the same approaches as for soil samples, that is, either using the PowerMax® Soil DNA Isolation Kit from MoBio Laboratories (see e.g., Haile *et al.* 2009) or the phosphate approach (see e.g., Giguet-Covex *et al.* 2014; Pansu *et al.* 2015b). However, as DNA might be much less concentrated in ancient sediments than in soil, the extraction must be carried out following guidelines for ancient DNA (Poinar & Cooper 2000) due to high contamination risks. A concentration step using Amicon Ultra-15 10K spin columns can be added, either just after the desorption when using the phosphate buffer approach, or at the end of the extraction when using the PowerMax® Soil DNA Isolation Kit (MoBio).

5.3 From litter

Several studies have already used leaf-litter to assess the local biodiversity of a target taxonomic group. Porazinska *et al.* (2010) isolated nematodes from litter, and then extracted DNA from bulk nematode samples to compare nematode diversity between soil, litter, and canopy in a tropical rainforest of Costa Rica. In the same way, Yang *et al.* (2014) first separated arthropods from litter using a Winkler litter extractor (www.entowinkler.at) for 36–48 h, before extracting DNA from the obtained bulk samples.

However, alternative solutions exist for avoiding the bulk sample step. England *et al.* (2004) proposed extraction of extracellular DNA using a sodium pyrophosphate buffer. It is thus possible to use the same approach for litter as that developed for large amounts of soil (see Section 5.1). For example, in a tropical rainforest, we collected litter over 0.5 m² into a large disposable autoclave plastic bag. We performed the desorption by adding one liter of saturated phosphate buffer (see Box 5.1, item 18) to the plastic bag, before mixing vigorously for 15 minutes. The volume of phosphate buffer to be added was adjusted depending on the humidity level of the litter. If the litter was very dry, a higher volume of phosphate buffer was added. After mixing, 2 mL of the liquid phase were transferred to a 2 mL tube with a cap using a disposable plastic Pasteur pipette and the remaining part of the extraction was done as described in Box 5.2 from step 8.

5.4 From fecal samples

Two strategies can be implemented for extracting DNA from feces. The first option is to use a kit dedicated to DNA extraction from feces (e.g., QIAamp DNA Stool Mini Kit from Qiagen or the PowerFecal Kit from MoBio). However, one should be aware that the QIAamp DNA Stool Mini Kit from Qiagen contaminates the DNA extract with potato DNA, and that might be a problem when analyzing herbivorous diets. A second alternative involves extracting extracellular DNA via the phosphate buffer approach (Taberlet *et al.* 2012c). The use of a dedicated extraction kit is particularly recommended when the objective is to both genotype individuals, and obtain the diet from the same DNA extract. Indeed, as intestinal cells are not subject to lysis when using the phosphate buffer, less host DNA is retrieved. Thus, while the phosphate buffer approach might be less effective to fulfill this genotyping/diet objective, it remains highly convenient for processing a very large number of samples, or when the sample is too large to be extracted with a classical extraction kit. We have already implemented large-scale DNA extractions from herbivore scats using a phosphate buffer. For large scats, such as those of elephants or rhinoceroses, we first homogenized a fecal pellet before subsampling 15 g and carrying out the extraction in 50 mL Falcon tubes as described in Box 5.2. This approach provided reliable results (unpublished data) for both the diet analysis (using the Sper01 primers amplifying Gymnosperms and Angiosperms; see Chapter 2 and Appendix 1), and the identification of the host (using the Mamm02 primers amplifying mammals; see Chapter 2 and Appendix 1).

5.5 From water samples

If the water sample is of moderate volume, has not been subjected to filtration, and is freshly sampled or preserved in alcohol (for 15 mL of water, 1.5 mL of sodium acetate 3M and 33 ml of absolute etha-

nol), the 50 mL tube is centrifuged at a high speed to pellet precipitated DNA and cellular remains (Valière & Taberlet 2000). It is important to not exceed the maximum speed specified for the tube. Then, the supernatant is discarded, and the pellet is finally subjected to a classical DNA extraction using a commercial kit (e.g., DNeasy Blood & Tissue Extraction Kit, Qiagen), starting with a digestion step using proteinase K.

However, the environmental DNA or microbial cells are often too diluted to work with small sample volumes. Instead, it is common to filter water samples (from ten to hundreds of liters) using pumps and filtration capsules (reviewed in Valentini et al. 2016; Zinger et al. 2012). Microbiologists have been using a wide variety of DNA extraction procedures from filtration capsules, including different commercial kits (see e.g., Eichmiller et al. 2016 and Hwang et al. 2012 for benchmark studies). The protocol published in Civade et al. (2016) is working very well for studying fish, and particularly for detecting rare species. The effects of water pH and proteinase K treatment on the yield of environmental DNA from water samples have been studied by Tsuji et al. (2017). It appeared that successful eDNA extraction can be carried out without any special attention to the water pH, and that proteinase K treatment is appropriate.

CHAPTER 6

DNA amplification and multiplexing

6.1 Principle of the PCR

The polymerase chain reaction (PCR) allows the *in vitro* amplification of short fragments of DNA, up to several kilobases (kb) long. Invented by Karry Mullis, PCR first involved the use of a thermosensitive polymerase (Klenow fragment of the polymerase I of *E. coli*; Saiki *et al.* 1985), which was later replaced by a thermostable enzyme, the *Taq* polymerase (Mullis & Faloona 1987). This polymerase was isolated from *Thermus aquaticus*, a bacterium living at 70–75°C in hot springs in Yellowstone National Park. DNA amplification by PCR really became accessible to molecular ecologists in 1989, with the availability of the first PCR machines and the seminal paper of Kocher *et al.* (1989) describing universal primers for animal mitochondrial DNA. PCR revolutionized our field, making the analysis of genetic polymorphism much easier than ever before, and reducing the prohibitive constraints of having to sample large quantities of tissues preserved in liquid nitrogen for allozyme studies, or large amounts of fresh tissues for mitochondrial DNA extractions. Since then, and before the rise of next-generation sequencing, virtually all papers analyzing DNA polymorphism in molecular ecology or conservation genetics were PCR-based.

In the context of DNA metabarcoding, the purpose of PCR is to synthesize a large number of copies of a homologous DNA fragment that will be subsequently sequenced on next-generation sequencers. It is important to properly understand the amplification mechanism in order to adjust the reaction in case of difficulty. Typically, a PCR reaction contains a tiny amount of template DNA, two primers (synthesized oligonucleotides), the four nucleotides, and a thermostable DNA polymerase, in an appropriate buffer (see Section 6.3).

A PCR amplification is realized by repeating 25–45 times three major steps (see Figs. 6.1 and 6.2 for details). The first step, denaturation, corresponds to the dissociation of the two DNA strands by heating the PCR mix at 94–96°C. It occurs instantly, as soon as the reaction mix reaches the target temperature.

The second step is the annealing of primers, which requires cooling the PCR mix down to 45–65°C. During this step, the primers move around due to Brownian motion. They constantly form hydrogen bonds with the single-stranded DNA present in the reaction. The more stable the bonds are, the longer they last. The most stable bonds occur when the primer matches the target sequence exactly, each A (adenine) binding to a T (thymine), and each C (cytosine) binding to a G (guanine). The polymerase can bind to this small piece of double-stranded DNA, and extend it from the 3'-end of the primer, by copying the template. This extension only occurs if the 3'-end of the primer matches the target DNA perfectly. The bond between the extended primer and the template DNA becomes stronger, meaning it will not break, even during the subsequent temperature increase toward the next step. The power of PCR comes from the amazing capability of the primers to find their exact complementary sequence among billions of base pairs of template DNA during the first few cycles.

The third step is extension (or elongation) at 72°C, which is the optimal working temperature of the polymerase. The polymerase binds to the small pieces of double-stranded DNA generated in the previous step, and extends a complementary strand by reading the template DNA from 3' to 5.'

By the end of the extension step, the quantity of target DNA fragments is roughly twice the amount present at the previous cycle. Therefore, during the

Environmental DNA for Biodiversity Research and Monitoring. Pierre Taberlet, Aurélie Bonin, Lucie Zinger, & Eric Coissac, Oxford University Press (2018). © Pierre Taberlet, Aurélie Bonin, Lucie Zinger, & Eric Coissac 2018.
DOI: 10.1093/oso/9780198767220.001.0001

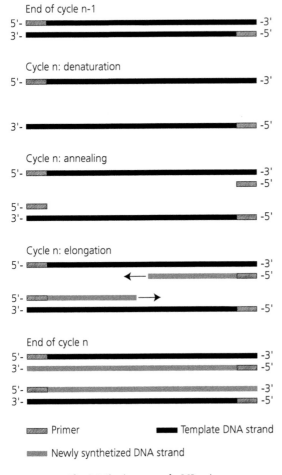

Fig. 6.1 The three steps of a PCR cycle.

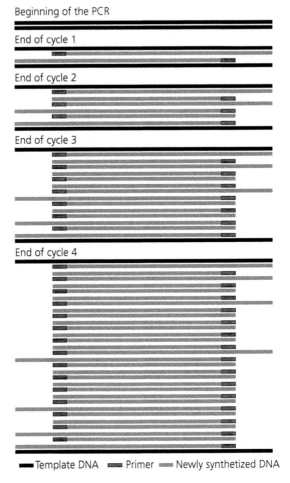

Fig. 6.2 Newly synthesized DNA strands generated during the four first cycles of a PCR.

amplification process, the DNA concentration increases exponentially, and then reaches a plateau. The number of cycles necessary to reach the plateau mainly depends on the initial concentration of target DNA in the template and on the efficiency of the PCR. A very small reduction of the efficiency, due for example to the presence of polymerase inhibitors, or to a hybridization temperature that is too high, strongly affects the number of cycles necessary to reach the plateau. After 20–30 cycles, the amount of initial template DNA in the PCR becomes negligible compared to the number of amplicons (newly synthesized DNA fragments). It is interesting to note that each extremity of the amplicons contains the primer sequences and its reverse complement, and not the initial sequence of the template DNA.

When designing a PCR experiment, the biggest constraint lies in our *a priori* knowledge of the sequences flanking the target DNA fragment to be amplified, to design the primers (see Chapter 2). Other limiting factors concern the length of the amplicons. When using non-degraded DNA as the template, it is easy to amplify up to 1,000 bp (1 kb), or even up to 4 kb in favorable circumstances (several commercial kits are dedicated to longer PCRs, up to 20–30 kb). However, if the template DNA is highly degraded, for instance when working with museum specimens, ancient DNA, or extracellular DNA from soil, the maximum size of the amplicons should not exceed 100–150 bp (excluding primers), if possible.

6.2 Which polymerase to choose?

It is now widely recognized that the amplification and sequencing steps of DNA metabarcoding produce many artifactual sequences, making it difficult to avoid overestimating the number of taxa, that is, the alpha diversity (Coissac *et al.* 2012). This overestimation led to the "rare biosphere" (Sogin *et al.* 2006) controversy, with at least part of the "rare biosphere" attributed to amplification/sequencing errors as shown by several papers (e.g., Huse *et al.* 2010; Quince *et al.* 2011; Reeder & Knight 2009). Unfortunately, even with improved algorithms for filtering the raw data, it is difficult to completely overcome this bias. Consequently, in addition to the bioinformatic treatment, several scientists have tried to use proofreading polymerases to limit the amplification errors (e.g., Mahé *et al.* 2015; Vermeulen *et al.* 2016). As well as their 5′→3′ polymerase activity, proofreading polymerases also have a 3′→5′ exonuclease activity. During the elongation step (see Fig. 6.1), if there is a mismatch (i.e., an error) at the 3′ end of the DNA strand being elongated, the proofreading polymerase removes the incorrect nucleotide, and replaces it with the correct one before continuing the replication. If the error rate of non-proofreading *Taq* polymerases is about one error per 10,000 nucleotides (Eckert & Kunkel 1990; Tindall & Kunkel 1988), the different proofreading polymerases have a 4–100 times lower error rate. It is thus tempting to use proofreading polymerases to reduce the overestimation of the alpha diversity. However, proofreading polymerases can also remove mismatches at the 3′-end of the primers, leading to non-specific amplifications, particularly when a complex template such as eDNA from soil is used. Such non-specific amplifications can be limited by the use of three to five phosphorothioate bonds between the nucleotides at the 3′-end of the primers. Indeed, if proofreading polymerases can break the normal phosphodiester bonds between nucleotides, they cannot easily break phosphorothioate bonds.

We tested a collection of non-proofreading and proofreading polymerases (Table 6.1; unpublished data). We observed that when using a proofreading polymerase with "normal" primers (i.e., only with phosphodiester bonds) and a complex template (such as eDNA extracted from soil), a very large number of non-specific PCR products is generated. Thus, switching from a non-proofreading to a proofreading polymerase without modifying the amplification protocol leads to an important loss of specificity associated with the removal of mismatches at the 3′-end of the primers. Two solutions exist to avoid this loss of specificity. The first solution, as indicated previously, involves using primers with three to five phosphorothioate bonds on the 3′-end. This significantly increases the cost of the primers, but reduces the synthesis of non-specific products to a level comparable to that obtained with non-proofreading polymerases. The second solution is the implementation of a two-step PCR, using an internal primer for the second step. In this case, the non-specific products produced during the first step are not amplified during the second step according to the specificity of the internal primer. Using such a protocol, Oliver *et al.* (2015) observed a reduction

Table 6.1 The different DNA polymerases tested for their performance in metabarcoding studies

	Polymerase	Company
Non-proofreading polymerases	AmpliTaq Gold® 360 Master Mix	Applied Biosystems™, Foster City, CA, USA
	Platinum® PCR SuperMix	Life Technologies™, Carlsbad, CA, USA
	TaqMan® Environmental Master Mix 2.0	Applied Biosystems™, Foster City, CA, USA
	Taq DNA polymerase	QIAGEN, Hilden, Germany
Proofreading polymerases	Q5® High-Fidelity DNA Polymerase	New England Biolabs, Ipswich, MA, USA
	AccuPrime™ Pfx SuperMix	Life Technologies™, Carlsbad, CA, USA
	PfuTurbo DNA Polymerase	Agilent Technologies, La Jolla, CA, USA
	PfuUltra II Fusion HS DNA Polymerase	Agilent Technologies, La Jolla, CA, USA

of 15% of the number of molecular taxonomic units (MOTUs) when using a proofreading polymerase.

For each polymerase, we carried out additional experiments using the Sper01 primer pair (Chapter 2 and Appendix 1) with a template composed of a mixture of genomic DNA from 16 plant species. In the mixture, the DNA of species A was two times more concentrated than the DNA of species B, which was itself two times more concentrated than that of species C, and so on, so that the most concentrated DNA was about 32,000 times more concentrated that the DNA with the lowest concentration. The PCR reaction was repeated eight times with each polymerase.

The results were highly reproducible for each polymerase, but were not fully consistent across the different polymerases. It is therefore important to give the name of the polymerase used in any metabarcoding experiment, because the final results can vary according to the polymerase. AmpliTaq Gold (AmpliTaq Gold® 360 DNA Master Mix, Applied Biosystems, Foster City, CA, USA) provided better results than proofreading polymerases concerning the quantitative aspect. It was interesting to note that the Accuprime Pfx systematically missed one of the rare species in the 16-species plant template. To conclude, if proofreading polymerases are chosen, it is important to use phosphorothioate primers. However, the use of these primers is much more expensive and only provides marginal improvement. It seems that hot start *Taq* polymerases represent a good compromise when considering the level of amplification artifacts, the quantitative aspects, and the cost. In a hot start PCR, a specific antibody is used to block the polymerase at low temperatures. An initial step at 95–96°C is required to remove the antibody from the active center of the polymerase. This prevents the elongation of unspecific DNA fragments during the initial temperature ramp, before the first denaturation step, and significantly reduces non-specific priming and the formation of primer dimers.

6.3 The standard PCR reaction

It is impossible to propose a versatile set of reagents and cycling parameters that will guarantee success in the wide variety of situations in which PCR is implemented. We deliberately chose to describe the standard PCR that we currently use in our laboratory when we develop a new metabarcode. This standard PCR is adapted to the *Taq* polymerase that we use (AmpliTaq Gold® 360 DNA Master Mix, Applied Biosystems), but different concentrations and parameters might be needed if another *Taq* polymerase is used. In any case, it is important to carefully read the manufacturer's instructions.

Typically, the PCR is done in a 20 µL volume, containing the following reagents:

(i) 10 µL of 2× PCR master mix (containing Tris-HCl, KCl, $MgCl_2$, each dNTP, and the polymerase);

(ii) 4 µL of DNA-free water;

(ii) 2 µL of a 2 µM solution of forward primer (final concentration: 0.2 µM);

(iii) 2 µL of a 2 µM solution of reverse primer (final concentration: 0.2 µM);

(iv) 2 µL of DNA extract (corresponding ideally to about 100 ng of template DNA);

(v) optional: 0.16 µL of Bovine Serum Albumin (Roche Diagnostics GmbH, Mannheim, Germany, Cat. No. 10 711 454 001) if we suspect the presence of PCR inhibitors (i.e., 0.04 µg per PCR reaction).

The cycling parameters involve:

(i) 10 minutes at 95°C to activate the polymerase (look at the polymerase manufacturer's instructions to see if this step is necessary and/or needs to be adjusted);

(ii) 35 cycles: denaturation at 95°C for 30 s, hybridization at 50°C for 30 s, elongation at 72°C for one minute;

(iii) storage at 4°C.

The hybridization temperature must be adjusted, usually 2–5°C below the salt-adjusted *Tm* of the primer with the lower *Tm* (estimated using OligoCalc: http://biotools.nubic.northwestern.edu/OligoCalc.html). This point is very important: for example, O'Donnell *et al.* (2016) showed a strong bias according to the sequence of the tag used on the 5'-end of the primers. Through careful reading of their protocol, it can be suspected that the bias was induced by a hybridization temperature that was too high. The elongation time must also be adjusted according to the length of the PCR product, at least one minute in

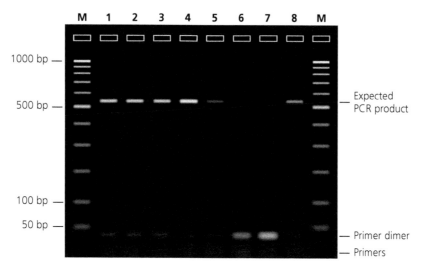

Fig. 6.3 Schematic results of a typical PCR experiment on an agarose gel. Lanes M, molecular weight markers; lanes 1 to 5, samples; lane 6, DNA extraction negative control; lane 7, PCR negative control; lane 8, positive control.

DNA metabarcoding, even if the amplified fragment is very short. This prevents occurrence of single-stranded DNA that interferes with the subsequent library preparation and sequencing, leading to artifactual sequences. The number of cycles must be adjusted according to the amount of template DNA, from 25 cycles, if about 10,000 target DNA copies are present in the template DNA, to 45 cycles if less than ten target DNA copies are present. Figure 6.3 shows the results of a typical PCR on an agarose gel.

6.4 The importance of including appropriate controls

Including multiple appropriate controls in any DNA metabarcoding experiment is crucial for monitoring potential contamination and for interpreting the results correctly. Five types of controls can be distinguished. In our typical experimental design, we systematically include 10–20% of controls.

6.4.1 Extraction negative controls

An extraction negative control corresponds to a mock DNA extraction carried out at the same time and using the same consumables as a regular extraction, except that the sample (i.e., the source of DNA) is omitted. Including such an extraction negative

control within each batch of extractions is highly recommended. The DNA extract originating from the mock extraction is then amplified, sequenced, and analyzed as regular DNA extracts. The results obtained from the extraction negative controls, when compared to the results of the PCR negative controls (see Section 6.4.2), allows for the identification of potential contaminants during the extraction process. When using universal primers to amplify Bacteria, eukaryotes, or plants (e.g., using primers Bact01, Euka02, or Sper01, see Chapter 2) and when pushing the PCR for the detection of highly diluted templates, it is difficult to avoid sporadic contamination as the consumables from the extraction or PCR commercial kits can contain a few molecules of contaminant DNA (Leonard *et al.* 2007; Willerslev & Cooper 2005). Common sporadic contaminants in DNA metabarcoding experiments come from animals (e.g., human, pig, cow, chicken) or from plants (e.g., potato, tomato, rice, banana).

6.4.2 PCR negative controls

A PCR negative control is a PCR where the addition of template DNA (i.e., DNA extract) is replaced by the addition of water (the same DNA-free water that is used for diluting the PCR reagents). This allows for the detection of potential contamination arising at the amplification step. However, negative results (i.e., no amplification) in the PCR negative controls do not necessarily mean that there are no contaminants (Kolman & Tuross 2000). This is due to the carrier effect: contaminant DNA in low concentration may be unavailable for amplification due to adherence to a plastic tube or to tips; if additional DNA is added (i.e., template DNA from the samples) the few contaminant molecules will be in competition with non-contaminant DNA for adherence and they become amplifiable (Cooper 1994; Malmström et al. 2005). At the data analysis stage, the filtering strategy can be adjusted to discard sequences observed in these controls (see e.g., De Barba et al. 2014 and Chapter 8).

6.4.3 PCR positive controls

In DNA metabarcoding studies, positive controls are often overlooked, even though they can greatly help at the data analysis stage (De Barba et al. 2014; Lopes et al. 2015). The ideal positive control should be comparable to the analyzed samples (i.e., it should have a similar DNA concentration and complexity). In practice, such a positive control is a mock community, preferably composed of species that do not appear in the actual study. For example, if the target species are eutherian mammals from Europe, the PCR positive control should include either marsupials, or eutherian mammals from other regions of the globe. At the data analysis stage, positive controls will greatly help to adjust the filtering strategy with the aim of limiting the impact of PCR and sequencing errors, and thus reducing the usual overestimation of the alpha diversity (Coissac et al. 2012). As negative controls, they can also be used to detect PCR contaminants as they are less sensible to the DNA carrier effect.

6.4.4 Tagging system controls

Due to the considerable number of samples involved in DNA metabarcoding studies, it is generally necessary to implement a tagging system where each sample displays a unique combination of forward and reverse forward or reverse tags. When several samples share the same forward or reverse tags (see Section 6.10.4 for more details), it is important to include a few unused tag combinations in the experimental design. By looking at the sequences attributed to these unused tag pairs during sequence analysis, it is possible to assess the occurrence of chimera among samples (i.e., tag jumps; Schnell et al. 2015a), and thus to filter the whole data set accordingly.

6.4.5 Internal controls

An internal control is a template DNA added into the PCR mix at a low concentration to produce a small percentage of the final PCR product, to act as a positive control (see e.g., Zhou et al. 2011). Alternatively, the template DNA can be added to the sample before DNA extraction. The added template must not interfere with the experiment (i.e., it should not generate a PCR fragment that can be confused with a true MOTU). Thus, it can be composed of genomic DNA from an organism occurring in another geographic area or of a PCR product containing the target sequences of both primers on the two extremities. Such a PCR product can be obtained either from genomic DNA, or from synthetic oligonucleotides (with an arbitrary sequence between the two primers). To date, internal controls have only been used marginally in metabarcoding studies.

Using internal controls is particularly useful for estimating the relative amount of target DNA in the different samples by comparing the ratio between target DNA and internal control, even if the amount of PCR inhibitors varies among samples. Furthermore, the impact of potential contaminants can be evaluated by comparing the ratio between contaminants and internal control in the PCR negative controls. This means that it might be advisable to run two types of PCR negative controls (i.e., with and without an internal control).

6.5 PCR optimization

Each new PCR experiment is likely to require optimization. Adjusting a PCR always involves a compromise between specificity and efficiency. When starting with the concentrations recommended by the manufacturer, an increase in magnesium, nucleotides, primers, or enzyme concentrations, as well as a reduction of the hybridization temperature, will favor efficiency of the reaction. Inversely,

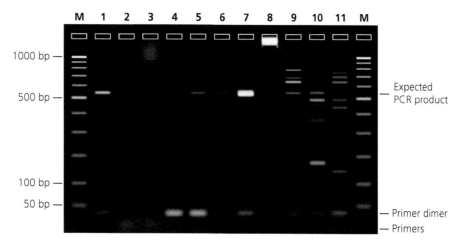

Fig. 6.4 Schematic results of PCR experiments illustrating some of the most common problems (see Table 6.2 PCR troubleshooting). Lanes M = molecular weight markers; lane 1 = positive PCR (presence of the expected PCR product); lane 2 = PCR failure (no PCR product, no primer dimer); lane 3 = PCR failure (only a light smear of high molecular weight); lane 4 = PCR failure (only primer dimer); lane 5 = PCR to be adjusted (primer dimer stronger than the expected PCR product); lane 6 = PCR to be adjusted (expected PCR product, but too faint); lane 7 = PCR to be adjusted (expected PCR product, but too strong); lane 8 = PCR to be adjusted (strong product of very high molecular weight); lane 9 = PCR to be adjusted (expected PCR product, but with unspecific bands of higher molecular weight); lane 10 = PCR to be adjusted (expected PCR product, but with unspecific bands of lower molecular weight); lane 11 = PCR to be adjusted (only presence of unspecific bands).

a decrease in magnesium, nucleotides, primers, or enzyme concentrations, along with an increase of the hybridization temperature, will favor specificity.

If the PCR products obtained look like those presented in Figure 6.3, no further adjustments are necessary. However, after a first experiment, the results might be different; there might not be any PCR products, or the products might be unspecific (see Fig. 6.4 for examples). In this case, the first thing to do is to repeat exactly the same experiments, in order to see if the results are repeatable. Indeed, many

PCR failures are due to missing reagents, pipetting errors, or errors in programming the PCR machine. Only then, one can begin to adjust the PCR by changing some parameters. We do not recommend altering more than one parameter at a time. The first two parameters to adjust are the annealing temperature and number of PCR cycles.

The potential causes leading to a PCR failure are numerous, and we do not expect to provide solutions for them all. Table 6.2 deals with the most common problems and their likely solutions. When

Table 6.2 PCR troubleshooting. Obtaining a clean PCR product might require several steps of adjustment. Here we provide the PCR parameter modifications that might solve the most common problems encountered when trying to adjust a new PCR, starting with the recommendations for the standard PCR protocol presented in this chapter. The potential problems can be identified after a careful examination of the agarose gel showing the PCR results

Problems	Solutions
No DNA visible at all on the agarose gel (but no positive control implemented, lane 2 of Fig. 6.4).	– Repeat the same experiment, but with a positive control (if possible, also with another combination of primer pair and template known to work under these conditions).
No DNA visible at all on the agarose gel, including the absence of PCR products in the positive control (a PCR positive control that worked previously with another primer/ template combination under the conditions used).	– Carefully check the cycling parameters of the PCR machine. – Repeat the experiment on another PCR machine. – Change the PCR reagents one by one, and replace them with new ones (dNTPs, primers, buffer, $MgCl_2$, template DNA, and so on).

(continued)

Table 6.2 *Continued*

Problems	Solutions
No DNA visible for the PCR being adjusted, but with the expected band in the positive control.	– Decrease the hybridization temperature by up to 10°C. – Additionally, increase the MgCl$_2$ by up to 2 mM.
No expected PCR product, no primer dimer, but a light smear of high molecular weight (lane 3 of Fig. 6.4).	– Dilute the template DNA by 10, 100, or even 1,000, and then repeat the experiment.
No expected PCR product, but with a clear primer dimer (lane 4 of Fig. 6.4).	– Decrease the hybridization temperature by up to 10°C. – Additionally, increase the MgCl$_2$ concentration by up to 2 mM.
Faint band corresponding to the expected product, but strong primer dimer (lane 5 of Fig. 6.4).	– Decrease the primer concentrations down to 0.2 μM.
Faint expected band (lane 6 of Fig. 6.4).	– Increase the number of cycles. – Decrease the hybridization temperature (by 1–5°C). – Increase the MgCl$_2$ concentration by up to 2 mM.
Very strong expected band (lane 7 of Fig. 6.4).	– Decrease the number of cycles. – Decrease the MgCl$_2$ concentration.
Strong product of very high molecular weight (lane 8 of Fig. 6.4).	– Decrease the number of cycles.
Expected band, but with presence of unspecific bands of higher molecular weight (lane 9 of Fig. 6.4).	– Increase the hybridization temperature. – Decrease the elongation time. – Decrease the MgCl$_2$ concentration.
Expected band, but with presence of unspecific bands of lower molecular weight (lane 10 of Fig. 6.4).	– Increase the hybridization temperature. – Increase the elongation time. – Increase the MgCl$_2$ concentration.
Absence of the expected band, but with presence of unspecific bands (lane 11 of Fig. 6.4).	– Decrease the hybridization temperature (this is counter-intuitive, but the problem might be because only a single primer is able to anneal; decreasing the hybridization temperature might allow the second primer to anneal). – Decrease the MgCl$_2$ concentration.
No possibility of adjusting the PCR with the solutions provided here.	– Modify the primer pair (sometimes, shifting one of the primers of a few base pairs can solve the problem). – Target another DNA fragment.

a problem occurs, the first step should be a meticulous observation of the PCR products on agarose gel. One of the most common mistakes when performing PCR experiments for the first time is adding too much template DNA to each reaction. Such a mistake will lead, in the best-case scenario, to the amplification of unspecific products, and in the worst case, to a complete inhibition of the PCR due to the presence of a high concentration of inhibitors.

The dilution of the extracts and the number of PCR cycles can be adjusted based on quantitative PCR (qPCR) experiments (see Chapter 9). Ideally, the dilution factor, as well as the number of PCR cycles could be adjusted differently for each extract, but this becomes very difficult in practice if there are

many samples. If a representative number of samples is tested by qPCR, it is usually possible to find the best compromise for the dilution and number of cycles (see Fig. 6.5 for an example of qPCR results).

6.6 How to limit the risk of contamination?

The risk of contamination is a serious threat to any PCR, as it can result in the production of false positives (Kwok & Higuchi 1989). This is due to the extreme sensitivity of this reaction: in some circumstances, it is possible to amplify a single target molecule. There are three types of contaminations. First, a

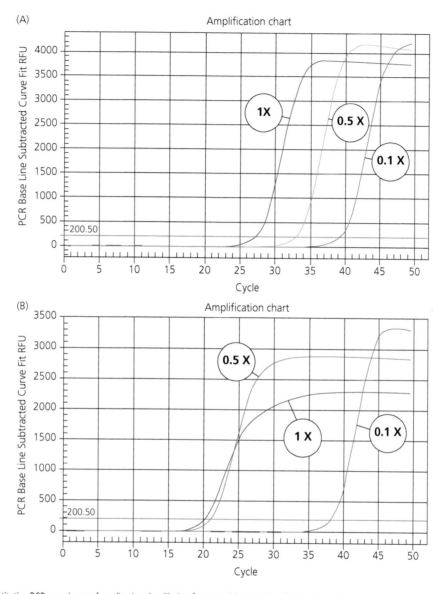

Fig. 6.5 Quantitative PCR experiments for adjusting the dilution factor and the number of cycles in metabarcoding experiments on soil samples extracted using a phosphate buffer approach (Taberlet *et al.* 2012c): A, example of results where both the initial amount of template DNA and the level of PCR inhibitors are low; B, example of results where both the initial amount of template DNA and the level of PCR inhibitors are high. Based on the A and B results, if the dilution and the number of PCR cycles must be the same for both samples, we would recommend a 10-times dilution (0.1×) and 45 cycles.

DNA fragment generated by a previous PCR can contaminate the reaction. This type of contamination has been termed "PCR product carryover" and represents the most common contamination type. Second, a false positive can be caused by a sample-to-sample contamination, if the target sample is more degraded than the contaminant sample, or if the primers match the contaminant better than the target sample. Finally, the contaminant can originate from the consumables used and is quite difficult to trace, as the contamination can be sporadic and requires many negative controls to be detected.

Generally, the best strategy for avoiding and monitoring contaminants is to assume that the obtained PCR product is a false positive, and to establish an experimental protocol that demonstrates that this is not the case. Such an experimental protocol must obviously involve strong measures for limiting contaminations. It must also allow for the detection of contamination when it occurs, and should be adjusted according to the risk. Amplifying DNA from bulk tissue samples does not require the same level of stringency as amplifying DNA from lake sediments or permafrost samples, however. The false positive risk strongly depends on the number of cycles that are performed, as more cycles increase the probability of amplification of a contaminant sequence.

Following some simple rules can help minimize the contamination risk. The complete physical separation (i.e., in different rooms) of the pre-PCR and the post-PCR areas is extremely important. Different laboratory coats and gloves must be used in these two areas. As a rule, nothing should go from the post-PCR area into the pre-PCR area. All reagents used for the PCR reaction must be stored in the pre-PCR area. The PCR machines can be either in the post-PCR area, or in a special room dedicated only to PCR machines. We advise setting up different pre-PCR areas if samples with varying levels of DNA quality and concentration are commonly used in the laboratory. For example, in our laboratory, we work with tissue samples, soil samples, hair or feces samples, fossil bones, lake sediments, and permafrost samples. Accordingly, we set up four different pre-PCR areas (tissue, soil, hair and feces, ancient DNA), each with a DNA extraction area, and a PCR set-up area. If the risks of contamination are high for a sample type (i.e., ancient DNA), working in the post-PCR area before extracting DNA or setting up a PCR reaction for these types of samples later in the day is not recommended.

Pipettes are the main source of contamination. Each set of pipettes must be restricted either to the pre-PCR area or to the post-PCR area. Even in the pre-PCR area, we recommend dedicating a pipette set to the DNA extraction procedure, another set to the preparation of the PCR mix, and a single pipette to the addition of the DNA extract to each tube. Despite the cost, we highly recommend

always using filter tips, not only when working with stock solutions, but also for the whole metabarcoding procedure, from the DNA extraction to the PCR set-up.

Good laboratory practices, such as changing gloves frequently, systematically centrifuging tubes before opening them, uncapping and closing tubes carefully to prevent aerosols and contamination of gloves, minimizing sample handling, and adding the DNA extract at the end of the PCR set-up, should minimize the contamination risk. Although PCR product carryover represents the main contamination source, cross-contamination between samples can also lead to problems, and is in a way even more difficult to detect. Therefore, all aspects of the sample handling, from the sampling in the field to the PCR set-up, must be carried out bearing in mind this cross-contamination risk. Finally, the systematic use of appropriate negative controls should allow for the detection of most contaminants (see Section 6.4).

6.7 Blocking oligonucleotides for reducing the amplification of undesirable sequences

There are several situations where it is suitable to reduce the amplification of some abundant sequences to reveal the rarest ones. This can be obtained via the addition of a PCR blocking oligonucleotide (Vestheim & Jarman 2008). For example, when studying the diet of carnivores using feces as a source of DNA, the primer pair often also amplifies the predator DNA, which can represent a very high proportion (up to 100%) of the obtained amplicons. In this situation, the use of a blocking oligonucleotide is common (e.g., Deagle et al. 2009; De Barba et al. 2014; McInnes et al. 2016; Shehzad et al. 2012a, 2012b, 2015). Another situation where it is useful to use blocking oligonucleotides is the amplification of highly diluted templates extracted from ancient material, such as lake sediments or permafrost samples (Boessenkool et al. 2012). In this case, the problem comes from the common contaminants in PCR reagents that can prevent the amplification of DNA from the sample. For example, if the primers used target mammals, adding blocking oligonucleotides for human, cow, and pig is recommended. Boessen-kool *et al.* (2012) clearly showed, using permafrost samples, that the application of blocking human oligonucleotides allows the amplification of three species (moose, reindeer, and woolly rhinoceros) that were not detected without the human blocker.

Figure 6.6 illustrates the design of blocking oligonucleotides. It is important that the blocking oligonucleotide overlaps with one of the amplification primers. We usually target an overlap of at least six nucleotides. It is better to put the blocking oligonucleotide on the extremity of the PCR fragment where there is more variation between the sequence of the species to be blocked and the sequences of the target species. According to our experience, it is not a problem if the blocking oligonucleotide is long (up to 40 bases). The optimization of the PCR involving a blocking oligonucleotide mainly concerns the concentration of the blocker. A concentration that is too high can lead to a complete failure of the PCR. In practice, we observed that a concentration 10 times higher than that of each primer is a good compromise (0.2 µM for the PCR primers, and 2 µM for the blocking oligonucleotide). To prevent the 3'-extension of the blocking oligonucleotide, the best solution is to add a C3 carbon spacer during the synthesis (Vestheim & Jarman 2008). This C3 carbon spacer does not influence the hybridization with the target sequence.

6.8 How many PCR replicates?

Besides biological replicates at the sampling step, and technical replicates at the extraction step, it is possible to perform additional technical replicates at the amplification step. Increasing the number of replicates leads to a much better evaluation of potential methodological biases, and reduces uncertainties during data analysis (see e.g., Prosser 2010; Searle *et al.* 2016). As the stochasticity of individual PCR reaction is well known, technical PCR replicates have the potential to improve the accuracy of the metabarcoding results.

The first objective when performing several PCR replicates is to reduce the impact of the PCR stochasticity. This could be done by pooling, before sequencing, the PCR products from several reactions based on the same DNA extract. Recovering the results of each individual PCR later on during data analysis will unfortunately no longer be possible with this procedure (see e.g., Schmidt *et al.* 2013). This commonly used strategy assumes that the PCR noise is unbiased, which is incorrect (Coissac *et al.* 2012). By considering PCR replicates independently, it becomes possible to assess the consistency of the PCR signal among replicates and to eliminate the divergent ones (see e.g., De Barba *et al.* 2014). This sometimes leads to the complete rejection of some samples because of the inconsistency among all their replicates (see details in Chapter 8).

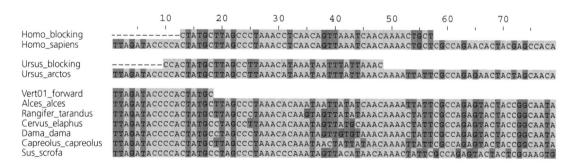

Fig. 6.6 Design of two blocking oligonucleotides for human and brown bear for the Vert01 primer pair (Chapter 2), allowing the amplification of target ungulate species. This example corresponds to the De Barba *et al.* (2014) study that analyzed brown bear diet using feces as a source of DNA. Homo_blocking, sequence of human blocking oligonucleotide (a C3 spacer must be added on the 3'-end). Homo_sapiens = human sequence. Ursus_blocking = brown bear blocking oligonucleotide (a C3 spacer must be added on the 3'-end). Ursus_arctos = brown bear sequence. Vert1_forward = sequence of the forward primer. *Alces_alces, Rangifer_tarandus, Cervus_elaphus, Dama_dama, Capreolus_capreolus,* and *Sus_scrofa* correspond to sequences of potential prey (moose, reindeer, red deer, fallow deer, roe deer, and wild boar, respectively).

The analysis of PCR replicates offers several other opportunities: that of increasing the detection probability in situations where the DNA extracts are highly diluted (Giguet-Covex *et al.* 2014; Pansu *et al.* 2015b; Valentini *et al.* 2016); that of assessing species presence probability (Ficetola *et al.* 2016); and that of estimating relative abundances by counting the number of positive replicates for each MOTU. When analyzing plants or vertebrates from lake sediments or from freshwater, we routinely run 8–12 PCR replicates per sample (Giguet-Covex *et al.* 2014; Pansu *et al.* 2015b).

6.9 Multiplexing several metabarcodes within the same PCR

To be multiplexed, the different metabarcodes should have comparable sizes, and their primers should have similar *Tm*, in order to perform the annealing steps of the PCR at the same temperature. Possible cross-hybridization among primers must be checked (see https://www-s.nist.gov/dnaAnalysis/primerToolsPage.do), to avoid generating primer dimers, and to ensure maximal efficiency of the amplification. The relative amount of each of the different markers multiplexed within the same PCR reaction can be adjusted by modifying the concentration of the different primer pairs. Increasing the concentration of a primer pair will increase the relative contribution of the relevant marker to the final PCR product. The suitable final concentrations of the different primers in the PCR reaction can vary from 0.1 to 0.5 μM.

Generally, we try to avoid multiplexing several metabarcodes within the same PCR as the adjustment phase can be complex, and some additional artifacts due to the multiplexing itself can be generated. Multiplexing can be useful when the amount of DNA extract is limited, and DNA concentration is low. In this situation, it is important to run many PCR replicates, and multiplexing helps obtain more information from the DNA extract. A successful multiplexing example can be found in De Barba *et al.* (2014).

6.10 Multiplexing many samples on the same sequencing lane

Here, we only discuss the multiplexing strategy for Illumina sequencers. However, the basic concepts presented can be easily adapted to other next-generation sequencing platforms. Note that next-generation sequencers have been initially designed for sequencing genomic DNA, therefore protocols are not optimized for amplicon sequencing. To properly understand this section, it is advisable to first read Chapter 7 dealing with next-generation sequencing.

6.10.1 Overview of the problem

Figure 6.7 illustrates the structure of a DNA fragment corresponding to an amplicon that can be sequenced on Illumina platforms. The amplicon itself is flanked by two different adapters. The external part of each adapter is complementary to oligos that are bonded to the sequencing surface of the flow cell. The internal part of each adapter is complementary to the two sequencing primers for reading the first and the second strands, and to the primer(s) for reading the indexes.

On Illumina sequencers, the smallest experiment that can be carried out produces eight million reads (mid-output kit on MiniSeq). Obviously, for DNA metabarcoding, it is not necessary to have such a high number of reads per PCR, and many samples can be combined per sequencing experiment to reduce costs. For example, a total of 800 PCRs can be analyzed on the same sequencing run using the mid-output kit on MiniSeq, with 10,000 reads per reaction. Different options can be implemented to load many PCR products within the same sequencing lane.

The first option involves taking advantage of the multiplexing system of the sequencer itself, based on different 6–8 base indexes that are inserted within the P7 adapter. The basic system consists of 24 such indexes. To load more than 24 PCR products within the same sequencing lane, it is possible to either combine this first index on P7 with another one on P5, or to use longer indexes leading to more possibilities (e.g., Caporaso *et al.* 2012; Hamady *et al.* 2008). With such a system, it is possible to properly analyze 96, or more, different PCRs together. A second option is to add different sequence tags on the 5'-end of the two amplification primers (Binladen *et al.* 2007; Valentini *et al.* 2009). This allows the multiplexing of several hundred, or even thousand, PCRs within the same sequencing lane. Finally,

Structure of a DNA fragment that can be sequenced on Illumina platforms

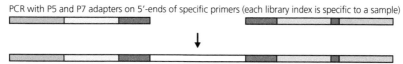

Strategy 1: single-step PCR with Illumina adapters

PCR with P5 and P7 adapters on 5′-ends of specific primers (each library index is specific to a sample)

Strategy 2: two-step PCR with Illumina adapters

Step 1: PCR with a linker on the 5′-end of the primers

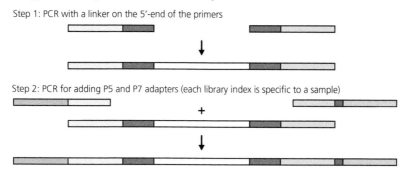

Step 2: PCR for adding P5 and P7 adapters (each library index is specific to a sample)

Strategy 3: single-step PCR with tagged primers

PCR with tags and spacers on the 5′-end of the primers (each tag combination is specific to a sample)

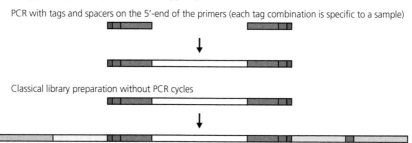

Classical library preparation without PCR cycles

Fig. 6.7 Structure of DNA fragments to be sequenced on Illumina sequencing platforms, with the three different strategies for multiplexing several samples on the same sequencing lane (see text for further explanations). Another index can be inserted on the P5 adapter.

these two options can be combined, resulting in the simultaneous sequencing of a very large number of PCR products, corresponding to the number of indexes of the first option multiplied by the number of tag combinations of the second option.

Amplicon sequencing imposes additional constraints when compared to sequencing of genomic DNA. First, the identification of the different clusters on the sequencing surface is based on the first four to five nucleotides sequenced (four nucleotides on HiSeq, five on MiSeq). Therefore, for reliable cluster identification, it is important that these first sequenced nucleotides are variable enough. This is not the case when sequencing amplicons, if the first four to five nucleotides are on the 5′-end of the two amplification primers. A second constraint is directly linked to the fact that the Illumina sequencers are designed for sequencing genomic DNA, and thus expect the same proportion of C versus G, and of A versus T, as well as the same CG/AT ratio during each sequencing cycle. Obviously, this is not the case when sequencing amplicons. During the raw data analysis, the program uses the first 25 first nucleotides (the first 25 sequencing cycles) for adjusting its base calling. Again, when sequencing amplicons, the fact that these ratios are different among these first sequencing cycles alters the reliability of the base calling algorithm. Fortunately, there are solutions to overcome these issues. To allow a better cluster identification, it is advisable to add a few variable nucleotides on the 5′-end of each amplification primer. To improve base calling, first up to 10–30% of PhiX can be loaded together with the amplicons to increase the diversity, and second, different numbers of variable nucleotides can be added to the 5′-end of the primers (Wu *et al.* 2015; see Section 6.10.4 for examples).

6.10.2 Strategy 1: single-step PCR with Illumina adapters

This first strategy consists of carrying out the DNA amplification with very long primers, containing the specific primers on the 3′-end, and the P5 and P7 adapters on the 5′-end (Fig. 6.7). Such primers are typically 75–95 bp long, depending on whether or not tags and spacers of variable length are added onto the 5′-end of the specific primers. The main advantage of this strategy is its simplicity (i.e., a single PCR with a single primer pair). Furthermore, the resulting fragment is ready to be loaded on Illumina sequencers after purification, and there is no possibility of tag jumps if tags are used on the 5′-end of the primers (Schnell *et al.* 2015a), because samples are amplified independently. Tag jumps are due to the artifactual production of chimeras during the library preparation step necessary prior to sequencing on next-generation sequencers (see Fig. 6.7), leading to tag exchange between two sequences originating from different samples (see Schnell *et al.* 2015a for detailed explanations). However, this strategy has several drawbacks. First, it is only easy to implement if the template DNA is relatively highly concentrated. When there is a low number of template DNA copies, the amplification process begins to have difficulties due to the much lower mobility of long primers when compared to short ones. Another drawback is the possibility of index jump during the sequencing itself. This refers to the assignment of a sequence read to the wrong index (i.e., to the wrong library) if several other indexes corresponding to the same metabarcode are present within the same sequencing lane. This occurs during the index sequencing when the signal from a cluster is particularly strong, and another nearby cluster exhibits a weaker signal. The sequence of the cluster with the weaker signal might then be incorrectly assigned to the index of the cluster with the strong signal. Usually, on Illumina sequencers, about 0.1% of the sequences are misassigned due to this phenomenon. Such an artifact might lead to wrong ecological conclusions if the presence of very few reads are interpreted as the presence of the associated taxon. It might be the explanation for the surprising results that "Central Park soils harboured nearly as many distinct soil microbial phylotypes . . . as we found in biomes across the globe (including arctic, tropical and desert soils" (Ramirez *et al.* 2014). According to the information provided in the Materials and Methods section of this paper, it is not possible to exclude that at least a part of such a prominent level of overlap between Central Park and the rest of the world does not come from index jumps. A potential solution for testing this artifact hypothesis would be to reload all the Central Park samples and all the samples from around the world separately, on different sequencing lanes to avoid index jumps among samples. Several variants of this first strategy have been implemented. Caporaso *et al.* (2012) used sequencing primers that correspond to the specific primers plus 12 nucleotides on the 5′-end. This significantly reduces the length of the long amplification primers (60 and 68 bp for the forward and reverse primers, respectively), and the forward specific primers used for the PCR is conveniently not sequenced, leading to longer reads within the metabarcode. However, this approach does not use the standard sequencing primers and thus the standard sequencing protocol. In this context, it is more difficult to outsource the sequencing. Another variant involves using the same index for all samples, but adding tags of different lengths on the 5′-end of the specific primers (Elbrecht & Leese 2015). However, with this solution, the cluster detection might not be optimal, as the diversity of the first four to five sequenced nucleotides is reduced when compared to random nucleotides.

6.10.3 Strategy 2: two-step PCR with Illumina adapters

This second strategy is recommended by Illumina for carrying out metabarcoding experiments. It involves two consecutive PCRs: a first with specific primers tailed on their 5′-end by a 24–25 bp linker (corresponding to the priming sites of the forward and reverse sequencing primers), and a second aiming to add both (P5 and P7) flow cell binding sites, as well as indexes (Fig. 6.7). In addition to the index on the P7 side, an optional index can also be added on the P5 side (double indexing) to increase the number of samples analyzed simultaneously within the same sequencing lane (see Fonseca & Lallias 2016 for a detailed protocol). When compared to the two other strategies, the cost of the primers is significantly reduced, as the same indexed primers can be used for different metabarcoding markers during the second PCR (provided that the 5′-tails of the specific primers are identical). When compared to the first strategy, the length of the primers for the first PCR is about 45 bp, corresponding to a significant length reduction, allowing for a much better amplification of low copy number templates. The disadvantage of such a two-step procedure is the

contamination risk with PCR products originating from the first PCR. Usually, such contaminations do not occur during pilot experiments or at the beginning of the implementation of such a two-step PCR, but later on during routine experiments. The relatively limited number of samples that can be run in the same sequencing lane represents another disadvantage, even when using a double-indexing approach. The last potential problem is the index jump artifact, as for the first strategy.

6.10.4 Strategy 3: single-step PCR with tagged primers

This approach involves a single PCR with relatively short primers of about 30 nucleotides, with a subsequent step consisting of the ligation of Illumina adapters (i.e., library preparation). The main advantage of the use of short tagged primers is its simplicity and the efficiency of the amplification, even when starting with low copy number templates. This strategy is particularly suitable when working with highly diluted templates, as can be the case with some types of eDNA samples. There are two alternative ways of tagging the different PCR products. The first possibility entails using all the possible tag combinations, with one tag combination per PCR product (Fig. 6.8). It is adapted particularly when the number of PCR products is high. For example, by using 36 different forward primers, and 32 different reverse primers, it is possible to theoretically tag, in the same library, the equivalent of twelve 96-plates (i.e., 1,152 PCR products or controls). Table 6.3 provides a list of 36 tags of eight nucleotides, with at least five differences between each of them. These five differences make the confusion of two tags due to sequencing errors almost impossible. Using all tag combinations is interesting not only due to the fact that it reduces the primer cost, but it also prevents the detection of tag jumps during the library preparation. Therefore, a careful design of the experiment is necessary to estimate the extent of the tag jump problem. Some tag combinations should not be used in the experiment (= blanks, but see Esling *et al.* 2015 for a more complex design). We suggest having at least one blank per forward tag, and one per reverse tag (i.e., one per line and one per column of the

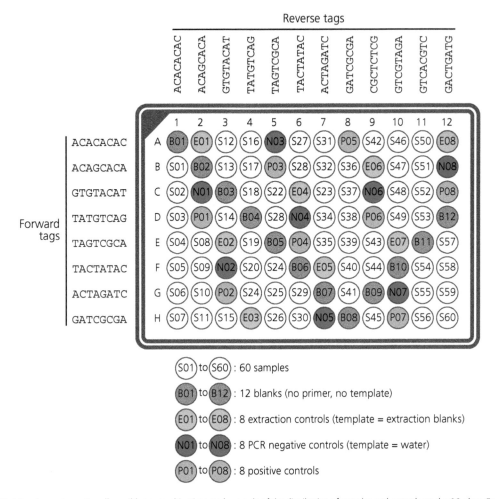

Fig. 6.8 A tagging system using all possible tag combinations and example of the distribution of samples and controls on the 96-plate. Each tag is distributed on the relevant line and column of the plate. For example, the S13 sample has the ACAGCACA forward tag and the GTGTACAT reverse tag. Table 6.4 gives an example of the 20 primers that can be ordered to tag a single 96-plate. The same tagging system can be used to tag up to twelve 96-plates using the 36 tags given in Table 6.3, with 32 forward primers, and 36 reverse primers. To properly assess the extent of tag jumps, it is advisable to not use some tag combinations (i.e., to perform blanks), with at least one such blank per line and per column (but see Esling *et al.* 2015 for a more complex design). It is also important to have several controls (extraction negative controls, PCR negative controls, and positive controls) that will help at the filtering stage of data analysis (see Chapter 8).

experimental design; Fig. 6.8). Table 6.4 illustrates the sequences of the Sper01 primers (Appendix 1) that can be used to tag a single 96-plate. The second possibility for tagging the different PCR products is more restrictive and corresponds to the particular case of the first possibility, with the forward and reverse tags used only once for a single PCR or control. This possibility is not appropriate when the number of samples is high, because the primer cost is prohibitive (192 primers must be ordered

for tagging a single 96-plate). Appendix 2 gives a list of 384 tags of eight nucleotides, with at least three differences between each of them. These tags can be used to label samples from a maximum of four 96-plates. Note that it is not necessary to leave blanks, as the sequence reads with an unused tag combination can be eliminated at the data analysis stage. However, even when using this restrictive tagging strategy, the assignment of a sequence read to a wrong sample is still possible, but at a

Table 6.3 List of 36 eight-nucleotide tags allowing the tagging of twelve 96-plates. Each tag differs from the others by at least five differences. A text file containing these tags can be downloaded from www.oup.co.uk/companion/taberlet

acacacac	gtcgtaga	acgacgag	atatagcg
acagcaca	gtcacgtc	catcagtc	ctatgcta
gtgtacat	gactgatg	atcagtca	tcgcgctg
tatgtcag	agactatg	tctactga	agcacagt
tagtcgca	gcgtcagc	gatgatct	tagctagt
tactatac	tgacatca	ctgcgtac	agtgctac
actagatc	acatgtgt	agcgacta	cgtataca
gatcgcga	gtacgact	tcagtgtc	cgagtcgt
cgctctcg	atgatcgc	actctgct	cacatgat

Table 6.4 Example of the tagged Sper01 primers (see Appendix 1) that can be ordered for tagging a single 96-plate, using the first 12 tags of Table 6.3 and corresponding to the design presented in Figure 6.8. The specific parts of the primers are in capital letters. The tags are in lowercase, and the 5′-end has two, three, or four random nucleotides to improve the detection of the different clusters on the sequencing surface and to break the sequencing frame for increasing the quality of the base calling. This primer design corresponds to the third strategy (Section 6.10.4). However, the P5 and P7 adapters must be added to the 5′-end of the forward and reverse primers for implementing the first strategy (Section 6.10.2)

Code	Primer sequence (5′–3′)
Sper01_F_01	NNacacacacGGGCAATCCTGAGCCAA
Sper01_F_02	NNNacagcacaGGGCAATCCTGAGCCAA
Sper01_F_03	NNNNgtgtacatGGGCAATCCTGAGCCAA
Sper01_F_04	NNtatgtcagGGGCAATCCTGAGCCAA
Sper01_F_05	NNNtagtcgcaGGGCAATCCTGAGCCAA
Sper01_F_06	NNNNtactatacGGGCAATCCTGAGCCAA
Sper01_F_07	NNactagatcGGGCAATCCTGAGCCAA
Sper01_F_08	NNNgatcgcgaGGGCAATCCTGAGCCAA
Sper01_R_01	NNacacacacCCATTGAGTCTCTGCACCTATC
Sper01_R_02	NNNacagcacaCCATTGAGTCTCTGCACCTATC
Sper01_R_03	NNNNgtgtacatCCATTGAGTCTCTGCACCTATC
Sper01_R_04	NNtatgtcagCCATTGAGTCTCTGCACCTATC
Sper01_R_05	NNNtagtcgcaCCATTGAGTCTCTGCACCTATC
Sper01_R_06	NNNNtactatacCCATTGAGTCTCTGCACCTATC
Sper01_R_07	NNactagatcCCATTGAGTCTCTGCACCTATC
Sper01_R_08	NNNgatcgcgaCCATTGAGTCTCTGCACCTATC
Sper01_R_09	NNNNcgctctcgCCATTGAGTCTCTGCACCTATC
Sper01_R_10	NNgtcgtagaCCATTGAGTCTCTGCACCTATC
Sper01_R_11	NNNgtcacgtcCCATTGAGTCTCTGCACCTATC
Sper01_R_12	NNNNgactgatgCCATTGAGTCTCTGCACCTATC

very low rate corresponding to two tag jumps, one on each side of the same DNA fragment. Note that tailor-made collections of tags can be designed using the `oligotag` software (Coissac 2012). When choosing the single-step PCR with tagged primers, the library preparation is not as straightforward as for genomic DNA, due to the tag jump problem (Schnell *et al.* 2015a). To limit this problem, along with the experimental design explained here, we recommend using a library preparation protocol that does not include any PCR cycles and to possibly omit the T4 DNA polymerase step to obtain blunt-end fragments (see Schnell *et al.* 2015a for details). The MetaFast protocol developed by Fasteris (https://www.fasteris.com/metafast) for building libraries from tagged amplicons strongly limits the occurrence of tag jumps. Finally, the index jump also represents a potential problem with this third strategy. However, it is easier to monitor, as many PCR products can be pooled within the same library (i.e., with the same index), and the solution is to load a single index for each metabarcode within the same sequencing lane. Thus, if many libraries have to be analyzed for the same metabarcode, it is advisable to load each of them into a different sequencing lane if possible. If it is not possible,

the level of index jump can be estimated by running two different metabarcodes per sequencing lane, and by looking at the second metabarcode in a library originally containing only the first metabarcode. Assessing the extent of index jump in the experiment allows the data to be filtered in an appropriate way, so as to limit false positives.

CHAPTER 7

DNA sequencing

The emergence of eDNA analysis is mainly linked to the development of next-generation sequencing. In this chapter, we will briefly provide basic information about the associated technologies. For a more detailed presentation, and as the technologies develop over time, we recommend consulting the websites of the corresponding companies. Note that many excellent reviews on next-generation sequencing are available and also provide precise comparative information (e.g., Glenn 2011; Metzker 2010; Schadt et al. 2010; Shokralla et al. 2012; van Dijk et al. 2014). Here, we will focus on the different aspects that are directly connected to the analysis of amplicons for eDNA studies.

7.1 Overview of the first, second, and third generations of sequencing technologies

First-generation sequencing is mainly based on the Sanger method (Sanger et al. 1977). To implement this method, the template DNA must be represented by several million identical fragments generated by cloning or by polymerase chain reaction (PCR). Starting from a sequencing primer, a DNA polymerase extends the fragment. A low percentage of chain-terminating dideoxynucleotides are randomly incorporated during this process, stopping the *in vitro* DNA replication at that position. The different fragments terminated by each of the four dideoxynucleotides are then separated by electrophoresis according to their length, and the DNA sequence is reconstructed by combining length information and the corresponding dideoxynucleotide. Initially, Sanger sequencing was based on polyacrylamide gels and radioactivity. Then, it relied on fluorescence on 96-capillary automated sequencers able to produce about one megabase per day, with routine read lengths of about 800 bp.

The second-generation sequencing technology is based on massively parallel sequencing. It encompasses the Roche 454 system, the PGM (Ion Torrent™), and Illumina sequencers. The first technology that was available on the market was the 454, which used pyrosequencing (Margulies et al. 2005), but it is no longer maintained by Roche. All these sequencers require the amplification of individual fragments prior to sequencing. This involves adding adapters to each extremity of the DNA fragments to be sequenced, and amplifying them, either by emulsion PCR on beads (454 and PGM), or by bridge PCR on the sequencing surface (Illumina).

Then, the nucleotide sequence is determined by an approach called "sequencing by synthesis" whereby the nucleotides are added during several sequencing cycles. A single nucleotide is added at each sequencing cycle for the 454 and the PGM, while for Illumina the four nucleotides are added together. This difference leads to sequences of variable length within the same run for the 454 (400 bp on average and up to 800 bp) and the PGM (up to 400 bp) and to sequences of the same length for Illumina (up to 250 and 300 bp for the HiSeq and MiSeq, respectively). Currently, the most powerful second-generation sequencer, the NovaSeq 6000, can produce up to 800,000 times more data than the most powerful first-generation sequencer. This implies that sequencing is no longer the cause of hold-ups in studies based on DNA metabarcoding. Illumina sequencers globally provide better data quality than 454 and PGM, which have more difficulties in properly sequencing homopolymers (i.e., in evaluating the exact number of times the same nucleotide is repeated).

Environmental DNA for Biodiversity Research and Monitoring. Pierre Taberlet, Aurélie Bonin, Lucie Zinger, & Eric Coissac, Oxford University Press (2018). © Pierre Taberlet, Aurélie Bonin, Lucie Zinger, & Eric Coissac 2018. DOI: 10.1093/oso/9780198767220.001.0001

The third-generation sequencers are characterized by their ability to sequence single DNA molecules, without any prior amplification. Two sequencers are now available in this category: the Sequel, which relies on the single molecule real-time (SMRT) sequencing technology (www.pacb.com/) and the MinION (nanoporetech.com) based on the nanopore approach (Jain *et al.* 2016). Both can sequence relatively long DNA fragments (up to 20 kb routinely). As these two technologies still have a relatively high error rate (up to 10%), they are not appropriate for analyzing eDNA at the moment. However, due to its portability, the Nanopore sequencer may be promising if the error rate becomes acceptable, eventually allowing for small-scale experiments to be conducted in the field.

7.2 The Illumina technology

At the onset of DNA metabarcoding, which was mainly used for the study of microorganisms, the 454 was the most popular approach. However, since Roche decided to shut down its 454 sequencing business in 2016, the 454 is no longer an option. As only very few metabarcoding studies are based on the Ion Torrent PGM (e.g., Deagle *et al.* 2013; Geml *et al.* 2014; Kraaijeveld *et al.* 2015), the clear majority of eDNA research now relies on Illumina sequencing. To date, it represents the only technology suitable for large-scale analysis with hundreds to thousands of samples. Therefore, hereafter we will only describe the sequencing process based on Illumina sequencers.

7.2.1 Library preparation

When using Illumina sequencers, the DNA fragments to be loaded must have the structure presented in Figure 6.7. The construction of these fragments is called library preparation. In the case of DNA metabarcoding, such a structure can be produced via PCR (see sections 6.10.2 and 6.10.3), or by protocols initially developed for sequencing genomic DNA (Adey *et al.* 2010; Head *et al.* 2014). Note that in this last case, the fragmentation step is skipped as the starting material is already at the right length.

The first and more conventional protocol (TruSeq® DNA library preparation kits, Illumina) includes the following steps (Fig. 7.1): (i) fragmentation of the genomic DNA via sonication to obtain fragments of about 300 bp long; (ii) end-repair of the fragmented DNA using the T4 DNA polymerase, the DNA polymerase I large (Klenow) fragment, and the T4 polynucleotide kinase to obtain blunt-end DNA fragments; (iii) addition of an "A" base to the 3'-end of each of the blunt-end DNA fragments using the Klenow fragment (3'→ 5' exo-) or a *Taq* polymerase; (iv) adapter ligation using the T4 DNA ligase, with several alternative strategies according to the kit used.

The second protocol ("Nextera™" kits, Illumina) is based on an engineered transposase that simultaneously cuts the DNA and adds specific adapters to both sides of the break (Fig. 7.1). This process is called "tagmentation." These short, added sequences are subsequently extended via a few PCR cycles in order to produce the appropriate fragment structure, including the index and flow-cell binding sites P5 and P7. The first advantage of the Nextera protocol is its simplicity, with far fewer steps than the conventional protocol. Another advantage is that it allows the library to be constructed with a small amount of initial genomic DNA (50 ng). However, these two advantages are counterbalanced by the fact that the tagmentation can introduce significant sequence-dependent biases when compared to the conventional protocol.

For both protocols, the final library product must be properly quantified in order to load the appropriate amount onto the sequencer (i.e., to have the expected fragment density on the sequencing surface). These two protocols are suitable for sequencing genomic or environmental DNA via a shotgun approach (i.e., to sequence random fragments of the DNA extract, without any PCR cycles before the library preparation). However, they can generate problems with metabarcoding amplicons. The two crucial steps are the end-repair and the PCR at the end of the library preparation. These two steps produce a significant number of chimeras among amplicons, leading to tag jumps (Schnell *et al.* 2015a) and other artifacts. When sequencing amplicons, it is thus highly advisable to use a PCR-free kit for library preparation, and to skip the end-repair step (Schnell *et al.* 2015a). We routinely use the MetaFast protocol (www.fasteris.com/metafast) that significantly limits the tag jump problem.

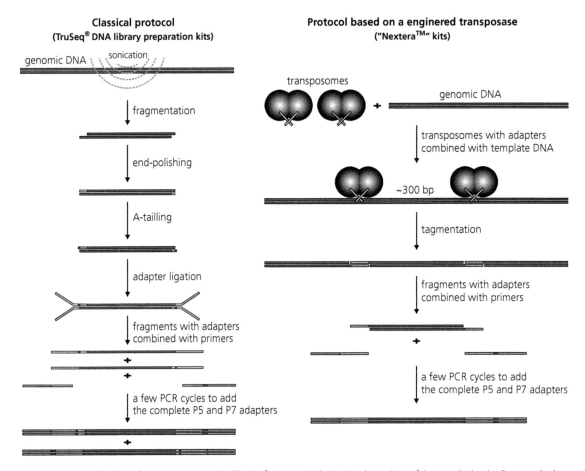

Fig. 7.1 Overview of methods for preparing sequencing libraries from genomic DNA. Note that variants of these methods exist. For example, the last step of the classical protocol (PCR) can be omitted, and the structure of the adapters to be ligated can change among the different library preparation kits.

7.2.2 Flow cell, bridge PCR, and clusters

The sequencing reaction is carried out on a flow cell, which is a glass slide with 1, 2, or 8 physically separated lanes, depending on the sequencing platform (Table 7.1). Each lane is coated with a lawn of two types of oligonucleotides, bound to the surface, and complementary to flow cell binding sites P5 and P7 (see Fig. 6.7).

Once the library is loaded on the flow cell, the first step is to amplify individual molecules via a process called "bridge PCR." The amplification of each initial molecule occurs on a very limited surface of the flow cell, using as primers the two oligonucleotides that are bound to the surface. During the annealing phase of this particular PCR, the DNA fragment forms a bridge between the two different types of oligonucleotides fixed on the surface. This is why this method has been called "bridge PCR." To obtain approximately one thousand copies of the initial DNA fragment, only a few cycles are performed, forming a cluster of identical DNA molecules (Fig. 7.2). Each well-delimited cluster (also called sequencing spot) on the flow cell produces a single forward (R1) sequencing read, and a single reverse (R2) sequencing read if a paired-end sequencing strategy is implemented (see Section 7.2.3).

Usually, up to 24 libraries can be loaded on the same sequencing lane when using a single index per library on the P7 side, provided that all these

Table 7.1 Number of flow cells, number of lane(s) available per flow cell, and maximum number of clusters for different Illumina sequencing platforms

Sequencing platform	Number of flow cells	Number of lanes per flow cell	Maximum number of clusters per lane (in millions)	Maximum length of the sequence reads (in b)
HiSeq 2500	1 or 2	8	250	150
HiSeq 2500 rapid run	1 or 2	2	300	250
HiSeq 1500	1	8	250	150
HiSeq 1500 rapid run	1	2	300	250
MiSeq	1	1	25	300
MiniSeq	1	1	25	150

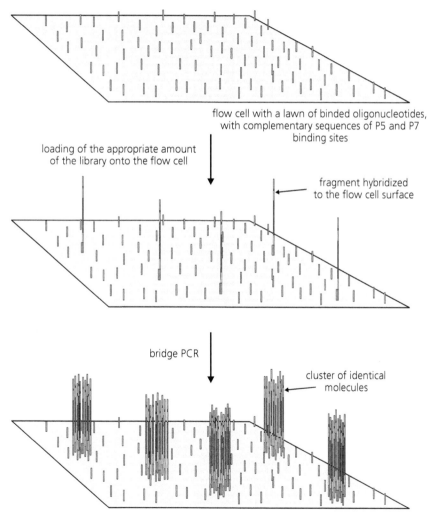

flow cell with a lawn of binded oligonucleotides, with complementary sequences of P5 and P7 binding sites

loading of the appropriate amount of the library onto the flow cell

fragment hybridized to the flow cell surface

bridge PCR

cluster of identical molecules

Fig. 7.2 Single molecule amplification via bridge PCR on the flow cell.

indexes are different. Up to 96 libraries can be routinely loaded on the same sequencing lane when using a dual indexing system (indexes on both the P7 and the P5 sides).

7.2.3 Sequencing by synthesis

This technology allows for the parallel sequencing of millions of clusters of identical fragments, and uses four fluorescently labeled nucleotides, each nucleotide displaying a different fluorochrome. During each sequencing cycle, a single fluorescently labeled nucleotide is incorporated in each newly synthesized DNA strand. The fluorochrome acts as a "reversible terminator" for polymerization. After the nucleotide incorporation, the fluorescent dye corresponding to each cluster is detected and identified through laser excitation. Then, the fluorochrome is enzymatically cleaved to allow for the addition of the next labeled nucleotide during the following sequencing cycle.

There are two strategies for the sequencing on Illumina platforms: single-read or paired-end sequencing. The single-read approach corresponds to the sequencing of a single strand, starting with the sequencing primer on the P5 adapter side, producing the R1 reads. The paired-end approach too sequences the strand starting from P5 (R1 reads), but it also sequences the complementary strand starting from P7 (R2 reads). The index that identifies the library is sequenced between the R1 and R2 reads (Fig. 7.3).

It is important to know that if several different libraries are loaded on the same sequencing lane, the Illumina technology assigns reads to the wrong library at a low rate (about 0.1%). This artifact can be termed "index jump" (see Section 6.10.2). This problem is important when working with eDNA, as common molecular taxonomic units (MOTUs) in some libraries can artifactually appear as rare MOTUs in other libraries. To solve this potential bias, the solution is either to estimate the rate of index jumps and to filter the rare MOTUs accordingly, or to load a single metabarcoding library that uses the same specific primers and the same tags per lane.

Another potential problem when sequencing amplicons is the lower diversity of the sequences

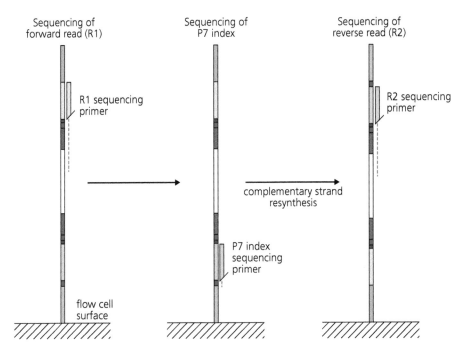

Fig. 7.3 Sequencing strategy on Illumina platforms when a single index is used. See Figure 6.7 for the structure of the fragment.

```
@HWI-M01126:144:000000000-A78T8:1:1101:14640:1680 1:N:0:TAGCTT

AGTACTAGATCGGGCAATCCTGAGCCAAATCCTATTTTCCAAAAGGAAGAATAAAAAAGGATAGGTGCAGAGACTCAATGGGATCTAGTCCCAGATCGGA

+

1>AAABDDF33AEECEFGGGGGEBGHHFFFGCGHFDGHHHFHDCGEHGGHGHHEGHHGGFHHHHHBF1FBFGFFGHHFD2@B0FC1BFFGBGG@GGHFG?
```

Fig. 7.4 Example of a FASTQ record.

compared to genomic DNA. A possible solution is to add a variable number of random nucleotides to the 5'-end of the primers (see Section 6.10.1 for more explanations).

7.2.4 Quality scores of the sequence reads

The output of Illumina sequencing platforms are large FASTQ files. For each library, a single file is produced if the sequencing strategy was to sequence only a single strand, whereas two files are produced for paired-end sequencing. In the second case, the two files contain the same number of records, the first sequence read of the first file and the first sequence read of the second file corresponding to the forward (R1) and reverse (R2) sequences of the same cluster, and so on. Figure 7.4 presents an example of a sequence record in the FASTQ format. Each sequence record has four lines. The first line, starting with a "@", is the sequence identifier

Table 7.2 Correspondence between quality scores and ASCII characters in FASTQ files

ASCII character	Quality score	Probability of error	ASCII character	Quality score	Probability of error
!	0	1	8	23	
"	1		9	24	
#	2		:	25	
$	3		;	26	
%	4		<	27	
&	5		=	28	
'	6		>	29	
(	7		?	30	0.001
)	8		@	31	
*	9		A	32	
+	10	0.1	B	33	
,	11		C	34	
-	12		D	35	
.	13		E	36	
/	14		F	37	
0	15		G	38	
1	16		H	39	
2	17		I	40	0.0001
3	18		J	41	
4	19		K	42	
5	20	0.01	L	43	
6	21		M	44	
7	22		N	45	

Table 7.2 Continued

ASCII character	Quality score	Probability of error	ASCII character	Quality score	Probability of error
O	46		g	70	0.0000001
P	47		h	71	
Q	48		i	72	
R	49		j	73	
S	50	0.00001	k	74	
T	51		l	75	
U	52		m	76	
V	53		n	77	
W	54		o	78	
X	55		p	79	
Y	56		q	80	0.00000001
Z	57		r	81	
[	58		s	82	
\	59		t	83	
]	60	0.000001	u	84	
^	61		v	85	
_	62		w	86	
`	63		x	87	
a	64		y	88	
b	65		z	89	
c	66		{	90	0.000000001
d	67		|	91	
e	68		}	92	
f	69		~	93	

while the second line is the DNA sequence itself. The third line is the quality score identifier line, consisting of a "+," and the fourth line contains the encoded quality score for each nucleotide.

The Phred quality score Q is related to the base calling error probability P according to the following formula:

$$Q = -10 \log_{10} P$$

For example, for a given nucleotide, a quality score of 30 (Q30) is equivalent to the probability of an incorrect nucleotide call of one out of 1,000. Quality scores of 20 (Q20) and of 10 (Q10) means a probability of correct calls of 99% and 90%, respectively. When the sequencing quality reaches Q30 for all the nucleotides of a particular sequence, it becomes

very unlikely that this sequence contains errors. Indeed, the probability of having an error is less than one out of 1,000.

In a FASTQ file, the quality scores from 0 to 93 are encoded using ASCII characters 33 to 126. Table 7.2 presents the relevant ASCII characters and their associated quality scores. In raw data reads, the quality score rarely exceeds 40, but higher scores are possible when different raw reads are assembled in subsequent analyses, leading to consensus sequences that can have very high scores. For example, when using a paired-end approach for a short metabarcode, the overlapping region between the forward (R1) and reverse (R2) reads has a much higher quality score than the two flanking fragments that were only sequenced once.

## CHAPTER 8

DNA metabarcoding data analysis

As in other fields of science, the big-data era has led to a dramatic shift in the allocation of working time and costs in ecology. Up until recently, the most constraining part of biodiversity surveys involved the compilation of species inventories in a sufficient number of sites or points in time, in order to be representative of a given ecological phenomenon. The previous chapters of this book showed that DNA metabarcoding, coupled with high-throughput sequencing, now allows this limitation to be circumvented by producing comprehensive biodiversity censuses with minimal effort. Today, the most important part of the next-generation molecular ecology work lies in processing huge quantities of DNA sequence data. In addition to the computational and storage resources it demands, one crucial aspect when working with DNA metabarcoding data is its noise. Improved sequencing coverage of samples has allowed for better descriptions of both intra and interspecific biological diversity, but also of all molecular artifacts that were previously invisible. The main challenge is, therefore, to filter out these artifacts and reduce the size of the data to produce a final output similar to classical taxonomic inventories. This can then be readily analyzed statistically to answer ecological questions.

This data curation process involves successive steps that are increasingly dependent on study design and questions (Fig. 8.1). These can be accomplished with several programs, which all have their pros and cons as well as their own format of data input or output. We will not recommend the use of one above the other, as they are sometimes complementary. However, hereafter we will mainly consider the OBITools program suite (Boyer et al. 2016), QIIME (Caporaso et al. 2010), and R (R Development Core Team 2016) to implement DNA metabarcoding analyses, but other solutions exist,

such as MOTHUR (Schloss et al. 2009), DADA2 (Callahan et al. 2016), and PipeCraft (Anslan et al. 2017). Given the success of the Illumina sequencing platforms and their paired-end technology, the recommendations provided next will focus on this type of data.

8.1 Basic sequence handling and curation

The first steps of sequence handling and curation are common to all metabarcoding studies, but might differ slightly in their philosophy depending on the multiplexing and sequencing strategies used.

8.1.1 Sequencing quality

Before going further in the data processing, the global quality of the sequencing run needs to be determined. If the overall quality is not sufficient, a resequencing of the samples should be considered, instead of relying on the part of the data set for which sequencing qualities are acceptable. A global overview of the quality of the sequencing run can be visualized with FastQC (Andrews 2014).

8.1.1.1 The pros and cons of read quality-based filtering

All sequencing platforms provide both the sequence and a quality score (often termed "phred quality score") associated with each base. This score is determined during the base calling process and provides information on the probability of miscalling at each sequenced position (see Section 7.2.4). Within a sequencing run, sequence quality is not homogeneous, with some sequencing reads having poorer average quality than others. Globally, within one sequencing read, sequence quality decreases

Environmental DNA for Biodiversity Research and Monitoring. Pierre Taberlet, Aurélie Bonin, Lucie Zinger, & Eric Coissac, Oxford University Press (2018). © Pierre Taberlet, Aurélie Bonin, Lucie Zinger, & Eric Coissac 2018. DOI: 10.1093/oso/9780198767220.001.0001

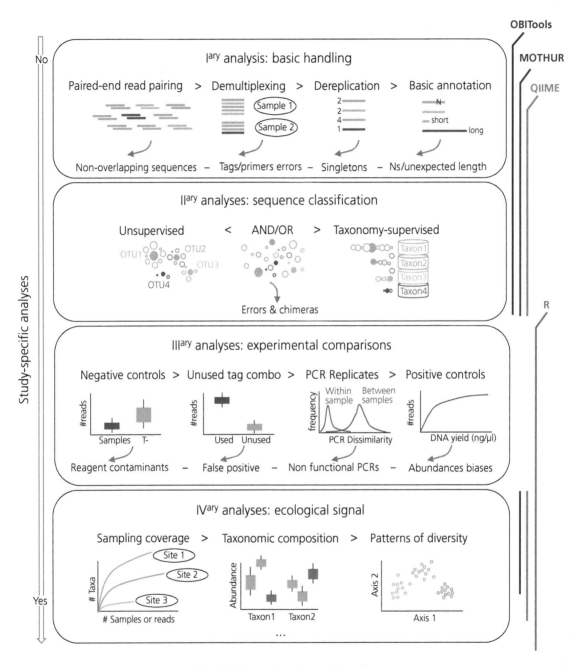

Fig. 8.1 DNA metabarcoding analysis workflow.

from the 5'- to the 3'-end. To ensure the quality of the data, it is often recommended to either remove reads with low quality or to trim the reads in their 3'-end to remove the low-quality part of the sequence (Bokulich *et al.* 2013). While this strategy has been shown to improve overall results quality, such data filtering may have a strong side effect that has never been considered thus far: quality is not independent of the sequence itself. For example, the presence of an homopolymer in a metabarcode sequence

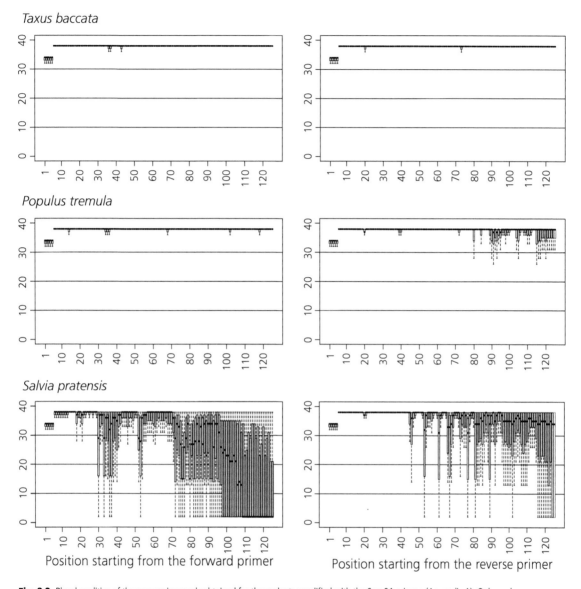

Taxus baccata

Populus tremula

Salvia pratensis

Position starting from the forward primer   Position starting from the reverse primer

Fig. 8.2 Phred qualities of the sequencing reads obtained for three plants amplified with the Sper01 primers (Appendix 1). Only reads without sequencing errors (*Taxus baccata*: atccgtattataggaacaataattttattttctagaaaagg; *Populus tremula*: atcctatttttcgaaaacaaacaaaaaaacaa acaaaggttcataaagacagaataagaatacaaaag; and *Salvia pratensis*: atcctgttttctcaaaacaaaggttcaaaaaacgaaaaaaaaaag) were considered for these graphs.

leads systematically to a decrease in the quality of the reads produced. Hence, it is possible that the average quality changes consistently from one species to another. To illustrate this phenomenon, we have extracted the reads corresponding to the three most abundant species of a mock plant community (i.e., *Taxus baccata*, *Populus tremula*, and *Salvia pratensis*), which were analyzed using the Sper01 me-

tabarcode (Taberlet *et al.* 2007; see also Appendix 1) and sequenced on a HiSeq 2000 machine. The distribution of quality scores per position and per species (Fig. 8.2) clearly differs: while the sequence qualities are excellent for all *Taxus baccata* reads, they are not optimal at the 3′-end of the sequencing reads for *Populus tremula*, when sequencing is initiated on the reverse primer. The average quality of *Salvia pratensis* reads is much lower. By filtering on the basis of reads quality, the data set will be depleted of most *Salvia pratensis* sequences and some of the *Populus tremula* sequences. It will therefore introduce biases in diversity estimates, or at least in the composition and structure of the community. Some sequencing centers provide only high-quality sequence reads by additional, stringent, in-house filtering without informing the customer. This systematically skews metabarcoding results.

8.1.1.2 Quality trimming software

Many programs propose trimming Illumina sequencing reads based on their quality (review in Del Fabbro *et al.* 2013). Most of them offer more than a quality trimming procedure, and are often able to remove the sequencing adapters. Some of them, like Flexbar (Dodt *et al.* 2012) and Skewer (Jiang *et al.* 2014), are able to deal with paired-end reads. Flexbar is one of the numerous programs proposing basic quality trimming options. Because the first few bases of a read usually have a slightly lower score, Flexbar trims a fixed number of bases on the 5'-end. A similar option is proposed for the 3'-end. There are two options that allow for actual trimming based on quality. The first one rejects reads containing a given number of ambiguous bases. The second option involves trimming from the 3'-end of the read until finding a base with a phred quality score higher than a given threshold. The Skewer program offers the additional possibility of filtering out reads based on their average quality. In QIIME, the filtering and trimming of low-quality reads can be performed with the command split_libraries_fastq.py (Bokulich *et al.* 2013). In OBITools, no command is especially dedicated to read quality trimming, but the user can implement its own quality filtering rules with the obigrep and obiannotate commands.

8.1.2 Paired-end read pairing

This step aims to align both the forward and reverse reads based on their 3'-end to reconstitute the full-length sequence. FLASH (Magoč & Salzberg 2011), COPE (Liu *et al.* 2012), and PANDASeq (Masella *et al.* 2012) are the three most popular algorithms used to accomplish this task. All of them use an ungapped alignment algorithm, as the probability of insertion or deletion errors on Illumina platforms is relatively low. The sequence quality scores are not considered during the alignment process in FLASH. However, one can expect an inflated number of mismatches in the aligned regions as the phred qualities are generally lower in the read 3'-end (Fig. 8.2). To take this property into account, COPE considers the quality scores at mismatched positions and keeps the base with the highest quality in the consensus sequence. COPE relies on a k-mer-based alignment procedure, which makes it 10 times faster than FLASH. PANDASeq is the only program from this trio that fully considers quality scores for the alignment scoring, and performs well even when the overlap between reads are short. However, it tends to inappropriately align reads when there is no overlap between the forward and reverse reads (i.e., when the metabarcode sequence is too long). This particular case is now handled by PEAR (Zhang *et al.* 2014). PEAR can also deal with the situation where a metabarcode is shorter than the read length, which requires an alignment algorithm based on the 5'-end of the reads. The QIIME join_paired_ends.py command is a wrapper on fastq-join (Aronesty 2011), which uses ungapped alignments and does not take into account quality scores. QIIME also wraps SeqPrep (https://github.com/jstjohn/SeqPrep), which considers the quality scores only at mismatch positions. The OBITools comprise the illuminapairedend algorithm that considers the quality scores at all positions during the alignment process. The consensus computation is implemented such that the quality of the final consensus is maximized. illuminapairedend uses an exact alignment algorithm (dynamic programming) that also considers possible gaps in the alignment. It is hence slower than the other software programs discussed here, but presents the advantage of being able to run without any prior knowledge of the paired-end reads overlap length. Like PEAR, it also considers the possibility that the sequenced metabarcode can be shorter than the read length, or too long for a paired-end read overlap.

Command 8.1

```
illuminapairedend -r reverse.fastq forward.fastq > paired.fastq
```

The command here merges the forward reads stored in the `forward.fastq` file with the reverse ones stored in the `reverse.fastq` file. The consensus sequences resulting from the pairing are stored in the FASTQ file `paired.fastq` with several annotations allowing for an easy filtering of reads based on the alignment quality. From our experience with such data, we recommend filtering out any consensus sequence with a score_norm <3.90 or a score <40.

8.1.3 Sequence demultiplexing

Usually, many polymerase chain reaction (PCR) amplicons are mixed together in a single sequencing library (see Section 6.10). Once the full amplicon sequence has been built from forward and reverse paired-end reads, each sequence must be assigned to its initial PCR according to the tag added at one or both extremities of the amplified fragment. Many tools perform this job, including QIIME and the OBITools. Both systems use a tabular text file to describe sample properties and associated tags. QIIME tools are better adapted to the tagging strategy proposed by Caporaso *et al.* (2011) and detailed in Section 6.10.2. The OBITools are better adapted to a tagging strategy where the tag is attached directly to the 5′-end of one or both primers, and where all tags used in the experiment have the same length. Sequence demultiplexing then involves detecting amplification primers and tags located at amplicon extremities, extracting the central metabarcode sequence, and assigning it to its corresponding PCR. The QIIME command performing this job is `split_libraries_fastq.py` (Caporaso *et al.* 2010), and the corresponding OBITools command is `ngsfilter` (Boyer *et al.* 2016). If specified, `ngsfilter` can keep only full-length metabarcodes and rejects partially sequenced ones. Both QIIME and the OBITools produce a sequence file in which each sequence is tagged with its sample name. If needed, the file can be split into subfiles, each corresponding to one sample using the `obisplit` (OBITools) or `split_fasta_on_sample_ids.py` (QIIME) commands. The procedure is not recommended at this stage of the analysis, as samples are often required to be treated all together in downstream analyses.

Command 8.2

```
ngsfilter -t sample_description.txt -u not_assigned.fastq paired.fastq \
> data.fastq
```

This command tags the sequences contained in the FASTQ file `paired.fastq` with their sample identifiers according to the description of samples provided in the `sample_description.txt` file. The results are stored in the file `data.fastq`. Sequences that cannot be assigned to a sample for a given reason (i.e., no primer and/or tag found) are saved in the `not_assigned.fastq` file.

8.1.4 Sequence dereplication

In a typical DNA metabarcoding data set, the same sequence occurs many times. Once each sequence has been associated with a sample, it is common practice to group all strictly identical sequences and their associated information all together to reduce the size of the data. This process is called sequence dereplication and can be viewed as a clustering operation with an identity threshold of 100%. In QIIME, this is done with the `pick_de_novo_otus.py` with specific parameters. In the OBITools, it is done with the `obiuniq` command.

Command 8.3

```
obiuniq -m sample data.fastq > data.uniq.fasta
```

Although this is not mandatory, the best strategy when using the OBITools is to keep all the data in a single file. The `ngsfilter` command run previously annotates each sequence header with an attribute named `count` containing the occurrence of the reads in the entire data set. This command, with the *-m* option, will dereplicate the sequence data set and keep the number of occurrences of each unique sequence in each PCR in the `data.uniq.fasta` file. It will be used later to build the contingency table for ecological analyses.

8.1.5 Rough sequence curation

The basic characteristics of sequences can be checked to remove obvious artifacts. This includes metabarcode sequences containing ambiguous bases (bases other than A, C, G, and T), or sequences with a length that differs from the expected metabarcode size, which can be evaluated by running an *in silico* PCR on GenBank (see Chapter 3). Singleton sequences are also commonly considered as spurious (e.g., Bálint *et al.* 2014; Zimmermann *et al.* 2015; Geisen *et al.* 2015). They correspond to sequences represented by one single sequence in the whole data set and are usually excluded. All of these obvious artifacts can be easily filtered out using the OBITools command `obigrep`. In QIIME, most of these filtering criteria can be specified during the assignment of reads to samples with the `split_libraries_fastq.py` command, and singletons can be removed with `filter_otus_from_otu_table.py`.

Command 8.4

```
obigrep -p 'count > 1' -l 8 -L 150 -s '^[acgt]+$' data.uniq.fasta > \
data.uniq.raw_curated.fasta
```

This command selects sequences occurring more than once in the data set, having a length ranging from 8 to 150 bp, and composed only of the four nucleotides A, C, G, T. Input data are stored in the file `data.uniq.fasta` and the results are pushed into the file `data.uniq.raw_curated.fasta`.

8.2 Sequence classification

In metabarcoding data, PCR and sequencing errors are numerous (see Chapter 6). Consequently, the number of unique sequences (i.e., dereplicated sequences or classes of sequences with an identity of 100%) far exceeds the true diversity for any taxonomic group (Huse *et al.* 2007; Schloss *et al.* 2011; see Fig. 8.3 for an example on plants). Apart from this experimental noise, many organisms can display intraspecific variability in the metabarcode sequence. Constructing groups of similar sequences (i.e., clustering) is therefore compulsory to minimize the artifactual inflation of biodiversity estimates caused by intraspecific polymorphism or experimental noise. Computer scientists roughly sort classification methods into two main groups: supervised and unsupervised methods. Supervised methods classify items into pre-existing classes defined by a model, which can consist of a set of example members. Unsupervised methods define the classes during the classification process. In metabarcoding data, the taxonomic assignment is a classification problem supervised by classes already defined by the taxonomy. On the other hand, the clustering of sequences without any other external data to define the group is an unsupervised classification process. It corresponds to the building of molecular operational taxonomic units (MOTUs).

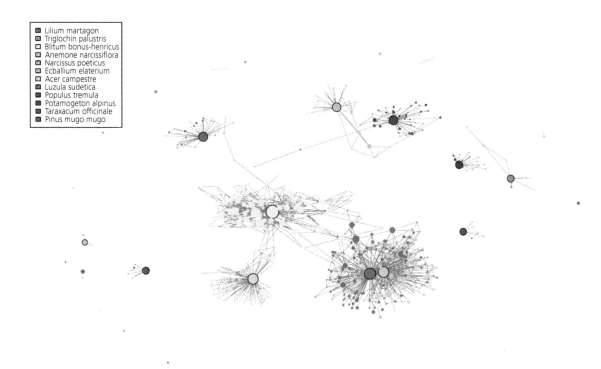

- ▣ Lilium martagon
- ▣ Triglochin palustris
- ▢ Blitum bonus-henricus
- ▢ Anemone narcissiflora
- ▣ Narcissus poeticus
- ▣ Ecballium elaterium
- ▢ Acer campestre
- ▣ Luzula sudetica
- ▣ Populus tremula
- ▣ Potamogeton alpinus
- ▣ Taraxacum officinale
- ▣ Pinus mugo mugo

Fig. 8.3 Structure of sequence variants in the metabarcoding data set of a mock community consisting of 12 plants. Colors used in this figure correspond to the plant species. In this graph, each node (circle) corresponds to a sequence, and edges link two sequences with only a single nucleotide difference. Node sizes are proportional to the log-transformed sequences abundances and black-framed nodes represent errorless sequences.

Whatever the approach, all the methods used in the classification of metabarcodes rely on the similarity between sequences and therefore on a sequence alignment procedure. The choice of the alignment method is very important, as it may have a strong impact on the classification. However, it is rarely considered by end-users, as it is often imposed by the classification software. For instance, multiple alignment algorithms are useful when reconstructing phylogenies of small sets of closely related sequences as their aim is to find homologous segments. However, they perform very poorly when it comes to aligning thousands of metabarcodes, highly variable in both composition and length, and that do not necessarily share common regions, as could be the case for an entire fungal or bacterial community (Sun *et al.* 2009). For such applications and to reduce the computational effort, pairwise sequence alignments are usually preferred. To further reduce computational time, it is also often necessary to rely on heuristic alignments (approximative approaches) that might yield

suboptimal alignments. Consequently, we encourage the users to have a good understanding of the main principles of their classification tools, including their advantages and limits, to ensure that they are the appropriate tools for their specific needs.

8.2.1 Taxonomic classification

Chapter 3 provides an overview of the different reference databases that are currently available or that can be constructed for various organisms. These contain metabarcode sequences that are already assigned to classes defined by taxonomists. Taxonomic assignment involves classifying sequences into these taxonomic classes. The relationship between a class and a sequence is based on the similarity between that sequence and reference metabarcodes. As similarity assessment can be time-consuming, we recommend performing the taxonomic classification after data set reduction through sequence dereplication and unsupervised classification (Section 8.2.2).

The simplest assignment method is called "nearest neighbor." It is implemented in QIIME with the command `assign_taxonomy.py` when used with the Basic Local Alignment Search Tool (BLAST) strategy. It involves comparing a query sequence against all reference metabarcodes and assigning this query to the closest reference. This method is efficient if the database is complete and if the metabarcode distinguishes all taxa. However, if the database is incomplete, which is almost always the case, the sequence belonging to a taxon that is absent from the database will be assigned to a closely related, yet incorrect, taxon. Furthermore, if the metabarcode marker does not distinguish all taxa, the sequence will be assigned randomly to one of the taxa sharing the same metabarcode.

A way to circumvent this problem is to use the lowest common ancestor (LCA) algorithm. Here, the query sequence is still compared to all reference metabarcodes. Then, one selects a set of references that are sufficiently similar to the query. The query is then assigned to the most specific taxonomic label that is shared by this set of references (i.e., the lowest common ancestor). As reference metabarcodes might contain taxonomic annotation errors, this rule can be relaxed by assigning the query sequence to the most specific taxonomic label shared by a subset of this set of references larger than a given quorum. The quorum is defined by the expected rate of erroneous taxonomic annotations in the reference database. To date, there are unfortunately no clear ways to estimate this error rate, which also depend on the taxonomic group studied. In practice, if the rate of erroneous taxonomic annotations is overestimated, this should lead to taxonomic assignments that are more precise than what they should be. Inversely, underestimating this rate leads to coarser taxonomic assignments. The size of the reference set used to estimate the LCA can be defined in two ways: either the reference set has a constant size, in which case the similarity of these sequences against the query is variable, or the reference set is defined against the query by a minimum similarity threshold, in which case its size is variable. The latter strategy is the most commonly used. The command `assign_taxonomy.py` implemented in QIIME is based on this approach when used with SortMeRNA, UCLUST, or RTAX.

The `ecotag` program from the OBITools also implements an LCA procedure, except that the size of the subset is neither defined as a constant, nor by an identity threshold. Instead, it first finds the nearest neighbor sequence S of the query sequence Q in the database. Second, it builds the set of reference sequences by selecting those that exhibit a similarity with S that is higher than the similarity between S and Q. The LCA is then computed on this subset of references. Note that `ecotag` uses an alignment procedure which assumes that both the queries and references are full-length barcodes.

Command 8.5

```
ecotag -d refDB -R refDB.fasta data.uniq.raw-curated.fasta > \
data.uniq.rcurated.tax.fasta
```

The `ecotag` command included in the OBITools annotates each sequence with several attributes related to the taxonomy. For instance, the field `genus_name` will correspond to the genus the sequence belongs to and the field *genus* to the corresponding taxid in the `refDB` taxonomy. The field `best_identity` corresponds to the similarity of the query sequence against its best match in `refDB.fasta`. The field `scientific_name` corresponds to the lowest common ancestor between the query sequence and the set of sequences in `refDB.fasta`, as already defined here.

The program `classifier` from RDPII (Wang *et al.* 2007) is also commonly used. In this approach, the similarities between the queries and references are estimated from the number of k-mers they share. A k-mer is a DNA word of length k ($k = 7$ nucleotides by default in this case). This measure of similarity allows for the use of simple probabilistic models and is computable in a highly time-efficient way. Note here that the number of shared k-mers remains a very rough approximation of the similarity between sequences. For each taxon of the taxonomy (i.e., including internal nodes), a probabilistic model of k-mer occurrence is estimated. It is referred to as the training set in `classifier`. The taxonomic assignment then uses a Bayesian probabilistic approach to evaluate the model that is most likely to have produced the query sequence. Therefore, the output of `classifier` provides a classification to each query sequence together with a probability of belonging to a given taxon from the genus to the phylum levels. As the training sets provided by `classifier` are limited to prokaryotic 16S rRNA genes and fungal ITS, its use is currently limited to these markers and clades.

Phylogenetic placement methods are now proposed as promising alternatives to the aforementioned approaches. The main argument in favor of these methods is that the position of the query sequence in a phylogenetic tree should provide more precise taxonomic information than a single taxon name. Several software programs have been developed recently for this purpose, for example, `pplacer` (Matsen *et al.* 2010) and EPA (Berger *et al.* 2011). Briefly, the input is a reference tree, a reference multiple alignment, and the query. The query is crudely assigned to a region of the tree based on its distance to the reference sequences. The placement of the query in the tree is then refined by locally recomputing the phylogenetic tree using maximum likelihood algorithms. Hence, for these methods, query sequences must contain enough phylogenetic signal (i.e., they should be sufficiently long and/or with moderate evolution rate) to be consistently placed within the tree. This is usually not the case for short metabarcodes that do not contain enough phylogenetic signal.

In summary, there is no optimal solution for solving the taxonomic classification problem, and current solutions are not universally applicable. Trade-offs between the various aspects addressed here should hence be considered when defining the strategy to be used when analyzing a given data set.

8.2.2 Unsupervised classification

Although taxonomy-supervised approaches would be preferable, they cannot be applied when most of the diversity remains unknown. For example, soils most likely harbor species that are undiscovered, taxonomically unclassified, or most simply undocumented in public sequence repositories, even if vouchers are available. In this context of partial knowledge, unsupervised approaches are a convenient alternative. There is a wide variety of unsupervised clustering algorithms that have been implemented to cluster sequences (or other objects, such as individuals in social networks). This section will certainly not provide an exhaustive review of this vast topic (Kopylova *et al.* 2016), but will instead aim to provide some clues to help understand the main principles of sequence clustering (see Fig. 8.4 for examples). Note that most clustering methods do require a sequence similarity threshold to be set, which should be evaluated and will depend on the metabarcode used.

Clustering algorithms can be roughly divided into four families, although some tools make use of two of them in their implementations. The most widely known is the family of ascendant hierarchical clustering algorithms, which are implemented, for example, in MOTHUR (Schloss *et al.* 2009). Ascendant hierarchical clustering algorithms initially consider each sequence as a class. The two closest classes A and B are merged into a new class N. The distances between the new class N and all the remaining classes C_i are then evaluated. These distances can be set as the smallest distances between $d(A,C_i)$ and $d(B,C_i)$—single linkage; or the largest distances between $d(A,C_i)$ and $d(B,C_i)$—complete linkage; or the average distances between $d(A,C_i)$ and $d(B,C_i)$—average linkage. This process is repeated until a single class persists. The final set of classes is defined by truncating the classification tree at a given distance threshold (often 3% of dissimilarity).

A second family of clustering methods are based on the graph theory. These include community detection algorithms such as MCL, Markov clustering (Van Dongen 2001; Enright *et al.* 2002), and Infomap (Rosvall & Bergstrom 2008). These consider a graph composed of nodes (here sequences) connected by edges, if they are for example more similar than a given threshold. If a random walker tends to be trapped in a subregion of the graph, this means that the nodes are highly connected, and this subregion is called a "community" or a cluster. Such

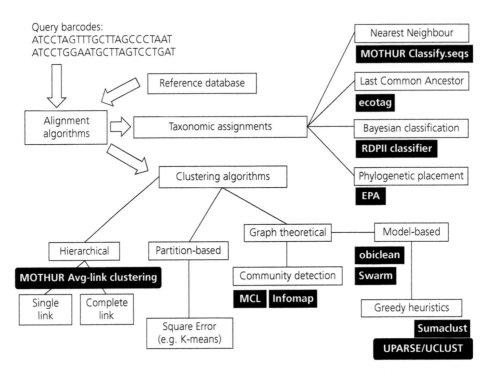

Fig. 8.4 Methods for supervised and unsupervised classifications. Classification types (white boxes) are provided along with examples of tools (black boxes).

approaches have already been proven to form reliable MOTUs (Zinger *et al.* 2009a; Kunin *et al.* 2010).

In the third family of clustering methods, the graph of sequences is built according to a model fitting a given process. For example, obiclean (Boyer *et al.* 2016) and Swarm (Mahé *et al.* 2014) build a graph to model the production of errors during the PCR. obiclean and Swarm differ in their implementation, but basically involve building a directed acyclic graph (DAG). In this DAG, nodes are sequences, and sequences differing at a single position are linked by an arrow, which is directed from the most abundant sequence to the rarest one. In this model, the less abundant sequence can be considered as being an erroneous variant of the most abundant one, resulting from a punctual PCR error. The representative sequences of clusters can then be defined by the top nodes of the DAG.

Such models can be simplified to reduce computational time. Greedy clustering algorithms such as sumaclust (Mercier *et al.* 2013) or UPARSE (Edgar 2013) define the most abundant sequence as the

first representative sequence of a cluster. All other sequences are then assigned to that cluster if they are similar enough to this representative sequence, depending on a threshold chosen beforehand. The same procedure is repeated to define other clusters on the remaining sequences until all of them are classified.

Model-based methods can also rely on biological hypotheses. For example, the ABGD method (Puillandre *et al.* 2012) assumes that metabarcode intraspecific variability is far lower than the metabarcode interspecific variability (i.e., the so-called "barcoding gap") to partition the sequence data sets into MOTUs.

Finally, partitioned-based clustering methods are based on the principle that there is a finite number of possible partitions of k classes and consist in classifying the objects in these k classes. Therefore, the number of classes should be known prior to the analysis, which is actually what one wants to estimate in most cases when working with environmental samples. This type of method is hence seldomly used in the field.

Next, we provide an example of a command line with the sumaclust algorithm. This tool is not implemented in the OBITools but can be viewed as a companion software (http://metabarcoding.org/sumaclust).

Command 8.6

```
sumaclust -r -t 97 -p 4 data.uniq.rcurated.tax.fasta > \
data.uniq.rcurated.tax.cl.fasta
```

In the example here, the clustering is done at a threshold *-t* of 97% sequence similarity. If the data set is large, the user can take advantage of multithreading computations with the *-p* option. The algorithm assigns each sequence to a given cluster of whose name corresponds to the most abundant sequence of the cluster. This information is stored in the field cluster, along with other information related to the clustering. In QIIME, unsupervised classification can be run with the command pick_de_novo_otus.py, which is a wrapper of UCLUST/UPARSE, sumaclust, and Swarm.

8.2.3 Chimera identification

Chimeras result from the recombination between abundant or closely related barcode sequences and may constitute up to 45% of the diversity in metabarcoding data sets (Ashelford *et al.* 2005; Quince *et al.* 2011; Haas *et al.* 2011). Significant efforts have been made toward the development of chimera detection algorithms. Tools like Chimera-Slayer (Haas *et al.* 2011) and DECIPHER (Wright *et al.* 2012) basically aim to determine whether the end-segments of a sequence correspond to the same reference in established chimera-free reference databases, whereas Perseus (Quince *et al.* 2011) and UCHIME (Edgar *et al.* 2011) carry out this task based on databases newly constructed from the data set under study. However, these tools are not easily applicable to current metabarcoding data as they perform poorly either on short sequences (<300 nucleotides) or for chimeras resulting from closely related parents (Haas *et al.* 2011; Wright *et al.* 2012). From our experience, these algorithms produce a high number of false positives corresponding to abundant taxa that are actually present in our experiment.

While the problem of detecting chimeras from closely related parents remains unsolved, there is a simple, viable alternative to identifying chimeras from divergent parent sequences. For such chimeras, similarity with reference sequences is much lower than for a true sequence, provided that the reference database covers most of the clade diversity. Figure 8.5 shows the distribution of sequences' maximum similarities against reference databases in a metabarcoding experiment on soils. The distribution is generally bimodal, with a gap observed at ~70–80% of similarity depending on the metabarcode used. Below this threshold, MOTUs are most likely to be putative chimeras or highly degraded sequences. For the fungal metabarcode (Fung01 in Appendix 1), the maximum similarities are distributed multimodally, and are overall much lower, probably due to the greater database incompleteness and variability of that marker. For the plant metabarcode (Sper01, Appendix 1), for which the reference database is more complete, most MOTUs seem to be chimeric sequences (although they represent a small fraction of total number of sequences). This suggests that the rate of chimeric fragments formation may be higher when amplifying low-complexity DNA templates. This contradicts previous studies reporting that chimera formation rate is higher in more genetically diverse samples (Wang & Wang 1997; Fonseca *et al.* 2012) and further analyses should be done to confirm this hypothesis. Note here that if the chimeras are formed at the initial stages of the PCR, they should also accumulate errors during the PCR. This filtering step can thus be performed after the clustering step to save computation time.

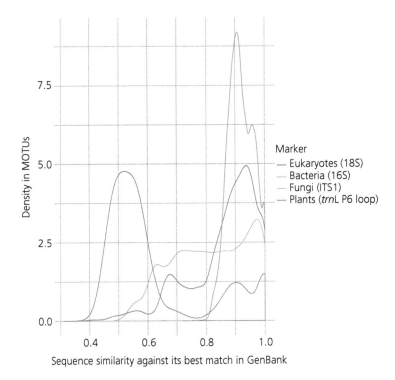

Fig. 8.5 Distribution of sequence similarities with their best match in GenBank (v197) for different markers.

8.3 Taking advantages of experimental controls

Sections 4.2.2, 6.4, and 6.8 discuss the different controls and replicates that should be implemented in a DNA metabarcoding experiment. These include extraction and PCR negative controls to monitor contaminations, tagging system controls to track tag jumps, positive controls, and technical as well as biological replicates. Although these are essential to ensure that the retrieved signal is not artifactual, they remain, thus far, seldom documented or even actually conducted in DNA metabarcoding studies. Consequently, the procedures implemented to correct the data based on these controls and replicates are also rarely documented. Here, we briefly describe how one can make use of such controls and replicates to further curate metabarcoding data sets for potential artifacts. We use R to conduct and automatize these filtering procedures, but other software can also be considered.

8.3.1 Filtering out potential contaminants

The sensitivity of PCR and high-throughput sequencing makes the metabarcoding approach prone to contamination, otherwise referred to as false positives (Ficetola *et al.* 2015, 2016). Consequently, the fact that negative controls in a metabarcoding experiment yield a small, or occasionally noticeable number of reads is a common observation, rather than an exception (see e.g., Fig. 8.6). Although this may suggest that the experiment has not been conducted in optimal conditions (see Chapter 6 for optimizations), one can take advantage of these experimental controls to curate the data from potential contaminants and hence improve the quality of the data set.

Negative controls usually contain several types of contaminants that can be separated into two main classes. First, one type can correspond to the pervasive exogenous contaminants occurring in the laboratory, or contained in some reagents such as certain bacterial strains (Salter *et al.* 2014), or DNA from animal or plants that are used in the production

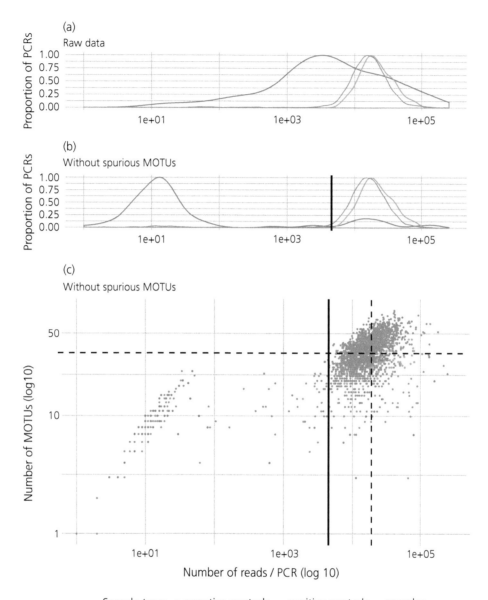

Fig. 8.6 Sequencing depth and sample richness before (a) and after (b–c) the removal of spurious MOTUs for a plant metabarcoding experiment on soils using the Sper01 metabarcode (Appendix 1). Spurious MOTUs correspond to chimeras and contaminants. These characteristics are shown for samples (blue), positive controls (green), and negative controls (red). (c) shows the relationship between sequencing depth and the observed richness in MOTUs. The axes are in log scale. The plain black line in (b–c) represents the proposed threshold below which PCRs are considered as not reliable. Dashed black lines in (c) correspond to the average sequencing depth and number of MOTUs.

of commercial reagents and kits (Chapter 6). Second, there could be internal contaminations, for instance, coming from the metabarcoding experiment itself through either the production of small aerosols when pipetting during the preparation of

the PCR plate, or caused by tag jumps at the library preparation/sequencing step (Schnell *et al.* 2015a; Esling *et al.* 2015). These latter contaminants hence have a biological meaning for the scale of the study, but can occur in PCRs where they are not supposed to be. Internal contaminants complicate the use of negative controls; for example, one cannot simply filter out any MOTU occurring in negative controls, as some of them actually correspond to MOTUs that have a high abundance in the biological PCRs studied.

Curating the data set from exogenous contaminants can nevertheless be relatively straightforward if there are some expectations on the taxonomic composition of biological samples. For example, detecting arctic plants in tropical soil samples strongly suggests that they are contaminants and the MOTUs corresponding to these allochthonous species can be confidently removed. However, this requires a good knowledge of the diversity in the studied environment, and this approach is hence not applicable in many cases, typically for microorganisms. A more objective alternative is to consider that any MOTU of whose frequency across the entire data set is maximal in negative controls, is an exogenous contaminant. The rationale behind this rule is that contaminants have no competitor fragments in negative controls and should hence be highly amplified. This criterion does not guarantee the exclusion of sporadic exogenous contaminants that occur in random PCRs, but not in negative controls. Nevertheless, it allows the pervasive exogenous contaminants present in the laboratory or in commercial reagents to be filtered out efficiently. Internal contaminants can result from sample cross-contamination in the field, through pipetting during the PCR plate preparation, or from tag jumps. In the latter case, they correspond to chimeras formed between DNA fragments from multiple PCRs during the library preparation or sequencing. This process generates new fragments for which the barcode sequence remains unchanged, but with a different tag combination that might correspond to one already used in the experiment (see Chapter 6). Hence, they may have dramatic impacts on the results, especially for species detection purposes. Whatever their origin, an internal contaminant in a given PCR should be of lower abundance than that of the original sequence occurring in another PCR. For internal contaminants resulting from tag jumps, the probability of their occurrence in other PCRs should be proportional to the abundance of

the genuine sequences in the amplicon multiplex. This implies that, in a given PCR, internal contaminants can still be more abundant than other rarer genuine sequences. Consequently, filtering out low abundance sequences/MOTUs (e.g., <10 sequences in each PCR) is inappropriate, and the abundance filtration threshold should be adjusted for each MOTU independently. For example, one can consider the distribution of the relative abundances of a given MOTU across all PCRs (i.e., ratio between MOTU abundance in the PCR and total MOTU abundance) and suppress that MOTU in PCRs where its relative abundance is too low (e.g., 0.03%). Tag jumps are also more likely to occur in PCRs sharing one tag, and the above-mentioned approach proposed here can use this property to refine the data curation procedure (Esling *et al.* 2015).

Finally, more generalistic frameworks allowing for the detection of false positives are now emerging, which rely on site occupancy modeling (Ficetola *et al.* 2015, 2016; Lahoz-Monfort *et al.* 2016). They use the occupancy of MOTUs in technical and biological replicates, from which one can infer the probability of a MOTU occupancy and detectability at each site with maximum likelihood techniques. Optimization of these models, taking into account the characteristics of eDNA data and their evaluations on various data sets and applications, will undoubtedly help in improving the quality of metabarcoding data. However, these may be not adapted to detect pervasive contaminants and reproducible PCR errors, and should be used in combination with the filtration procedures just mentioned here. Note that for all the proposed data curation procedures, the efficiency can be evaluated and adjusted by comparing the composition of negative and positive controls prior to and after the filtering process (e.g., Fig. 8.6).

8.3.2 Removing dysfunctional PCRs

When sampling biodiversity, the observed richness is capped by the richness of the sampled environment and by the sample size (i.e., the number of individuals sampled/identified for the census). Estimating biodiversity from DNA metabarcoding is done via a series of nested samplings; from the sample collected in the field, to the microliters of amplified DNA loaded on the sequencer, to the final set of sequence reads. A biased subsampling at any level of the process will impact the subsequent steps.

An amplicon can therefore yield a small number of sequences for many distinct reasons. For example, the quantity of extracted DNA may be too low, the PCR may have been inhibited, or the PCR product may have been underrepresented in the mix of PCR products in the sequencing library. Usually such amplicons represent a small fraction of the experiment and display a number of MOTUs lower than that of most other amplicons. They thus behave similarly to negative controls (Fig. 8.6b–c; De Barba *et al.* 2014). For these PCRs, MOTU richness is bounded by the sequencing depth, which corresponds roughly to the 4,500 reads in Figure 8.6b–c. They might actually be artifacts, and it is important to remove these amplicons from the data set.

The stochasticity of the PCR has been well documented (Polz & Cavanaugh 1998; Best *et al.* 2015). A dysfunctional PCR can produce amplicons with low sequencing depth, which are excluded with the criterion just explained here. However, it can also

produce PCRs with a sufficient number of reads, but where some MOTUs are over or underamplified, irrespective of the primers or experimental conditions. Such dysfunctional PCRs should be excluded from the data set as they may affect the ecological conclusions. To our knowledge, none of the available pipelines implement a procedure that allows for the removal of dysfunctional PCRs, but we next propose a simple approach to this problem.

In a repeatable metabarcoding experiment, the PCR replicates of a single sample should be more like each other than to PCR replicates obtained for another sample. PCR replicates that do not fulfill this condition can be considered as dysfunctional. This simple rule relies on the following procedure. First, one needs to evaluate whether the distance d_w of each PCR replicate to the centroid of their corresponding sample (i.e., average MOTU abundances across PCR replicates) is less than the distance between all centroids (hereafter d_b). This can be viewed with the distributions of d_w and d_b values (Fig. 8.7). If these distributions display an intersection I, the corresponding d_I can be viewed as a dissimilarity threshold, above which a PCR replicate is

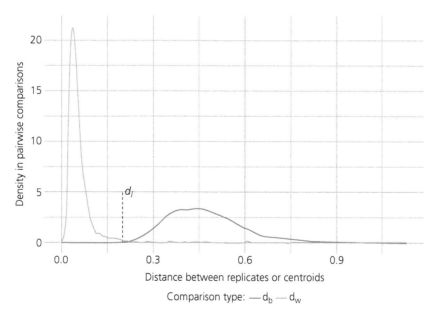

Fig. 8.7 Distributions of the distances between PCR replicates (blue) and between samples (red) obtained for a metabarcoding experiment on soil Fungi, using the ITS1 metabarcode (Fung01). A sample is assumed to be the average of its replicates. The dashed line indicates the lowest point between these two distributions and corresponds to the threshold above which a PCR is considered as dysfunctional.

too distant from its centroid to be reliable. PCR replicates above this threshold are therefore excluded from the analysis. As the sample centroids and distribution d_w and d_b will be affected by this exclusion, the whole process needs to be repeated iteratively until there is no more intersection between the distribution of d_w and d_b. If a sample is represented by a single PCR replicate during the iteration, it must be excluded as well. At the end of this procedure, the remaining PCR replicates are combined for each remaining sample. Although relatively simple to implement, this procedure may be too conservative, as it may exclude a noticeable number of replicates and even samples.

8.4 General considerations on ecological analyses

The ecological analyses of the curated metabarcoding data are largely driven by the initial scientific question. Our aim here is not to review all the available methods, and we will refer the reader to the literature dedicated to community ecology (Magurran & McGill 2011; Borcard *et al.* 2011; Escalas *et al.* 2013; Buttigieg & Ramette 2014), species detection (e.g., MacKenzie *et al.* 2006), ecological indicators (e.g., Baker & King 2010), trophic networks (e.g., Saint-Béat *et al.* 2015), and so on. Instead, we will discuss some considerations that are common to these analyses and their potential implications when drawing conclusions from metabarcoding data.

It is important to remember that most theories and metrics in ecology rely on the concepts of species and number of individuals, whereas many metabarcoding studies make use of MOTU and read counts. In the metabarcoding context, a MOTU corresponds to a set of sequence variants merged together in a single taxonomic unit based on a given similarity threshold. Considering a MOTU as a group of sequences allows PCR and sequencing errors to be dealt with, as well as intraclade variability (see Section 8.2.2). However, no explicit evolutionary rule supports the boundaries of a MOTU. For example, sequence dissimilarity of the ITS region varies from 0 to ca. 58% (2.5% on average) within fungal species, while the interspecific variability of that marker varies from 0 to 71% (30% on aver-

age; Schoch *et al.* 2012). In Bacteria, one single cell may harbor several versions of the 16S rRNA gene (Acinas *et al.* 2004). Therefore, MOTUs defined at a given similarity threshold remain taxonomically vague and can correspond to a species, a genus, a family, or even a group that does not fit with the current taxonomy, such as two genera within a family, and so on. In this context, what does it imply to transpose species-based ecological models, and their associated formulas, to analyses based on MOTUs? What does it mean to draw conclusions from a set of MOTUs where some are assimilable to species and others to families? Whether the analysis should be conducted at the MOTU level or at higher taxonomic levels remains an open question.

Likewise, how does one interpret a sequence count in metabarcoding studies? Does it correspond to species abundance in classical ecology? This is obviously not the case. Metabarcoding data only provides the relative proportions among the different MOTUs, while the number of individuals is an absolute value. Furthermore, the sequence counts of the different MOTUs depend on the presence of mismatches on the priming sites, the amplifiability of the considered metabarcode for the different MOTUs, and the number of copies of the target metabarcode per cell (Acinas *et al.* 2004; Vétrovský & Baldrian 2013), and per individual. One potential solution to reduce this problem would be to spike each DNA extract with the same quantity of a positive control in order to transform the number of sequences from a relative to an absolute value (Smets *et al.* 2016; Saitoh *et al.* 2016; Ushio *et al.* 2017). However, even in this situation, the direct correspondence between classical ecology and metabarcoding is not solved: metabarcoding data appear to be more related to the turnover of biomass than to the actual number of individuals (Yoccoz *et al.* 2012). To circumvent this problem, it is sometimes recommended to use presence/absence data, rather than the MOTU actual number of reads (or relative abundance). However, this approach gives too much weight to the errors that could remain in the data set, even after the most conservative data curation process. In this context, read abundance might be indicative of the signal reliability, and hence be considered as a probability of the MOTU presence in the sample.

It is clear that metabarcoding data contain pertinent ecological signals, but their interpretation following the classical ecological models should be made with caution. Based on these considerations, whether classical ecological theories can be tested with metabarcoding data is a legitimate question that is only beginning to be evaluated (e.g., Sommeria-Klein *et al.* 2016).

8.4.1 Sampling effort and representativeness

A first question that must be asked when dealing with biodiversity data is whether the system studied has been correctly sampled. This question remains in metabarcoding studies. In Chapter 4, we defined several sampling levels: the studied area is divided into sampling units, themselves composed of one or several samples, for instance, biological replicates (see Fig. 4.3). Furthermore, each sample is analyzed through one or several extractions and PCRs (technical replicates).

8.4.1.1 Evaluating representativeness of the sequencing per PCR

Even if PCRs with a very low number of sequences have been removed from the data set at this stage of the analysis (Section 8.3.2), the sequencing depth may vary among PCRs up to one or two orders of magnitude. It is, therefore, important to check if all PCRs have been sufficiently sequenced to be representative of their diversity in MOTUs. Several analyses commonly used in ecology can help to answer this question (see Magurran & McGill 2011; Chao & Jost 2012 for reviews), and some of them can be applied to DNA metabarcoding data. Rarefaction curves allow the rate at which the number of MOTUs increases with the sequencing depth for a given PCR to be determined (Fig. 8.8a, b). If these curves reach a plateau, one can get an idea of the representativeness of the data set for this PCR. They can be obtained in R with the functions `rarecurve` or `rarefy` of the vegan R package (Oksanen *et al.* 2007), the `iNEXT` function of the recent iNEXT R package (cran.r-project.org/web/packages/iNEXT), or the function `make_rarefaction_plots.py` in QIIME. This analysis is often used in microbial ecology to decide if

the PCR representativeness is sufficient. A common index used to evaluate PCR representativeness is Good's coverage estimator (Good 1953; see also Chao & Jost 2012 for a review),

$$Good's\ coverage\ estimator = 1 - \frac{F_1}{N}$$

where F_1 is the number of MOTUs that are singletons in a PCR (i.e., MOTUs of sequence count equal to 1), and N the total number of sequences obtained for the PCR. This estimator can be computed using the `iNEXT` function of the iNEXT R package. Originally, Good's estimator relies on the idea that the number of species observed only once in a sample is indicative of the number of undetected species and, therefore, of the sampling completeness. In metabarcoding studies, species become sequences, and sequences observed only once are numerous and usually considered as noise. Therefore, the suitability of Good's coverage estimator to evaluate whether the sequencing is representative of biological diversity can be questioned in metabarcoding. In the best case, it can be considered as an estimator describing the sequencing representativeness of the diversity within the PCR tube, including artifactual diversity.

8.4.1.2 Evaluating representativeness at the sampling unit or site level

Once the representativeness of each PCR has been assessed, it becomes possible to assess the sampling representativeness at the levels of the sampling unit or study site. Do the samples correctly describe the sampling unit? Are the sampling units representative of the studied area? Two methods can be implemented according to the number of samples considered. If many samples are available, the rarefaction curve (also referred to as the "accumulation curve" in this case) can be drawn, except here it is samples that are subsampled, and not sequences. The curve displays the accumulation of the number of MOTUs when considering an increasing number of samples. This curve also provides some ecological insights into species' spatial and temporal aggregation patterns if applicable to the study design. Note that sample-based rarefaction curves can be sensitive to the sequencing depth if the PCR

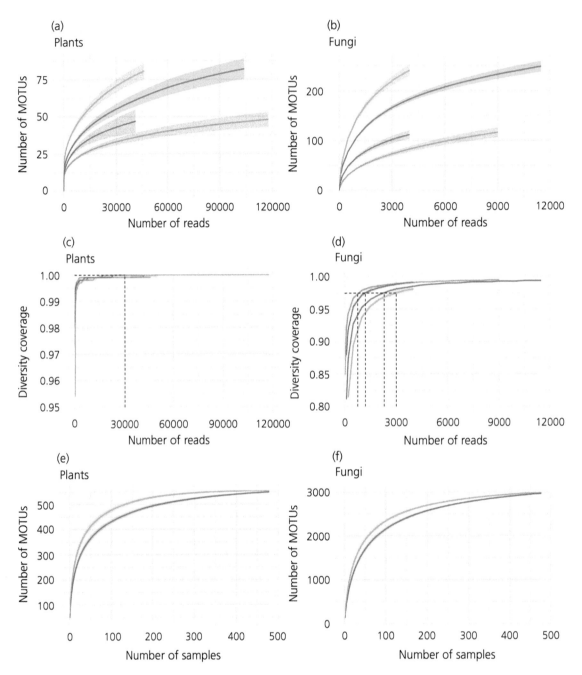

Fig. 8.8 Sampling and diversity coverage characteristics of metabarcoding data sets obtained with the plant Sper01 (a, c, e) and fungal Fung01 (b, d, f) metabarcodes. (a–b) Rarefaction curves. (c–d) Diversity coverage curves assessed with Good's coverage estimator. In (a–d), colors correspond to different samples of varying sequencing depths and MOTU richnesses. The dashed lines in (c–d) correspond to the sequencing depth for which the diversity coverage is equivalent among samples. (e–f) Sample accumulation curves for non-rarefied data (blue) and rarefied data (red) of the total data set. Data were rarefied using the criterion determined in (c–d).

diversity is not well covered. In this case, they should rather be plotted using the number of reads of the given set of samples considered, rather than the number of samples in that set (Gotelli & Colwell 2011). Sample-based rarefaction curves can be produced with the function `specaccum` of the vegan R package. If only two samples (i.e., biological replicates) are available, the only way to estimate the completeness of sampling is to evaluate the relative size of the intersect between the two samples: the larger it is, the better the representativeness.

8.4.2 Handling samples with varying sequencing depth

In metabarcoding data sets, when sequencing depth increases, more and more new MOTUs are detected, and more and more spurious sequences are also observed. For these two reasons, the observed MOTU richness tends to correlate with the sequencing depth within a PCR (Fig. 8.6). This feature is also observed for traditional biodiversity inventories, but only in the first case. Consequently, sampling and sequencing efforts have an influence on most diversity and distance metrics. Standardizing MOTUs' abundances by dividing by the sequencing depth (i.e., working with MOTUs' relative frequencies) might correct this phenomenon in some analyses, but remains inappropriate when estimating or comparing richness among samples.

A frequent practice for dealing with this issue is to standardize the sequencing depth among samples by rarefying the data set (i.e., by subsampling a given number of reads at random). This can be done with the `rrarefy` function of the vegan R package, or with dedicated options of the `single_rarefaction.py` script in QIIME. Most studies using this standardization procedure define a unique sequencing depth for the whole data set, which is chosen on the basis of the smallest sample. This convenient criterion is, however, not justified in many cases and the rarefaction procedure is currently under debate (McMurdie & Holmes 2014). The main argument against data rarefaction is that a significant volume of information is lost during this process, resulting in higher uncertainties in MOTUs' relative abundances. Detecting differences

between samples has therefore much less statistical power when the data are rarefied. To correct for differences in sequencing depth and avoid the rarefaction step, McMurdie and Holmes (2014) report different tools that enable stabilizing the variance between samples using mixture models, such as `calcNormFactors` in edgeR R package (Robinson & Smyth 2008), and `estimateSizeFactors` in DESeq (Anders & Huber 2010) R package. Nevertheless, if the sampling representativeness is high, the loss of information due to rarefaction should be negligible. In this case, fixing a subsampling size should be a compromise between maximizing the representativeness of the sample diversity while minimizing the number of samples excluded from the analysis, because their sequencing coverage is too low. For example, the plant MOTU diversity of all samples in Figure 8.8c is fully covered and the data can be confidently rarefied at a sequencing depth of ca. 30,000 reads, or even below. In contrast, fungal MOTU diversity in Figure 8.8d is unequally covered. To consider all samples in the analysis, one should standardize the data so that the diversity coverage is equal for all samples (e.g., equal to 97.5%). This therefore implies rarefying these communities at different sequencing depths.

For diversity estimations and comparisons, one can also use extrapolation estimates of species' richness that are able to deal with differences in sampling effort and diversity coverage. These estimates have been used for a long time in ecology (e.g., the Chao1 index, reviewed in Gotelli & Colwell 2011) and can be easily obtained with the `specpool` or `estimateR` functions implemented in the vegan R package, the `iNEXT` function of the iNEXT R package, and the `alpha_diversity.py` script implemented in QIIME. For metabarcoding data, it remains uncertain whether their use is appropriate as they often rely on singletons or rare species, which might be artifactual, as already explained in this chapter. Although efforts are currently being made toward developing estimates that reduce the impact of spurious singletons (Chiu & Chao 2016), they remain sensitive to artifactual MOTUs that are not singletons and that could persist in the data set, even after a careful curation procedure.

8.4.3 Going further and adapting the ecological models to metabarcoding

Metabarcoding studies have been shown to provide patterns of diversity (e.g., latitudinal gradients of diversity, distance decay of similarity) usually coherent with theoretical predictions and observations based on traditional sampling methods. However, as already outlined, there is still a need for methods specifically adapted to metabarcoding data. For example, the traditional diversity metrics should be refined to better comply with both the unclear nature of the MOTU concept and the uncertain meaning of sequence abundances. Hill numbers, which correspond to an "effective number of species" under a theoretical model (here the equal abundances of all species) are promising for that purpose. They are increasingly used in ecology as they offer many advantages over classical diversity indices (see Chao *et al.* 2014 for a review). A Hill number depends on a q parameter specifying the importance given to rare species. The more q increases, the less rare species have weight in the estimated diversity. Interestingly, Hill numbers provide a unified framework for the three most popular diversity indices in ecology: species richness; the Shannon entropy index; and the Gini-Simpson index, corresponding to $q = 0$, 1, and 2, respectively. In the context of DNA metabarcoding, by tuning this q parameter, one can lower the impact of artifactual sequences. Decomposition of diversity into alpha, beta, and gamma components can also be carried out within the Hill numbers framework. As such, they have connections with species' compositional similarity indices such as the Jaccard, the Sørensen, or the Morisita–Horn indices. This allows for both presence/absence and abundance data within a unified framework to be studied, an issue often debated in metabarcoding analyses. Along the same lines as the Rao's quadratic entropy index (Rao 1986), Hill numbers can be computed based on a species distance matrix (phylogenetic, taxonomic, or functional distances). In that case, an extra parameter allows the impact of closely related species on the diversity to be considered. Therefore, the simultaneous adjustment of both q and this extra parameter can help to lower the impact of PCR and sequencing errors, which produces close sequence variants, usually in low frequencies. Whatever the real meaning of read frequencies and how they relate to the actual abundances of species, diversity indices considering abundances seem to give more robust results by lowering the impact of rare MOTUs, as observed by Haegeman *et al.* (2013).

## CHAPTER 9

Single-species detection

On the spectrum of eDNA-based analyses, single-species detection lies at the opposite end to DNA metabarcoding. Indeed, while the latter provides diversity information on a wide taxonomic group without a prior knowledge of the list of taxa to be found, the former approach focuses on a restricted and predefined taxon, usually a single species (Lawson Handley 2015; Taberlet *et al.* 2012a). Historically, eDNA-based single-species detection has been considerably successful in biomonitoring invasive, elusive, or threatened animal species, mainly in aquatic environments (e.g., Bohmann *et al.* 2014; Lawson Handley 2015; Mahon *et al.* 2013). For example, eDNA-based surveys of the endangered great crested newt (*Triturus cristatus*) in the United Kingdom had a much higher detection rate than conventional methods and did not generate false positives (Biggs *et al.* 2015). The high sensitivity of the technique and the possibility of remote detection are the main reasons for the attractiveness of the eDNA approach (Thomsen & Willerslev 2015). In the case of invasive species, eDNA assays targeting a particular species are especially interesting, as they can detect low density populations (e.g., in the early stages of invasion). They can also detect any life stage, even those for which morphological identification is unreliable or time-consuming, such as larvae (Lawson Handley 2015).

Amplification of a specific barcode from a complex mixture lies at the heart of eDNA-based single-species detection, either via traditional polymerase chain reaction (PCR) (Ficetola *et al.* 2008; Jerde *et al.* 2011), or more often via quantitative PCR (qPCR; e.g., Biggs *et al.* 2015; Takahara *et al.* 2012; Thomsen *et al.* 2012b). Traditional PCR and qPCR can both reveal the presence of a target DNA sequence in an environmental sample. However, qPCR is more informative than its traditional counterpart. It can monitor in real time the number of target copies present in the mix (Higuchi *et al.* 1993), whereas simple PCR can only measure their final concentration, which is not necessarily proportional to their original concentration (Kim *et al.* 2013b; Valasek & Repa 2005). For biomonitoring purposes, qPCR can thus provide information on the abundance of the species of interest in the sampled environments (Lawson Handley 2015). For example, several studies observe a positive correlation between species density and/or biomass and eDNA concentration monitored using qPCR (Pilliod *et al.* 2013; Takahara *et al.* 2012; Thomsen *et al.* 2012b).

The rest of this chapter will address the topic of single-species eDNA detection from the point of view of qPCR. Compared with DNA metabarcoding, the technical constraints of qPCR are limited, as this approach does not involve sequencing and the associated bioinformatic analyses. That being said, there are a number of pitfalls that should be avoided to ensure validity of the results, and these will be further discussed next.

9.1 Principle of the quantitative PCR (qPCR)

9.1.1 Recording amplicon accumulation in real time via fluorescence measurement

The key aspect of a qPCR is the real-time recording of accumulating DNA sequences during the amplification process (Higuchi *et al.* 1993). This is generally achieved through the continuous measurement of a signal emitted during the incorporation of a fluorescent probe in the amplified DNA (Kim *et al.*

Environmental DNA for Biodiversity Research and Monitoring. Pierre Taberlet, Aurélie Bonin, Lucie Zinger, & Eric Coissac, Oxford University Press (2018). © Pierre Taberlet, Aurélie Bonin, Lucie Zinger, & Eric Coissac 2018. DOI: 10.1093/oso/9780198767220.001.0001

2013b). Several detection chemistries exist for this purpose, among which the SYBR green assay (Wittwer *et al.* 1997) and the TaqMan probe system (Heid *et al.* 1996; Holland *et al.* 1991) are the most popular. SYBR green is a fluorescent dye with an affinity for double-stranded DNA. During the amplification process, it binds to the produced amplicons in a non-specific way, and emits 200-fold more fluorescence in this bound state than when free (Zhang & Fang 2006). The TaqMan probe system relies on a particular oligonucleotide called the TaqMan probe (20 to 40-base long), which can specifically hybridize to a region within the DNA fragment of interest, and which is labeled with a fluorescent reporter and a quencher on its 5'- and 3'-ends, respectively. In such a configuration (i.e., when the reporter and quencher are on the same oligonucleotide), the signal from the reporter is canceled out by the quencher through a phenomenon called fluorescence resonance energy transfer (Valasek & Repa 2005). During template extension, the reporter is cleaved out by the DNA polymerase due to its 5'→3' exonuclease activity (Holland *et al.* 1991), and it starts to fluoresce. For both the SYBR green and the TaqMan probe assays, the level of fluorescence at any given time is thus proportional to the amount of amplified DNA in the reaction tube (Kim *et al.* 2013b; Zhang

& Fang 2006). Their advantages and disadvantages are discussed in Section 9.2.1.

9.1.2 The typical amplification curve

When measuring the fluorescence during a qPCR, the signal (and thus the number of DNA copies) follows a typical amplification curve (Fig. 9.1). At first, there is a lag phase, during which only a few amplicons are produced and the signal is indistinguishable from the background noise. After several cycles Ct, the amount of PCR product exceeds a detection threshold value and efficiency of the PCR increases progressively up to its maximum, which corresponds to the linear part of the curve. When the reactants start to become limited, amplification efficiency slows down, until reaching a plateau phase during which very few new amplicons are produced.

9.1.3 Quantification of target sequences with the Ct method

The number of cycles Ct to pass the detection threshold (Fig. 9.1) is central in qPCR quantification, as it is inversely proportional to the number of target sequences initially present in the reaction tube (Heid *et al.* 1996). As a result, for an un-

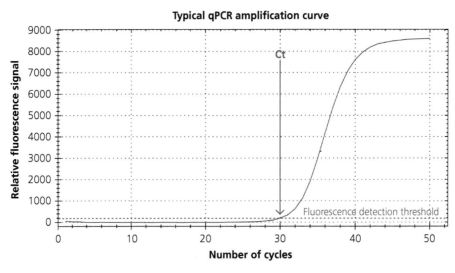

Fig. 9.1 Typical qPCR amplification curve. This figure shows the evolution of the fluorescence signal (y-axis) according to the number of cycles (x-axis) during a qPCR. The Ct (cycle threshold, i.e., the number of cycles necessary to detect a fluorescent signal) is determined when the signal exceeded a given detection threshold (dotted red line). In this example, with a threshold set at 200, Ct corresponds to 30 cycles.

known sample, the initial concentration in target sequences can be estimated using a linear regression of Ct values versus the log of concentrations for a set of standards (Fig. 9.2; Valasek & Repa 2005; Wong & Medrano 2005). The sensitivity of the Ct method is remarkable, as it can detect concentrations as low as 0.5 DNA copy/μL (Ellison et al. 2006).

9.2 Design and testing of qPCR barcodes targeting a single species

9.2.1 The problem of specificity

From the description of the two main chemistries used for qPCR fluorescence detection in Section 9.1.1, one can easily gather that the specificity of the primers used for amplification is crucial, especially in the SYBR green method where the fluorescent dye intercalates between bases in double-stranded DNA regardless of the sequence (Valasek & Repa 2005). As a result, emission of fluorescence will increase even if amplification is non-specific, or if primer dimers are produced. To reduce the risk associated with amplification and detection of unwanted amplicons, a melting curve analysis can be carried out after the SYBR green assay (Zhang & Fang 2005). This analysis relies on the fact that different DNA fragments will have different melting temperatures (Tm), that is, temperatures at which 50% of the two strands of DNA dissociate (Ririe et al. 1997). By monitoring denaturation of the PCR products and fluorescence levels over a temperature gradient, it is thus possible to draw the melting curve, where the presence of distinct peaks suggest that the PCR amplicons are heterogeneous, and/or possibly mixed with primer dimers (Valasek & Repa 2005). Despite its lack of specificity, the SYBR green assay is still largely popular as it is cheap and simple to use (Smith & Osborn 2009; Zhang & Fang 2006). The TaqMan probe method possesses an additional safeguard to the detection of non-specific amplification, as the fluorescent probe itself is specific to the fragment of interest. However, this imposes a supplementary experimental constraint, as there should be a third region specific to the species of interest in the amplified fragment, in addition to the two priming regions. Notwithstanding this need for a specific probe and the higher cost associated with the TaqMan system (Zhang & Fang 2006), its success remains unfailing in our community (Kim et al. 2013b).

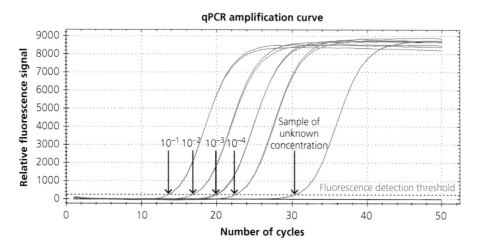

Fig. 9.2 Quantification of the initial concentration in target sequences using the Ct method. The red lines represent the amplification curves for standards of known concentrations: 10^{-1} ng/μL, 10^{-2} ng/μL, 10^{-3} ng/μL, and 10^{-4} ng/μL (in duplicates). The blue line represents the amplification curve for a sample whose unknown concentration in target sequences can be estimated at 2.6×10^{-7} ng/μL using a linear regression of standard Ct values versus a log of standard concentrations.

9.2.2 qPCR primers and probe

There are some similarities between the philosophies of primer design for DNA metabarcoding and for qPCR (see Chapter 2). Both qPCR primers should target sequences which are at the same time highly conserved for the studied species (no intraspecific polymorphism) and variable compared with the other taxa that may occur in the area. This is really important on the 3′-end of the primers (i.e., which is extended by the *Taq* polymerase). Primer specificity is even more of a major concern, if the objective is to survey a rare species in an environment where close relatives are abundant, and could lead to false positives (Wilcox *et al.* 2013). As with the ideal metabarcode, the ideal qPCR barcode should be as short as possible to be amplifiable from possibly degraded eDNA, especially since qPCR is more efficient with short fragments. However, unlike in DNA metabarcoding, sequence variability within the fragment is irrelevant as there is no sequencing involved in the qPCR process. Such variability could even be detrimental if a TaqMan probe must be designed.

The use of flawed or biased qPCR primer sets can have a tremendous impact on the final results, generating false positives, or even false negatives if target and non-target templates compete for the primers and an internal probe is used (Wilcox *et al.* 2013; Yu *et al.* 2005). It can also skew the Ct values, and thus affect estimation of template concentration (Wilcox *et al.* 2013). It is thus crucial to validate the qPCR assay sensitivity and specificity *in silico* (Tréguier *et al.* 2014), on a sequence database containing sequences of the species of interest and taxa that could occur in the same environment. This validation also needs to be done empirically, using for example samples from mesocosms experiments (Thomsen *et al.* 2012b).

9.2.3 Candidate qPCR barcodes

Non-coding, highly polymorphic loci, which are well represented in public sequence databases, are good candidate qPCR barcodes. Such loci include, for example, the chloroplastic *trnL* (UAA) intron in plants (Taberlet *et al.* 1991), or the animal mitochondrial control region (Boore 1999).

The mitochondrial cytochrome c oxidase subunit I (COI) gene (Hebert *et al.* 2003b) could also work for animals as it displays high interspecific variability and there is an abundance of reference sequences. However, special care must be taken that primers do not fall in a region showing intraspecific variation on the third nucleotide of each codon.

9.3 Additional experimental considerations

9.3.1 General issues associated with sampling, extraction, and PCR amplification

qPCR single-species detection has a lot in common with DNA metabarcoding with regards to the first steps of the experimental process (sampling, extraction) and even PCR amplification. Much of the information gathered in Chapters 4, 5, and 6 is thus also relevant to qPCR analysis. In particular, the use of proofreading *Taq* polymerases must be considered carefully: indeed, as explained in Chapter 6, their attractive low error rate is counterbalanced by the fact that they can trim the 3′-end of the primers due to their 3′→5′ exonuclease activity, and hence lead to non-specific amplifications.

9.3.2 The particular concerns of contamination and inhibition

As it can detect very small quantities of starting DNA, qPCR is even more sensitive to contaminants than classical PCR. It is therefore even more important to take precautions, both in the field and in the lab, to limit potential contamination. It is also highly recommended to monitor the occurrence of contaminants using a number of negative controls sufficient to overcome any stochastic effect. It should be noted, however, that qPCR analysis has the major advantage of not requiring post-PCR handling, because fluorescence data acquisition takes place directly during the reaction without opening the tube (Heid *et al.* 1996; Valasek & Repa 2005; Zhang & Fang 2006). As a result, in a lab dedicated exclusively to qPCR, levels of ambient contamination by PCR products are expected to be low.

A wide array of molecules can interfere with DNA isolation and/or amplification, either by hindering cell lysis during extraction and thus access to DNA, by degrading or capturing DNA molecules, or by inhibiting the *Taq* polymerase activity (Wilson 1997). These inhibitors have the potential to greatly skew qPCR assays (Hargreaves *et al.* 2013), which are highly dependent upon a perfect relationship between the DNA amount originally present in the sample and ultimately amplified at each PCR cycle. It is thus wise to consider inhibition in the qPCR experimental process (Jane *et al.* 2015). A widespread practice for this purpose is to check inhibition prevalence using template dilution series (Hargreaves *et al.* 2013; Wilson 1997).

CHAPTER 10

Environmental DNA for functional diversity

There is now increasing evidence that natural populations and species are declining at unprecedented rates (Barnosky *et al.* 2011). To which extent species loss will impact ecosystems functions and the multiple services they provide has fueled intense research over the last 20 years. While many experimental and empirical assessments clearly show that biodiversity has a positive effect on ecosystem function and stability (Cardinale *et al.* 2012; Naeem *et al.* 2012), predictions for the future remain hampered by the variety of threats to biodiversity, and by the idiosyncrasies and complexity of each natural ecosystem. It is therefore necessary to augment current models with more data collected across taxa and ecosystems, and at multiple spatial and temporal scales (Urban *et al.* 2016). Environmental DNA can potentially make an important contribution to this research area by accelerating the collection of multitaxa data, which could be used in turn to test, calibrate, or refine current predictive models.

However, looking at biodiversity only through the taxonomic lens might not always provide meaningful information on the link between biodiversity and ecosystem function (Díaz & Cabido 2001): species can be functionally redundant, or might instead show intraspecific functional differences. There is now a growing consensus that we need more data quantifying and qualifying how species acquire and use resources to produce biomass. This will allow biodiversity to be better related to the functioning of ecosystems and improve current predictions, particularly under scenarios of non-random species loss (Díaz & Cabido 2001; Naeem *et al.* 2012; Urban *et al.* 2016). Here, we will discuss the extent to which DNA metabarcoding, and more generally eDNA,

can provide such information (Fig. 10.1; see also Vandenkoornhuyse *et al.* 2010 for a review). Most of the following is particularly appropriate when unraveling the functions of microbes, but we will provide, when possible, some examples of application for macroorganisms.

10.1 Functional diversity from DNA metabarcoding

10.1.1 Functional inferences

DNA metabarcoding primarily provides a taxonomic perspective of the community studied. To which extent it can inform on the functional properties of communities depends on the actual ecosystem functions being investigated, as well as on the *a priori* knowledge of local species' functional characteristics. This knowledge usually involves coarsely defined functional groups (e.g., woody, leguminous, graminoid plants; shredders or decomposer soil organisms; pathogenicity or decomposition role of certain microorganisms). Thus far, such coarse functional information, when inferred from taxa identified with DNA metabarcoding, have been assigned manually, and have already provided valuable insights into the structure of food webs (i.e., diet analyses) or the potential role of microorganisms on the performance of their host (see Chapters 16 and 17).

However, functional annotations of molecular taxonomic units (MOTUs) could be refined by using specific functional information provided by ecophysiological traits. For example, it is now well established that the specific leaf nitrogen content

Environmental DNA for Biodiversity Research and Monitoring. Pierre Taberlet, Aurélie Bonin, Lucie Zinger, & Eric Coissac, Oxford University Press (2018). © Pierre Taberlet, Aurélie Bonin, Lucie Zinger, & Eric Coissac 2018.
DOI: 10.1093/oso/9780198767220.001.0001

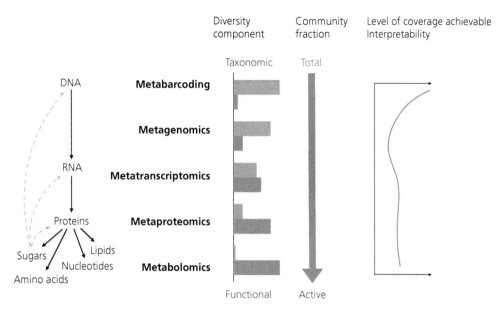

Fig. 10.1 The "omics" toolbox.

(or area) per unit mass and stem specific density reflect plant strategies for resource acquisition and use, respectively (Díaz *et al.* 2016). Based on a list of vascular plant taxa obtained with DNA metabarcoding, such ecophysiological traits can be retrieved from functional reference databases, such as the TRY Plant Trait Database (Kattge *et al.* 2011). Compared to plants, the functional classifications and appropriateness of functional traits in animal and decomposer organisms are less established. Existing databases are taxon-specific and still need to be compiled and unified (reviewed in Briones 2014; but see Gravel *et al.* 2016). Nevertheless, one consensus is that organism body size or mass is a relatively simple functional trait that encompasses many biological properties of species such as their metabolic rates, the extent of their spatial niche, as well as their broad trophic position (Woodward *et al.* 2005). As it concerns all living forms, body size can therefore be considered as a universal trait that may provide information on the broad spatial scaling and functional characteristics of multitaxa communities (e.g., top-down or bottom-up control in aquatic and terrestrial below-ground or above-ground food webs). Using this information has already provided a better understanding on soil

organism community assembly and spatial scaling (e.g., Bahram *et al.* 2016; Zinger *et al.* 2017). This could constitute a first step toward a functional understanding of multitaxa inventories compiled with universal DNA markers, until more complete functional databases are available.

Functional classifications are even less well determined when it comes to microscopic organisms, for which taxonomic diversity are also partially characterized. Nevertheless, several meta-analyses have suggested that some complex functional characteristics (e.g., photosynthesis, methane/sulfate-related functions) constitute key evolutionary innovations that are conserved at high taxonomic ranks in Bacteria (Martiny *et al.* 2015; Philippot *et al.* 2010) or Fungi (Chagnon *et al.* 2013; McGuire *et al.* 2010). Building on this observation, several inference tools based on current knowledge of microorganism taxonomy and ecology have been proposed. For instance, FUNGuild (Nguyen *et al.* 2016) is a database that compiles coarse ecological descriptions for many fungal taxa (e.g., saprotroph, ectomycorrhiza, plant pathogens, and so on). It also includes a functional annotation tool from MOTUs' taxonomic assignments, for example, with the UNITE fungal rDNA sequence database (Kõljalg *et al.* 2005). Similarly, PiC-

RUST (Langille *et al.* 2013) and Tax4fun (Aßhauer *et al.* 2015) are annotation tools that predict bacterial communities' functions from Greengenes or SILVA taxonomic annotations, and reference bacterial genomes or metabolic profiles. The emergence of such DNA-function coupled databases for other organisms in the near future will probably facilitate the functional annotation of multitaxa inventories obtained with DNA metabarcoding.

However, one inherent limitation of these inference methods is the extent to which the assumption of trait conservatism is met. If some functional traits are conserved within lineages, this feature is not necessarily true for many traits important to nutrient cycling. For example, many different bacterial clades can consume simple carbon sources (Martiny *et al.* 2015). In the same way, similar ecophysiological traits are present in different clades of the angiosperm phylogeny (Díaz *et al.* 2016). Therefore, some common functional properties have been acquired independently and repeatedly during evolution.

As DNA metabarcoding may produce MOTUs that are either undescribed in reference databases, or unresolved at the species level, inferring functional characteristics should be done carefully. In that case, FUNGuild functional entries are all annotated with a confidence ranking and the assignment tool includes a procedure excluding MOTUs with annotations higher than the genus level (Nguyen *et al.* 2016). PiCRUST strongly relies on the assumption of a phylogenetic trait conservatism. It involves (i) constructing a phylogenetic tree with environmental 16S MOTUs and reference databases; (ii) inferring the ancestral states of genome traits from the phylogenetic tree; and (iii) predicting contemporary states for undocumented MOTUs. This tool also does not model gene gain or loss, or horizontal transfers (Langille *et al.* 2013). Tax4fun makes such inference only from the nearest 16S sequence, but remains based on the assumption of phylogenetic trait conservatism (Aßhauer *et al.* 2015). Furthermore, it should be kept in mind that the two latter tools only infer potential functions (i.e., contained in genomes, that are not necessarily expressed). In that line, a trait value for one single animal or plant species may vary depending on the environmental conditions or depending on the genomes and epigenomes of individuals. Consequently, functional inferences from molecular taxonomic inventories should always be carefully evaluated by experts, and when possible, validated with experimental approaches measuring functional or life history trait values of community members.

10.1.2 Targeting active populations

For most questions related to the relationship between biodiversity and ecosystem functioning, it is commonly accepted that the retrieved functions should correspond with organisms that are metabolically active. When making use of eDNA, this assumption is not necessarily met due to the presence of dormant propagules (e.g., spores, seeds, pollens) as well as the persistence of extracellular DNA in the environment, sometimes over long time periods. Note that the following considerations are applicable for both metabarcoding and metagenomics (see Section 10.2).

In the microbial ecology literature, it is often suggested that using ribonucleic acid (RNA) instead of DNA as starting material would ensure that active populations are targeted (reviewed in Blazewicz *et al.* 2013). This hypothesis relies on the correlation repeatedly observed between total rRNA content and growth rates in Bacteria cultures. On this basis, several studies targeting active bacterial communities have extracted RNA from different matrices to construct cDNA libraries of the 16S rRNA gene. Comparatively, this approach has been far less used for metazoans (but see e.g., Guardiola *et al.* 2016). However, it has several limitations. First, RNA is a short-living and unstable molecule that requires specific extraction methods optimized for the environmental matrix studied (see Section 10.2). This makes its use more constraining and renders cross-experiment comparisons difficult. Second, this hypothesis does not hold true in many cases, as dormant propagules may actually contain large numbers of ribosomes (reviewed in Blazewicz *et al.* 2013).

For soil Bacteria or Fungi, it has been recently suggested that extracellular DNA (or "relic" DNA) may distort treatment effects or spatiotemporal patterns and should be excluded (Carini *et al.* 2016). Several methods have been proposed to maximize the capture of active populations and exclude extracellular DNA. For example, molecules such as propidium monoazide (Nocker *et al.* 2007), propidium iodide (Luna *et al.* 2002), or ethidium monoazide (Soejima *et al.* 2007) are DNA intercalating molecules that cannot penetrate cells, unless membranes are compromised, such as in dead organisms. When added to environmental samples, they complex with extracellular DNA only, hence rendering this fraction of the eDNA pool inaccessible for polymerase chain reaction (PCR) amplification. However, using these approaches might not always be appropriate. A substantial proportion of intact microbial cells are actually dormant in the environment: they represent about 40% in aquatic ecosystems and up to 80% in soils (Lennon & Jones 2011). On the other hand, microbial active biomass continuously releases DNA into the environment through cell turnover. It is hence possible that microbial extracellular DNA may mainly correspond with active populations (Zinger *et al.* 2016), provided that the remobilization of eDNA by microbial or plant biomass is relatively fast in matrices such as the rhizosphere. In this context, excluding extracellular DNA may have exactly the opposite effect to the one desired, by giving more weight to dormant communities. The short- to mid-term persistence of extracellular DNA in the environment could also be an advantage when the ecosystem process of interest occurs at larger temporal scales than the dynamics of the community studied. For example, the composition of microbial communities can change during the day (Gunnigle *et al.* 2017). Relating such punctual observations to processes occurring at the scales of weeks or months (e.g., productivity, carbon storage, and so on) might hamper the understanding of the relationship between biodiversity and ecosystem functioning. This bias may be potentially lower when making use of total or extracellular DNA as starting material, because extracellular DNA expands the spatiotemporal window of observation. Of course, these considerations are yet to be carefully explored and are likely to be strongly context-dependent. It is hence crucial that more efforts are made toward better characterizing the rate and mechanisms involved in extracellular DNA cycling prior to drawing any misleading conclusions.

Given the uncertainties on the meaning of DNA versus RNA, or total/extracellular DNA versus intracellular DNA, functional interpretations of molecular-based taxonomic inventories remain difficult, as they do not guarantee that all the detected taxa use resources and produce biomass. When one is interested in short-term processes, a feasible alternative to avoid this limitation is to probe the biomass, and therefore the newly produced DNA. Such probing can be performed using DNA stable-isotope probing (DNA-SIP; see Neufeld *et al.* 2007 and Radajewski *et al.* 2000 for detailed protocols), where stable-isotope-labeled ^{13}C or ^{15}N sources are added to the environmental sample and assimilated by the active biomass. Labeled nucleic acids are then added to a CsCl density gradient and can be isolated through ultracentrifugation. This approach is so far mainly used to identify the food web involved in the cycling of a given substrate. However, it requires high concentrations of isotopically labeled substrate, as well as isotope-related facilities. In addition, it cannot be used to monitor the biological transformation of complex natural organic compounds that are often encountered in nature. Another, less constraining approach involves incubating samples with bromodeoxyuridine (BrdU), which is a thymidine analog that can be incorporated into the DNA of newly synthesized cells (Urbach *et al.* 1999). Labeled DNA can be then retrieved by immunocapture. Although a few microbial isolates have been reported to not incorporate BrdU (Urbach *et al.* 1999), this approach seems a viable alternative to assess the active fraction of microbial diversity for high-throughput applications. It does not require specific nutrient sources either, and has already allowed identifying differences in carbon use between soil and aquatic bacterial or microeukaryotic taxa (Goldfarb *et al.* 2011; Hanson *et al.* 2008; Tada & Grossart 2014).

10.2 Metagenomics and metatranscriptomics: sequencing more than a barcode

As mentioned in Section 10.1, an inherent limit of DNA metabarcoding to retrieve organisms' functions lies in its limited taxonomic resolution for certain clades. Other approaches from the "omics" toolbox (Fig. 10.1) allow this limitation to be avoided by giving access to genes, transcripts, proteins, and metabolites contained in an environmental sample. Here, we will focus only on shotgun sequencing of environmental DNA (metagenomics) and RNA (metatranscriptomics), and will refer interested readers to reviews dedicated to metaproteomics and metabolomics for further information (e.g., de Bruijn 2011a).

Metagenomics and metatranscriptomics basically involves extracting DNA and RNA from an environmental matrix, respectively. Then, the retrieved nucleic fragments are sheared, converted into cDNAs when starting from RNAs, and subjected to high-throughput sequencing (often referred to as shotgun sequencing; see Chapter 7). This apparently simple approach hides complex methodological considerations and procedures. In particular, characterizing a given metagenome or metatranscriptome requires high sequencing coverage.

10.2.1 General sampling constraints

As sequencing cost remains an influential factor of metagenomics or metatranscriptomics analysis, it will generally determine the number of samples that can be analyzed. An appropriate sequencing coverage of one single metagenome or metatranscriptome depends on the number of species, on the identity of these species, which differ in their genome size, as well as on the structure of the community of interest (i.e., species relative abundances). This requires having a good prior knowledge of the studied community. If this is not the case, a pilot experiment should be performed. These considerations have been recently implemented in several methodological procedures that aim to guide researchers in estimating the sequencing coverage required for their samples in bacterial metagenomics applications (e.g., Ni *et al.* 2013).

10.2.1.1 Optimization of the number of samples

Having a good idea of the number of samples that can be processed will allow the sampling design to be optimized for the question of interest (Chapter 4) while keeping an acceptable level of biological replication (Prosser 2010). For example, the sampling

design could be set with a prior knowledge of the environmental gradient being studied in order to minimize the number of environmental conditions sampled. Alternatively, pilot experiments using DNA metabarcoding could be conducted to guide the selection of the optimal number of spatial, temporal, or biological units that would maximize the community differences in the field, or in experimental studies.

10.2.1.2 Enrichment in target organisms

Depending on the question and taxa of interest, it might be necessary to reduce the complexity of the studied metagenomes or metatranscriptomes. Bacteria are largely dominant in terms of biomass and diversity in the environment and bacterial genomes will overwhelm metagenomics or metatranscriptomics data in general, which is obviously undesirable if one is interested in fungal or metazoan communities, for example. In addition, metagenomics and metatranscriptomics require the retrieval of large amounts of clean nucleic acids, which might be difficult with environmental samples. Several relatively simple protocols enable enriching samples in the organisms of interest, while reducing the number of undesired molecules (humic acids, polyphenols, and so on) at the same time.

It is possible to divide the sample into different fractions through sifting or filtering at a given mesh size, in order to retain the organisms of interest based on their body size. For example, seawater samples from the Tara expeditions were successively filtered on sieves of decreasing mesh size to separate viruses, and pico, nano, micro, and mesoplanktonic communities (Karsenti *et al.* 2011). Similar approaches have been applied to marine benthic eukaryotic communities (Leray & Knowlton 2015). To target metazoans living in sediments, samples are commonly sieved with 1 mm (Aylagas *et al.* 2016) or 0.45 µm mesh (Fonseca *et al.* 2010). Such bulk samples can be constituted for aboveground insects using classical capture methods, such as light, pitfall, or malaise traps, as well as nets (Ji *et al.* 2013; Zhou *et al.* 2013). Epigeic and/or endogeic metazoan communities from litter or soil can be isolated using Berlese-based approaches, such as the flotation-Berlese-flotation protocol recently proposed for DNA metabarcoding and metagenomics applications (Arribas *et al.* 2016). Soil or sediment Bacteria can be separated from particles and eukaryotic cells by subjecting suspensions of samples to Nycodenz density gradient centrifugation (Bakken 1985; Delmont *et al.* 2011). This procedure relies on differential isopycnic centrifugation, where cells move at different rates depending on their density, and stop moving when reaching the zone of the gradient at which their density is the same. Each zone of the gradient corresponds with coarse size classes that can be each further subjected to successive filtrations to refine size fraction classes (Portillo *et al.* 2013). To our knowledge, no attempt has been made to determine the composition of the eukaryotic fraction retrieved with this approach. These different approaches are good candidates to enrich metagenomes with the target organisms. However, for most of these methods, large size fractions may still contain small organisms adsorbed on large particles or associated with larger organisms, and small size fractions may still contain extracellular DNA belonging to larger organisms (e.g., Leray & Knowlton 2015). New developments in flow cytometry will undoubtedly improve size-fractionation of microbial communities (Frossard *et al.* 2016), but require optimization for samples rich in organic matter.

When one is interested in host-associated organisms, metagenomics and metatranscriptomics data will be spoiled by the host genome. Fractionation or selective cell lysis may here again minimize the amount of host DNA retrieved, and different methods have been proposed for that purpose. Surface-associated microbes (e.g., microorganisms associated with the rhizoplane, the phyllosphere, or the insect cuticle) can be detached from their living substrate through surface washing with detergents (e.g., Tween-20) with or without mechanical treatments (e.g., shaking with beads or sonication), followed by centrifugation or filtration to concentrate microbial biomass (Birer *et al.* 2017; Richter-Heitmann *et al.* 2016; Ushio *et al.* 2015). When focusing on endogenous microbial communities, a pre-treatment of tissue surfaces may be required. This can be done by applying a bleach treatment (Richter-Heitmann *et al.* 2016), although it does not seem to be a determinant step when studying the overall endogenous communities

in insects (Hammer *et al.* 2015). Host tissues need to be scarified or disrupted in sterile saline solution, and subjected again to shaking to release endogenous microbes (Fieseler *et al.* 2006; Sessitsch *et al.* 2012). Alternatively, host tissues can be disrupted chemically through enzymatic treatments (Burke *et al.* 2009). The resulting solution is then centrifuged or filtered with a sufficiently small enough pore size to make sure that most of the host cell fragments are excluded. Unfortunately, there is currently no standard protocol for these applications and the isolation efficiency seems to be strongly dependent on the architecture of the tissues considered. Pilot experiments should be conducted to ensure optimal enrichment protocols.

10.2.1.3 Enrichment in functional information

Metagenomes, and to a lesser extent metatranscriptomes, comprise vast amounts of non-protein-encoding regions, which represent 5–25% of prokaryotic genomes and 25–99% of eukaryotic genomes (Mattick 2004). In metatranscriptomes, these non-protein-encoding regions mostly correspond to ribosomal RNAs and related introns (e.g., ITS, IGS). Ribosomal regions yield useful taxonomic information and are often used as classical barcodes. It is possible to take advantage of this property to refine the taxonomic identification of community members (Andújar *et al.* 2015; Coissac *et al.* 2016; Papadopoulou *et al.* 2015), and hence their potential functional characteristics. However, these regions often dominate metagenomes or metatranscriptomes and might hide more relevant functional information (i.e., from protein-encoding genes). The other non-protein-encoding regions, like tRNA or regulatory and structural DNA motifs, are rarely considered as their presence is difficult to link to pertinent ecological functions.

Metatranscriptomics is more adapted to focus on the community's functional properties, but can be difficult to implement due to the significant logistical constraints associated with RNA handling (see Section 10.2.2). When applicable, it might be necessary to reduce the amount of rRNAs. In eukaryotes, it is common practice to selectively target mRNAs through polyA enrichment approaches, which basically involve capturing polyadenylated RNAs using polyT probes. This procedure is not relevant for prokaryotes, whose transcripts are not poly-adenylated. In this case, the two most popular alternatives are either subtractive hybridization, or digestion with an exonuclease. The subtractive hybridization involves capturing rRNA fragments with rRNA-specific probes. Its efficiency will inherently rely on the appropriateness of the probes to the target community, as well as the quality of the RNA extracted, as low-quality RNA fragments are less likely to hybridize on probes. Digestion with an exonuclease preferentially targets rRNAs, tRNAs, and fragmented mRNAs through their 5′-monophosphate end. It seems to be efficient on any taxonomic group, except on Archaea and certain Actinobacteria (Wang *et al.* 2012).

10.2.2 General molecular constraints

Retrieving high quality DNA or RNA is a prerequisite for obtaining reliable functional data (Chapter 5; Bowers *et al.* 2015; Thomas *et al.* 2012). Furthermore, to be representative of the ecosystem under study, it is recommended that nucleic acids are extracted from substantial amounts of soil, water, or sediments. One should keep in mind that sample fractionation, nucleic acid extraction, and storage procedures yield different pictures of the studied community (Birer *et al.* 2017; Deiner *et al.* 2015; Delmont *et al.* 2011; Philippot *et al.* 2012), and should therefore be appropriately evaluated before running a full experiment. This consideration is even more important when working with RNA, as these molecules are highly unstable in environmental samples, particularly due to extracellular RNase activity. If RNA cannot be extracted directly after sample collection, it is advisable to store samples at -80°C, or in stabilizing solutions such as RNAlater (Ambion, Austin, TX) or RNAprotect (Qiagen, Valencia, CA), although these are also known to provide different pictures of the studied community composition (McCarthy *et al.* 2015 and references within). Compared to DNA, RNA is also more strongly adsorbed by particles due to the presence of an additional hydroxyl group, as well as its single strand structure, which can form hydrogen bonds. Its isolation suc-

cess is strongly affected by the soil clay and humic acid content (de Bruijn 2011a; Wang *et al.* 2012) and extraction protocols must be optimized. When possible, studying metagenomes and metatranscriptomes in parallel provides insightful information on active versus potential functionality of the community. In this case, it is better to co-extract RNA and DNA from the same samples, using a bead mill homogenization and organic extraction procedure, which is followed by a digestion of RNA or DNA on different aliquots (Griffiths *et al.* 2000).

The preparation of sequencing libraries is also an important part of the metagenomic and metatranscriptomic pipelines to consider as it may introduce biases in downstream analyses. When working with low DNA amounts, the risk of contamination is high, which can jeopardize the quality and assembly of sequence reads (Bowers *et al.* 2015). In this case, one can perform whole metagenome amplification based on multiple displacement amplification (MDA), which is one of the current isothermal and high-fidelity alternatives to PCR (Fakruddin *et al.* 2013). Briefly, it involves using random hexamers to initiate genome amplification, which is carried out by a highly processive and strand-displacing enzyme. However, the amplification remains biased toward certain groups and such biases seem context-dependent (Yilmaz *et al.* 2010). There have been many uncertainties on whether this method is acceptable for metagenomic studies (Fakruddin *et al.* 2013; Thomas *et al.* 2012). However, recent library preparation kits enable work to be carried out with much lower amounts of DNA (few picograms to nanograms) and seem to minimize the aforementioned biases (Bowers *et al.* 2015).

10.2.3 From sequences to functions

A good overview of existing methods for analyzing metagenomics data sets is presented in Sharpton's study (Sharpton 2014). Here, we will only discuss some basic key points, focusing on metagenomics, although many of the principles can also be applied to metatranscriptomic data.

When examining a PCR-based DNA metabarcoding data set, only the taxonomic diversity is assessed because only a single marker is amplified and sequenced. A metagenome data set is far more complex. It contains a collection of short sequences with a size usually ranging from 100 bp to a few hundred base pairs. These sequences are randomly sampled from an eDNA extract and therefore simultaneously represent the taxonomic and the gene (i.e., function) diversities present in the original sample. It is therefore crucial to simplify metagenomic data prior to further analyses. Two main strategies should be considered for this purpose: assembling reads to produce longer sequences, and classifying the sequences according to their type (mainly rDNA and protein-encoding sequences).

10.2.3.1 Assembling (or not) a metagenome

Sequencing the genome of a single species by a shotgun approach requires the assembly of overlapping reads to produce longer sequences (i.e., contigs). To ensure the reliability of the assembly procedure, the sequencing effort must be much larger than the size of the genome, as this increases the probability of drawing overlapping reads.

A metagenome contains sequence reads from all the different members of the community, with each member providing a number of reads roughly depending on its abundance in the community and its genome size. Another characteristic of metagenomes is that two similar reads can belong to different loci or different species. These two features imply that only deeply sequenced metagenomes can be properly assembled. It is therefore highly recommended to run assemblers specifically designed for metagenomes, rather than for single genomes. For example, MetaVelvet (Namiki *et al.* 2012), and Meta-IDBA (Peng *et al.* 2011) are two assemblers specifically dedicated to metagenomic data, and which can deal with the unevenness of sequencing coverage among community members.

Nowadays, it is possible to cover the metagenomes of simple bacterial communities such as those found in acid mine drainage or ocean oil spills relatively well (reviewed in Howe & Chain 2015). However, when considering highly complex metagenomes, such as those from soils (Howe & Chain 2015), one may ask whether an assembly-based approach is appropriate as these metagenomes remain difficult to assemble with current methods and computing capacities.

10.2.3.2 Sorting contigs or reads in broad categories

Sorting sequences in broad categories is a key step in the data analysis of metagenomes. It mainly aims at extracting sequences corresponding to rDNA genes and those corresponding to protein-encoding genes. This classification can be carried out on the contigs produced by the assembly procedure, or directly on the reads if no assembly has been performed. In this latter case, an assembly can be done for each category after the read classification.

Several programs are useful to classify reads into categories by matching them against a reference database, but here we will present only two robust and efficient ones. SortMeRNA (Kopylova *et al.* 2012) allows selecting sequences matching rDNA genes, by using a sliding window to search for short similarity regions between each sequence and an rRNA database. It is very fast and classifies sequences into broad categories (i.e., LSU and SSU), and in large taxonomic groups (i.e., Archaea, Bacteria, and Eukaryota). Diamond (Buchfink *et al.* 2015) allows nucleic sequences to be aligned against a protein database. The alignment is performed at the protein sequence level, which increases the sensibility of the algorithm. The two programs can be used in a single pipeline, where reads that are not identified as rDNA genes by SortMeRNA are sent to Diamond to test if they correspond with protein-encoding genes. However, one drawback of this approach is its dependency on available references, which precludes identifying new genes, and therefore new functions absent from public protein databases. *Ab initio* gene prediction methods can be adapted to deal with metagenome data. These algorithms seek biases in the sequence composition to infer the likelihood of a sequence to encode for a protein. Among these tools, we can cite Glimmer-MG (Delcher *et al.* 1999) and MetaGeneMark (Zhu *et al.* 2010), which are two adaptations of famous programs Glimmer (Delcher *et al.* 1999) and GeneMark.hmm (Lukashin & Borodovsky 1998), classically used to annotate bacterial genomes.

10.2.3.3 Extracting functional information via taxonomic inferences

Taxonomic information retrieved from metagenomes can be used to make indirect functional inferences. This approach is similar to the one used for DNA metabarcoding (see Section 10.1.1 for details).

Basically, the sequence of a gene depends on two components: its function and its taxonomic origin. These two components do not impact the sequence with the same strength, the influence of the function being far stronger than the influence of the taxonomic origin. In other words: two genes encoding two proteins having the same function but belonging to two different species are far more similar than two genes from the same species encoding two different functions. DNA barcode sequences contain a strong taxonomic signal as they all belong to the same gene (e.g., COI gene for animals; Hebert *et al.* 2003a). Analyzing the subset of reads corresponding to a barcode is therefore the straightest and most computationally effective way to retrieve taxonomic information from metagenomes.

Another strategy often coined "binning" can also be used to classify reads into taxa. Binning algorithms usually examine sequence composition biases to classify sequences. These biases can be simple ones, such as the guanine-cytosine content, or more often short DNA words like tetramers with differential frequencies in genomes according to the taxonomic origin. `PhyloPythia` and `PhyloPythiaS` (McHardy *et al.* 2007; Patil *et al.* 2012) rely on support vector machine (SVM) to classify reads according to these biases. The advantage of binning algorithms is that they favor the taxonomic message over the functional message by examining only intrinsic and taxon-specific properties of the sequences, independent from their functions. However, the taxonomic information retrieved via binning algorithms is not as reliable as that contained in a barcode. The models used by these algorithms are learnt from reference databases constituted of genomic sequences.

10.2.3.4 Functional annotation of metagenomes

Sequences are typically annotated functionally, based on their similarity with proteins of known function. It is commonly accepted that two proteins sharing at least 30% of their identity have similar biochemical functions. Sequence similarity can be estimated by comparing metagenome sequences with public protein databases like UniProtKB/trEMBL, or the less comprehensive, but manually curated UniProtKB/Swiss-Prot (The UniProt Consortium 2017). Sequences can also be compared with protein domain databases like Pfam (Finn *et al.* 2016), which is a collection of protein domains built using conserved regions observed across the UniProtKB proteins. Compared to full-length protein sequences, a domain has the advantage of usually being associated with a single function. The UniProtKB and Pfam entries are annotated by cross-references to functional ontology databases like Gene Ontology (Ashburner *et al.* 2000) or the Kyoto Encyclopedia of Genes and Genomes (KEGG; Kanehisa 2000). Gene Ontology is made up of three distinct ontologies describing the cellular components (i.e., where gene products are localized within or outside the cell, the molecular function of gene products, and the biological process in which the gene products are involved). The two last classifications are the most important for annotating metagenome data. KEGG is a metabolic database describing the pathways themselves, as well as the associated biochemical reactions and the enzymes catalyzing them.

Publicly available metagenomic pipelines represent a good starting point when initiating a metagenomic analysis. Many tools have been developed, but the two most famous ones are the European Bioinformatics Institute (EBI) metagenomics pipeline (https://www.ebi.ac.uk/metagenomics; Mitchell *et al.* 2016) and the MG-Rast pipeline (http://metagenomics.anl.gov; Meyer *et al.* 2008), both available through a web interface. However, there are no established rules for the analysis of metagenomic data. This difficulty has recently led to the CAMI challenge ("Critical Assessment of Metagenomic Interpretation"; http://www.cami-challenge.org), an initiative aiming to improve the recovery of the complex information encoded in metagenomes.

## CHAPTER 11

Some early landmark studies

In the last couple of decades, several technological barriers have been broken down in molecular biology. This has undoubtedly fueled the ever-growing appeal for environmental DNA that can be observed in ecological research over the same period (Fig. 11.1). A turning point was reached in 2005, with the breakthrough of next-generation sequencing technologies. Whereas many earlier eDNA projects were based on tedious cloning and low-throughput Sanger sequencing, next-generation sequencing made it realistic to exhaustively characterize complex communities using DNA metabarcoding and metagenomics. However, long before sequencing throughputs came close to being what they now are, several scientists anticipated the potential of eDNA and the range of scientific questions it could help address. This chapter is an opportunity to revisit several seminal papers that paved the way for today's eDNA studies.

11.1 Emergence of the concept of eDNA and first results on microorganisms

The expression "environmental DNA" was coined during the 1980s, in an article explaining how to isolate and purify nucleic acids from microbial origins found in various types of sediments (Ogram *et al.* 1987). The first attempts to analyze this environmental DNA involved either DNA-DNA hybridization, or separating the different molecules by high-resolution gel electrophoresis or cloning prior to sequencing (Olsen *et al.* 1986; Pace *et al.* 1986). Quickly, these early studies began to lift the veil on an unsuspected microbial diversity that morphology- or cultivation-based methods had failed to reach. At this time, it was estimated that 80% of existing Bacteria still remained to be discovered

(Wayne *et al.* 1987), and that a gram of soil, for example, could contain up to 10,000 undescribed prokaryote species (Torsvik *et al.* 1990).

In twin papers published in 1990 in *Nature*, Giovannoni *et al.* (1990) and Ward *et al.* (1990) brought the characterization of microbial communities to a higher level. For this purpose, they combined PCR amplification (Giovannoni *et al.* 1990) or synthesis of complementary cDNA from 16S rRNA templates (Ward *et al.* 1990) with cloning and sequencing to obtain 16S rDNA metabarcodes. A common feature of these two studies was their focus on two constrained habitats: the Sargasso Sea and the cyanobacterial mat of Octopus Spring in Yellowstone Park. Indeed, the Sargasso Sea displayed extremely low nutrient conditions, while in Octopus Spring, water temperature ranges from 90°C near the source to 55°C on the edges, and the water pH is alkaline. Giovannoni *et al.* (1990) analyzed the sequences of twelve 16S rDNA bacterioplankton clones obtained from the Sargasso Sea and identified two main bacterial groups. The first one had never been previously described, even though it represented a considerable proportion (15%) of the total prokaryotic DNA found in the Sargasso Sea and could also be detected in other waters. The second group branched within oxygenic phototrophs (i.e., Cyanobacteria, prochlorophytes, and chloroplasts) and included three distinct but closely related lineages. However sequence homology was rather low with known phototrophs.

In the cyanobacterial mat of Octopus Spring, microscopy and bacterial cultures had mainly pointed at the role of the cyanobacterium *Synechococcus lividus* and of the photosynthetic bacterium *Chloroflexus aurantiacus* in the production of the mat. However, this mat also harbored several other Bacteria species

Environmental DNA for Biodiversity Research and Monitoring. Pierre Taberlet, Aurélie Bonin, Lucie Zinger, & Eric Coissac, Oxford University Press (2018). © Pierre Taberlet, Aurélie Bonin, Lucie Zinger, & Eric Coissac 2018. DOI: 10.1093/oso/9780198767220.001.0001

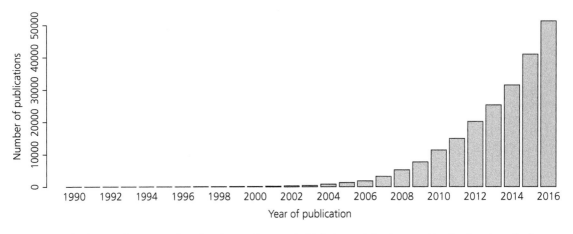

Fig. 11.1 Results of a Web of Science search conducted on March 10, 2017 with the keywords "environmental DNA" or "metabarcoding" or "metagenomic*".

(Woese 1987). Interestingly, a previous analysis of membrane lipid components suggested the presence of other undescribed microorganisms (Ward *et al.* 1985). In the Ward *et al.* (1990) study, 16S rDNA sequences were obtained from 16S rRNA fragments retrieved in the Octopus Spring mat. Eight sequence "types" (i.e., MOTUs, although this concept did not exist at that time), probably corresponding to eight different taxa, could be distinguished. No strict match was observed between any of these eight sequences and 16S rDNA reference sequences obtained from microorganisms previously isolated in this mat or from similar habitats. The highest homology was observed with bacterial, rather than archaeal, references. Overall, the Giovannoni *et al.* (1990) and Ward *et al.* (1990) results corroborated the then widespread view that only a small fraction of the Earth microbial diversity had been cataloged.

11.2 Examining metagenomes to explore the functional information carried by eDNA

In the late 1990s, microbiologists began to explore whether they could extract valuable functional insight from eDNA samples, on top of the taxonomic information these could provide. Handelsman *et al.* (1998) published an article introducing, for the first time, the notion of soil "metagenome", when speak-

ing of the "collective genomes of soil microflora". This paper described how the genomes of uncultured soil microorganisms could be accessed by cloning and performing a functional analysis of the soil metagenome. The underlying objective was to identify new synthetic pathways for biomolecules of interest. At that time, functional analysis was limited to screening the clones for biological activity and production of new molecules, instead of sequence analysis or annotation. Venter *et al.* (2004) pushed metagenomics into the big data era. This was quite an achievement. They filtered surface water collected from four sites in the Sargasso Sea with filters targeting microorganisms. After DNA extraction, insertion into plasmid vectors, and cloning, the clones were sequenced from both ends using the Sanger technology. The huge data set obtained included 1.66 million reads (average length 818 bp), which after assembly resulted in more than one billion base pairs of non-redundant sequences. Sequence homology of rRNA genes retrieved in the data revealed approximately 150 bacterial phylotypes, which had never been described before. Based on sequence depth coverage of whole genomes, the number of species was estimated to at least 1,800 species. Functional annotation allowed identification of over 1.2 million previously unknown genes. These included at least 782 new rhodopsin-like photoreceptors, potentially belonging to a non-chlorophyll-based pathway for harvesting light energy. Eventually, these findings highlighted the importance of energy metabolism in nutrient-poor environments such as marine ecosystems.

In another pioneer study, Dinsdale *et al.* (2008) used a metagenomics approach to characterize the functional potential of microbial and viral communities in nine biomes, including subterranean, hypersaline, marine, freshwater, coral-, mosquito-, microbialite-, fish-, and terrestrial animal-associated biomes. A total of 45 microbiomes and 42 viromes were analyzed by pyrosequencing. This resulted in approximately 15 million sequences significantly

similar to functional genes. In microbiomes, an important proportion (more than 30%) of the identified genes was involved in carbohydrate or protein metabolism, while an additional 15% was associated with respiration and photosynthesis. The most represented functions in viromes were nucleic acid metabolism and virulence. Most functional pathways were identified in all the environments, but each displayed a few dominant pathways. This was further confirmed by multivariate analyses (Fig. 11.2), which detected most of the variance between environments (79.8% for microbiomes,

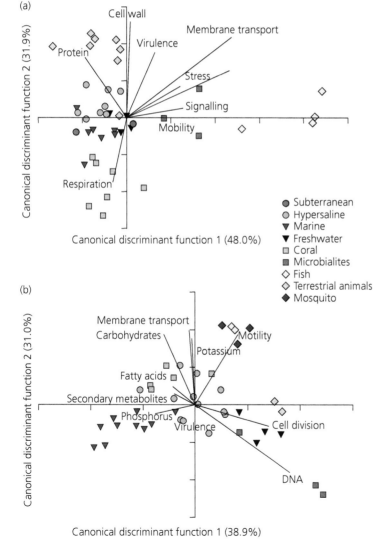

Fig. 11.2 Canonical discriminant analysis of the metagenome of 45 microbiomes (a) and 42 viromes (b) belonging to nine distinct biomes. The symbols differentiate the biomes, and the vectors indicate the metabolic function significantly influencing the separation of the metagenomes, with the length of the vector being proportional to the strength of this influence (from Dinsdale *et al.* 2008).

and 69.9% for viromes). Altogether, these results showed that there is a unique functional signature for each biome, and that this signature was coherent with the biological processes known to be at play in the corresponding biome. These findings agreed with those of Tringe *et al.* (2005), obtained on a more limited set of terrestrial and marine nutrient-rich environments.

11.3 Extension to macroorganisms

As outlined in Section 11.2, microbiologists had understood as early as the 1990s that intracellular DNA extracted from environmental samples could help resolve the functional and taxonomic diversity of microorganisms found in these habitats. Was it then realistic to try to obtain useful information on macroorganisms based on the extracellular DNA traces, gametes, or mucus they left behind in their environment? Researchers began to address this question in the 2000s, in two very different contexts: (i) to detect contemporary species that are difficult to observe at certain periods of time or developmental stages; and (ii) to reconstruct past communities, by taking advantage of the tremendous persistence of DNA under particular environmental conditions.

For example, Ficetola *et al.* (2008) explored the reliability of eDNA-based assessment of species presence in freshwater habitats. For this purpose, they focused on the American bullfrog *Rana catesbeiana*, a species listed among the 100 worst invasive alien species (Lowe *et al.* 2000). First, they conducted experiments in controlled environments, by keeping bullfrog tadpoles at different densities in 3-liter aquariums (with six replicates each time) for 24 h. DNA was extracted from water sampled in these aquariums, and used as a template to amplify a short fragment of the mitochondrial cytochrome *b* gene with primers highly specific of the bullfrog. Amplification was successful whatever the tadpole density, and systematically failed in empty aquariums. Second, this approach was extended to natural habitats (1,000–10,000 m² ponds) displaying different population densities (no bullfrog, low bullfrog density, high bullfrog density, with three field replicates each time). Again, DNA amplification was only possible if bullfrogs were present in the pond, and this was statistically significant.

Similarly, amplification success was significantly higher in ponds with higher population densities than in ponds with lower densities. Pyrosequencing confirmed that all amplification products perfectly matched the bullfrog *cyt-b* sequence, once artifacts commonly associated with this type of sequencing (e.g., in homopolymer stretches) had been removed. This work reports the first evidence that eDNA-based species detection is a sensitive method for aquatic habitats, even at low population densities. It also opened new avenues for the study and management of biodiversity, by demonstrating that eDNA species surveys could be an interesting alternative to traditional census techniques, especially in the case of elusive, threatened, or invasive species.

When adsorbed to a substrate, like clay or humic substances, and kept in stable and favorable environmental conditions, eDNA molecules can stay largely unaltered for millennia (see Chapter 15 for more details). Willerslev *et al.* (2003) exploited this feature in one of the first studies involving the analysis of ancient eDNA released by macroorganisms. At that time, paleoecology and paleontology relied mostly on the study of hard and soft tissues from preserved plant or animal specimens. However, due to the shortage of exploitable macrofossils, the reconstruction of past environments was very fragmented, even in areas with many specimens. Willerslev *et al.* (2003) assessed whether ancient DNA extracted from permafrost and cave sediment cores could help reconstitute Holocene and Pleistocene plant and animal communities. For this purpose, they analyzed five permafrost cores collected in Siberia and dated from today up to 1.5–2 million years ago, and three cave and coastal sediment cores collected in New-Zealand covering the time period from 0.6 to 3 thousand years ago. Ancient eDNA was extracted from different horizons along these cores. To study the past flora, a 130-bp fragment of the chloroplast ribulose-bisphosphate carboxylase (*rbc*L) gene was amplified, cloned, and sequenced. Out of 290 clones, 274 gave sequences that could be assigned at least to the class level by a Basic Local Alignment Search Tool (BLAST) search. This led to the identification of 11 classes (or subclasses), 23 orders, and 28 families of angiosperms, gymnosperms, and mosses across all samples. Interestingly, the plant sequences recovered from the 300,000 to 400,000-year-old permafrost horizon represented the oldest attested ancient DNA at that time. Permafrost samples showed dramatic changes in plant taxonomic diversity and composition, with a ratio of herbs to shrubs decreasing over time and possibly correlated with the extinction of megafauna. Cave sediment sequences were more diverse and typical of a pre-human environment.

The past vertebrate fauna was examined by amplifying, cloning, and sequencing several fragments (100–280 bp) of mitochondrial DNA (i.e., 16S rDNA, 12S rDNA, cytochrome *b* gene, and control region). Eight different vertebrate taxa were identified in Siberian permafrost: woolly mammoth, steppe bi-

son, horse, reindeer, musk ox, brown lemming, hare, and an unidentified bovid. Cave and coastal sand sediment samples contained DNA from three moa species, of which two are extinct, and one parakeet species which did not occur in the area anymore. Overall, the presence of extinct taxa in the samples, as well as the sudden changes in taxonomic composition along the cores, suggested that the results were sound and that the stratigraphic integrity had been preserved. In summary, Willerslev *et al.* (2003) had demonstrated that DNA from permafrost and temperate sediments was a good source of information on past ecosystems, and a reliable alternative to ancient pollen and macrofossils.

## CHAPTER 12

Freshwater ecosystems

The first report of eDNA from surface water concerns a test for detecting potential fecal pollution originating from human, cattle, sheep, and pigs via the amplification of mitochondrial DNA using specific primers (Martellini *et al.* 2005). Overall, freshwater ecosystems represent one of the most active targets for eDNA studies on macroorganisms. Four reasons can explain this keen interest. First, the landmark paper of Ficetola *et al.* (2008) stimulated further research on freshwater by showing that the detection of a frog species was possible using eDNA from ponds (see Chapter 11). Second, biomonitoring freshwater ecosystems is imposed by law in more and more countries (e.g., European Council 2000) and the potential of eDNA for this purpose has been identified relatively early (Baird & Hajibabaei 2012; Hajibabaei *et al.* 2011). Third, large efforts have been made to identify invasive fish species in North America (e.g., Jerde *et al.* 2011; Mahon *et al.* 2011, 2013). Finally, the greater sensitivity of eDNA versus traditional approaches is now well recognized (e.g., Civade *et al.* 2016; Jerde *et al.* 2011; Pilliod *et al.* 2013; Valentini *et al.* 2016; Wilcox *et al.* 2016).

12.1 Production, persistence, transport, and detectability of eDNA in freshwater ecosystems

The detectability of eDNA in freshwater ecosystems depends on its production, degradation, transport, and diffusion rates. Figure 12.1 presents a conceptual model summarizing the influence of various biological and environmental factors affecting the detectability of eDNA.

12.1.1 Production

According to the published studies, it seems that fish (e.g., Civade *et al.* 2016; Jerde *et al.* 2011; Takahara *et al.* 2012; Thomsen *et al.* 2012b), amphibians (e.g., Dejean *et al.* 2012; Pilliod *et al.* 2013; Valentini *et al.* 2016), and mollusks (Clusa *et al.* 2016; Goldberg *et al.* 2013) release a relatively large amount of eDNA into their environment. Furthermore, eDNA from arthropods seems more difficult to detect, probably due to their exoskeleton that reduces the desquamation of epidermal cells: Tréguier *et al.* (2014) had difficulties detecting crayfish eDNA at low abundances (but see Section 12.6 for opposite published results). In addition to the eDNA of animals living in the freshwater ecosystem, some allochthonous eDNA can be integrated into the system, leading to false positives when focusing on freshwater communities (Fig. 12.1). Such contaminants might come, for example, from droppings of domestic or wild animals, or from effluents of sewage.

12.1.2 Persistence

Dejean *et al.* (2011) performed several experiments to assess the persistence of eDNA in aquariums (with bullfrog tadpoles), and in small ponds (with sturgeons). They demonstrated that eDNA can persist from a few days to a few weeks and, in many cases, less than one month. They conclude that such a relatively short persistence time opens new perspectives in conservation biology by allowing access to the presence/absence of endangered or invasive species. Subsequent studies confirmed this time frame for eDNA persistence (Dunker *et al.* 2016; He *et al.* 2015; Piaggio *et al.* 2014; Pilliod *et al.* 2014;

Environmental DNA for Biodiversity Research and Monitoring. Pierre Taberlet, Aurélie Bonin, Lucie Zinger, & Eric Coissac,
Oxford University Press (2018). © Pierre Taberlet, Aurélie Bonin, Lucie Zinger, & Eric Coissac 2018.
DOI: 10.1093/oso/9780198767220.001.0001

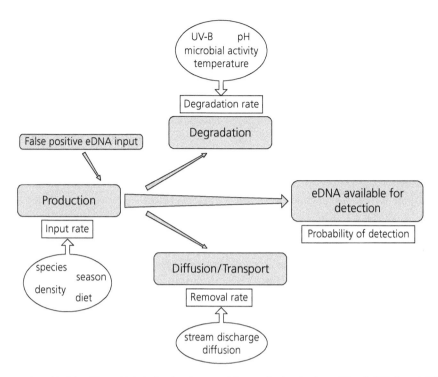

Fig. 12.1 Conceptual model of the different biological and environmental factors affecting the detectability of eDNA (from Strickler *et al.* 2015). Reprinted by permission from *Biological Conservation*, 183. Elsevier. Copyright (2015). Katherine M. Strickler, Alexander K. Fremier, and Caren S. Goldberg, Quantifying effects of UV-B, temperature, and pH on eDNA degradation in aquatic microcosms.

Thomsen *et al.* 2012b). The influence of some environmental and biological parameters on eDNA persistence has been assessed. It appears that water temperature is an important factor, with better persistence occurring at lower temperatures (He *et al.* 2015; Lacoursière-Roussel *et al.* 2016b; Pilliod *et al.* 2014). Additionally, sunlight and the absence of particles promote eDNA degradation (Gutiérrez-Cacciabue *et al.* 2016; He *et al.* 2015). Finally, a high level of microbial activity also increases the degradation rate.

12.1.3 Transport/diffusion distance

There are no publications, as yet, that study the eDNA diffusion rate in water for lentic ecosystems. Thus, in lakes and ponds, one solution for taking into account the diffusion problem is to systematically obtain several samples per sampling unit. In lotic ecosystems, despite the fact that a few papers have been published, it is difficult to draw

sound conclusions. Pilliod *et al.* (2014) found that eDNA from stream-dwelling salamanders is not detectable at more than 5 m downstream. Two other studies tried to detect lake-dwelling species in outflowing rivers. Deiner *et al.* (2014) looked at *Daphnia longispina* and *Unio tumidus* eDNA in the River Glatt outflowing from Lake Greifensee in Switzerland, using polymerase chain reaction (PCR) with specific primers and subsequent sequencing. They sampled 12 localities up to 12.3 km downstream of the lake. They found *D. longispina* eDNA in all localities, but they did not find any *U. tumidus* after 9.1 km. However, we cannot exclude that, for a small organism such as *D. longispina*, whole individuals have been transported downstream, making any inference about eDNA transport problematic. The second study, by Civade *et al.* (2016), targeted fish eDNA using a metabarcoding approach in the outlet of Lake Aiguebelette (France). They sampled five localities up to 6.9 km downstream. They concluded that the distance of eDNA detection is around 2–3 km (i.e., three to six times less than previous estimates). However, it is not surprising that different studies provide different results, which are influenced by the initial concentration of eDNA, the speed of the current, and by the detection sensibility, among others.

12.1.4 Detectability

In freshwater, eDNA mainly comes from cells or cell remains, at least for fish (Turner *et al.* 2014). Additionally, a small fraction can come either from free DNA or from DNA adsorbed on small particles. In this context, the best way to obtain eDNA from water is via a filtration of a relatively large volume of water (up to 100 L; Valentini *et al.* 2016) that is much more efficient that the initial strategy based on alcohol precipitation (Ficetola *et al.* 2008). Obviously, the detectability depends on the sampling strategy, on the volume of water, and on the way the filtration is performed. Currently, there is no consensus among the different research groups about the filtration approach. The volume of filtered water varies from 300 mL (e.g., Deiner *et al.* 2016) to 50–100 L (Civade *et al.* 2016; Valentini *et al.* 2016), and different filtering devices are used (from classical vacuum filter cups to capsules with large effective filtration area, up to 1,300 cm^2). In the same way, the type of membrane and the pore sizes are different among the different published studies. Obviously, in addition to these technical aspects, eDNA production, degradation, and transport/diffusion strongly influences the detectability.

12.2 Macroinvertebrates

Macroinvertebrates are widely used for assessing the quality of freshwater ecosystems (e.g., Cortes *et al.* 2013; Rosenberg & Resh 1993; Usseglio-Polatera *et al.* 2001). Consequently, it is tempting to use DNA metabarcoding for their taxonomic identification (Baird & Hajibabaei 2012). Two approaches can be implemented. The first one consists of collecting macroinvertebrates in the field, mixing them all together, extracting DNA from this bulk sample, and then using DNA metabarcoding only for the taxonomic identification. This approach is presented in Chapter 18. The second possibility relies on di-

rectly sampling eDNA, usually via water filtration. Thomsen *et al.* (2012b) were the first to demonstrate the possibility of obtaining eDNA from macroinvertebrates, targeting a dragonfly (*Leucorrhinia pectoralis*) and a crustacean (*Lepidurus apus*).

Bista *et al.* (2017) pushed the macroinvertebrate eDNA analysis a step further by looking at time series in a lake (16 time points over one year, and for each time point, 1 L of water collected in two sites of the lake). Two cytochrome c oxidase subunit I (COI) markers were used, including the classical Folmer region of 658 bp (Folmer *et al.* 1994) and a much shorter marker of 235 bp (modified from Carew *et al.* 2013). Despite the amplification of many invertebrates, they focused on an ecologically important group, the Chironomidae (Diptera). As expected, the shorter marker worked much better than the longer one, and the analysis of Chironomidae results showed that eDNA can reveal the temporal dynamics over the year. Thus, eDNA metabarcoding of macroinvertebrates represents a powerful tool for tracking seasonal variation at the ecosystem scale.

Two other studies have targeted invasive species. The first tracked mosquito species using both qPCR with internal probes and metabarcoding with primers targeting Culicidae (Schneider *et al.* 2016). Water samples were collected in 46 natural water bodies (3 × 15 mL of water per sampling site). The results were congruent between the two molecular methods, and the detection probabilities were, in most cases, higher for eDNA-based approaches than for traditional surveys. The second study described the distribution of an invasive gammarid species (*Gammarus fossarum*) in the UK using a metabarcoding approach (Blackman *et al.* 2017). Thus, eDNA-based techniques are efficient and reliable for the unambiguous detection of invasive macroinvertebrate species.

12.3 Diatoms and microeukaryotes

Because of their specific response to particular ecological conditions, diatoms are powerful bioindicators of water quality in streams and rivers (reviews in Lobo *et al.* 2016 and Stevenson *et al.* 2010). Two approaches can be implemented to explore their biomonitoring potential: the first is based on the species composition of diatom communities and on the ecological preferences of the different taxa; the second relies on diatom diversity as a general indicator of ecosystem health. For both approaches, DNA metabarcoding has the potential of being very helpful for the ecological assessment of rivers and streams, as classical identification based on morphology is relatively difficult. In this context, several DNA markers have already been developed, mainly targeting the nuclear 18S rRNA gene. Using primers amplifying the V4 region (Zimmermann *et al.* 2011), DNA-based identification was found more efficient than the classical morphological method for distinguishing the different diatom species (Zimmermann *et al.* 2015) and for assessing water quality (Visco *et al.* 2015). However, the whole V4 region is about 400 bp long. Primers also targeting the V4 region, but with a shorter fragment of 150 bp long, can be designed (see primer pair Baci01 in Appendix 1). Using a shorter marker will inevitably lead to a loss of taxonomic information, but this might not be critical for water quality assessment (Keck *et al.* 2016), and it leads to much more robust results. Both primer pairs (the long and the short) targeting the V4 region also amplify many other eukaryotes, which might provide additional informative data. In this context, it should be more appropriate to use universal eukaryote primers (e.g., Euka02 and Euka03 in Appendix 1). More recently, Apothéloz-Perret-Gentil *et al.* (2017) demonstrated that a taxonomy-free approach is possible when implementing an eDNA approach for assessing water quality via the diatom index. They deduced this index directly from the relative abundance of the different molecular taxonomic units (MOTUs) obtained, instead of trying to first give a scientific name to the MOTUs.

12.4 Aquatic plants

Of the relatively few papers dealing with eDNA from aquatic plants, all used a single-species approach based on qPCR with internal probes. Scriver *et al.* (2015) were the first to develop several markers for identifying 10 species of aquatic macrophytes, and to show that macrophyte detection is possible from eDNA extracted from water. Fujiwara *et al.* (2016) confirmed this result. They surveyed

23 ponds in Japan for the presence of an invasive aquatic plant, *Egeria densa*, originating from South America. Matsuhashi *et al.* (2016) also investigated the geographic distribution of another aquatic plant species (*Hydrilla verticillata*) in 21 ponds in Japan by using three different approaches: eDNA, visual observation, and past records. The past records indicated the presence of *H. verticillata* in all these 21 ponds. The visual observation recorded its presence in only two ponds. eDNA had a better detection probability and recorded the presence in five ponds, including the two where the plant had been spotted with visual observation. Additionally, these authors conducted aquarium experiments using *H. verticillata* and *E. densa* to assess the relationship between eDNA concentration and plant biomass, which they found to be positive, but not always significant. It might be interesting to implement DNA metabarcoding for the identification of aquatic plants from eDNA extracts, for example using the primer pair Sper01 (Appendix 1), which is probably resolutive enough, and to compare it with the species-specific approach.

12.5 Fish, amphibians, and other vertebrates

12.5.1 Species detection

Several studies now support the fact that fish and amphibian detection is more efficient with eDNA than with traditional approaches, using either a species-specific method (e.g., Jerde *et al.* 2011; Pilliod *et al.* 2013; Wilcox *et al.* 2016), or DNA metabarcoding (e.g., Civade *et al.* 2016; Hänfling *et al.* 2016; Shaw *et al.* 2016; Valentini *et al.* 2016). Usually, the aim is to detect invasive species (e.g., Dunker *et al.* 2016; Jerde *et al.* 2011; Mahon *et al.* 2013; Takahara *et al.* 2013), to draw the precise distribution map of a species (e.g., Laramie *et al.* 2015), or to assess the biodiversity of a target taxonomic group in the area under study (e.g., Hänfling *et al.* 2016; Shaw *et al.* 2016; Valentini *et al.* 2016). Obviously, the detection probability depends on the volume of filtered water (Shaw *et al.* 2016) that is itself conditioned by the filtration device. The most commonly used devices for filtration in the field are Sterivex (0.22 μm Sterivex-GP filters, Merck Millipore) and tangential filtration capsules (e.g., Envirochek HV 1 µm, Pall Corporation, Ann Arbor, MI, USA).

Civade (2016) carried out a thorough study in different French rivers to estimate the sampling effort necessary to obtain a substantial proportion of the total fish diversity. A single filtration capsule (Envirochek HV) revealed between 70–93% of the fish diversity, whereas two to eight capsules were necessary to obtain 95% of the total diversity.

In addition to fish and amphibians, it is also possible to detect other vertebrates from water eDNA. In Florida, the elusive Burmese python (*Python bivittatus*) is an invasive species. This semi-aquatic snake was identified in five field localities using species-specific PCR on eDNA (Piaggio *et al.* 2014). Mammals can also be detected from drinking water, as demonstrated by Rodgers and Mock (2015) for coyotes and by Ishige *et al.* (2017) for six endangered species from the forests of Borneo, including the Asian elephant (*Elephas maximus*), banteng (*Bos javanicus*), Sunda pangolin (*Manis javanica*), sambar deer (*Rusa unicolor*), bearded pig (*Sus barbatus*), and orangutan (*Pongo pygmaeus*).

12.5.2 Biomass estimates

For a given fish or amphibian species, it seems obvious that the eDNA concentration released in water is linked to the biomass of this species. This has been experimentally verified by quantitative PCR in aquariums, mesocosms, and experimental ponds (Evans *et al.* 2016; Lacoursière-Roussel *et al.* 2016b; Takahara *et al.* 2012), in lakes (Lacoursière-Roussel *et al.* 2016a), and in streams and rivers (Baldigo *et al.* 2017; Doi *et al.* 2017; Jane *et al.* 2015; Pilliod *et al.* 2013). Thus, the relationship between the amount of eDNA and species abundance should allow integration of the quantitative aspect of eDNA within fisheries management plans. However, it seems that fish biomass estimates also depends on the water temperature (Lacoursière-Roussel *et al.* 2016b).

If biomass estimates based on qPCR seems relatively straightforward, the situation is more complex when targeting many species using DNA metabarcoding. Several factors might influence biomass estimates from eDNA. First, it is not proven that for a given biomass the same amount of eDNA

is released by the different species. Second, the number of sequences obtained for each species is directly linked to the quality of the match between the primers and their targets in the different species. Finally, the number of sequence reads might also depend on the metabarcode sequences. Consequently, the metabarcoding approach can only provide a rough estimate of the biomass of the different species. Despite these potential difficulties, a few studies have already mentioned that DNA metabarcoding also provides some quantitative data on fish or amphibian biomass in mesocosms (Evans *et al.* 2016), and in lakes (Hänfling *et al.* 2016).

12.6 Are rivers conveyer belts of biodiversity information?

Several papers have already demonstrated that lake sediments contain eDNA from the whole catchment, including aquatic and terrestrial species (Alsos *et al.* 2016; Giguet-Covex *et al.* 2014; Pansu *et al.* 2015b; Pedersen *et al.* 2016). The most likely transfer of terrestrial DNA to the lake sediment is via rivers and streams, which are therefore conveyer belts of biodiversity. However, is it possible to analyze eDNA from rivers to have access to the total biodiversity of the whole catchment? Deiner *et al.* (2016) claimed that "environmental DNA transported in river networks offers a novel and spatially integrated way to assess the total biodiversity for whole landscapes and will transform biodiversity data acquisition in ecology." The next paragraph is a detailed analysis of the methods and results reported in this article.

Deiner *et al.* (2016) collected 1 L of water from eight localities along the same river system in Switzerland. They carried out three independent filtrations (using glass fiber filters with 0.7 µm pores) and DNA extractions (phenol-chloroform isoamyl protocol) per sample (i.e., per liter), and then mixed the three extracts. Three PCRs were performed per resulting extract (i.e., a total of 24 PCRs) with 35 cycles, using the classical COI primers amplifying a 658 bp fragment (Folmer *et al.* 1994). The three PCRs per locality were combined before library preparation, and the eight libraries for sequencing were prepared using the Nextera XT DNA 96 kit (Illumina Inc., San Diego, CA, USA). This protocol cleaves the amplicons into random fragments of about 300 bp in length and adds a short sequence on both sides of each fragment, in a process called tagmentation (see Section 7.2.1). Then, the tagmented DNA is amplified via 12 PCR cycles. During this amplification step, the P5 and P7 adapters are extended and indexes are added (see Chapters 6 and 7). The eight libraries were loaded on a MiSeq sequencing platform (paired-end sequencing, 2 × 250 bp). The resulting forward and reverse sequence reads were merged, provided that they overlapped by at least 25 bp. Non-overlapping sequences were discarded. Only merged sequences fulfilling the four following conditions were considered in the subsequent steps: (i) length of at least 100 bp; (ii) homology of at least 90% with any COI sequence; (iii) taxonomic assignment at least to the family level; and (iv) presence of this family in Switzerland (or in neighbor countries). This filtering protocol led to 96% of the initial number of reads being discarded: only 240,340 paired-end reads were considered out of 6,006,959 originally, and resulted in the identification of a

total of 296 eukaryotic families (with 54 families identified by a single sequence read). These families included both aquatic and terrestrial organisms.

Unfortunately, this paper suffers from several problems. First, there are no biological, nor technical replicates (see Prosser 2010 for the consequences). Second, amplifying a long marker of 658 bp is not optimal in this context as shorter markers are more efficient in revealing the diversity (e.g., Bista *et al.* 2017). Third, the use of the Nextera library preparation with 12 PCR cycles might lead to the production of a large number of chimeras. Fourth, a homology of 90% with any COI sequence is not restrictive enough to prevent false identifications and could lead to higher false positive rates. Finally, the taxonomic identifications were performed only at the family level, with only a limited number of sequences showing a 100% match in the reference database, out of more than six million paired-end sequences initially. In this context, the results of this study must be considered with extreme caution, until replications have been carried out with appropriate protocols.

## CHAPTER 13

Marine environments

The marine environment is the largest realm of life on Earth. It harbors a tremendous diversity of habitats and life forms, and constitutes what seems to be a limitless source of novel genes, enzymes, and metabolites. Fifty years of biodiscovery in this environment have considerably fueled biotechnological developments in both research and industry, leading notably to thermostable enzymes (reviewed in Heidelberg *et al.* 2010). One noticeable example is the proofreading polymerase *Pfu* isolated from *Pyrococcus furiosus*, an extremophile Archaea living in deep-sea hydrothermal vents. Marine microbiologists have also pioneered the use of environmental DNA for biodiversity research and revealed the existence of previously unknown microbial lineages (DeLong 1992; Fuhrman *et al.* 1992; Giovannoni *et al.* 1990), as well as novel functional genes or metabolic pathways (Venter *et al.* 2004). These findings have stimulated the launch of global-scale initiatives such as the Sorcerer II Global Ocean Sampling expedition (Rusch *et al.* 2007; Yooseph *et al.* 2007), the International Census of Marine Microbes (Amaral-Zettler *et al.* 2010; Zinger *et al.* 2011), and the Tara expeditions (Brum *et al.* 2015; De Vargas *et al.* 2015; Karsenti *et al.* 2011; Sunagawa *et al.* 2015). These projects have considerably improved, and are still improving, our understanding of marine microbial diversity and associated biogeochemical cycles in the oceans. Although eDNA-based research on marine macroorganisms has long lagged behind microbiology, its current expansion holds great promises for conservation biology and biomonitoring applications. We will here briefly review how eDNA research has contributed to our improved understanding of marine biodiversity.

13.1 Environmental DNA cycle and transport in marine ecosystems

The marine realm is characterized by a wide variety of environments (e.g., the oligotrophic ocean, the deep-sea, coastal, and estuarine environments) that can be each subdivided into different habitats (e.g., pelagic and benthic zones, coral reefs, hydrothermal vents, and methane seeps). These habitats exhibit contrasted physico-chemical characteristics, material transport, biomass, and biological activity that will inherently determine the fate of eDNA and, hence, its detectability. Although distinct, these habitats are strongly connected: life in the sunlit open ocean is sustained by autotrophic organisms that fix CO_2 and other elements in the food web. In estuarine and coastal zones, this biomass can be amended with terrestrial or freshwater inputs, as well as with benthic organisms due to up- and down-welling currents that may resuspend sediment particles and larvae in the water column (Levin 2006). The resulting pool of organic matter is then either recycled in the food web, or exported deeper through sinking, to ultimately reach the sediments (Buchan *et al.* 2014; Torti *et al.* 2015).

The sinking of organic matter to sediments is key to the benthic life. In particular, DNA from the surface of the ocean is an important allochthonous source of phosphorus for the benthos, which stimulates the production of heterotrophic bacterial biomass and makes these environments unlimited in phosphorus (Dell'Anno & Danovaro 2005). Total DNA usually represents dozens of micrograms per gram of wet sediments, of which an important part is extracellular (Torti *et al.* 2015). This latter pool contributes to 3% of the total phosphorus pool. It has a

Environmental DNA for Biodiversity Research and Monitoring. Pierre Taberlet, Aurélie Bonin, Lucie Zinger, & Eric Coissac,
Oxford University Press (2018). © Pierre Taberlet, Aurélie Bonin, Lucie Zinger, & Eric Coissac 2018.
DOI: 10.1093/oso/9780198767220.001.0001

relatively short residence time (of about 10 years) in the top sediment layer, despite their anoxic and cool temperatures that would otherwise constitute excellent conditions for long-term preservation (Dell'Anno & Danovaro 2005). Hydrothermal vent sediments also display large amounts of extracellular DNA that are important for the local food web and results from an intense viral predation of prokaryotic communities (Corinaldesi et al. 2014). To our knowledge, DNA transport in sediments has not been studied thus far, but should mainly arise from bioturbation or up-/down-wellings.

The water column of the sunlit open ocean contains comparatively much less DNA (0.5–10 µg/L; Torti et al. 2015) than its benthic counterpart, as life is highly diluted and the conditions are relatively oligotrophic. DNA degradation rate in seawater is very fast (between 10 hours to several days), which translates into relatively short distances of detection for animals (Thomsen & Willerslev 2015). Coastal, estuarine, and coral reef waters should display similar characteristics, with higher amounts of DNA and biomass due to the higher nutrient availability (Torti et al. 2015).

13.2 Marine microbial diversity

As outlined in Section 13.1, marine microbes fuel biogeochemical cycles of the global ocean and have received considerable attention in the last three decades through eDNA analysis, either by DNA metabarcoding, or metagenomics and metatranscriptomics (Heidelberg et al. 2010). The DNA metabarcoding of marine microbes is usually carried out using primers to target the V6 (Sogin et al. 2006) or V4 regions (Caporaso et al. 2011; but see Parada et al. 2016 for a modified version) of the 16S rRNA gene for prokaryotes and the V9 region of the 18S rRNA gene for protists (Amaral-Zettler et al. 2009). Fungi are usually assessed with either the primer pair dedicated to protists, which actually amplifies also metazoans, or classical internal transcribed spacer (ITS) primers (Richards et al. 2012). The rRNA fragments retrieved in metagenomics/ metatranscriptomics are also used to produce taxonomic profiles of the marine microbial diversity (see Chapter 10; Logares et al. 2014). This presents the advantage of avoiding spurious OTUs

introduced by polymerase chain reaction (PCR) errors, but remains constrained by the incompleteness of references databases. Finally, viruses, which are an important component of marine ecosystems in terms of both diversity and function, lack universal markers and their community structure is now usually assessed by defining viral protein clusters from metagenomes (Brum et al. 2015; Yooseph et al. 2007). There is now a huge body of literature on these systems, and we will only briefly review how eDNA has permitted identification of new taxa and revealing patterns of biodiversity in marine microbes.

Marine (and more generally aquatic) ecosystems are expected to be subjected to more physical mixing (e.g., currents) compared to other environments, which would favor a large, or even unlimited distribution of many phytoplankton species, in particular microbial species (Finlay 2002). Accordingly, some groups of Bacteria have been reported to be cosmopolitan (Fuhrman 2009). Whether these groups represent actual "species" or higher taxonomic levels remains uncertain, but this apparent cosmopolitan distribution could result from adaptive radiations, as suggested for the ubiquitous SAR11 clade (Brown et al. 2012).

Despite an apparent large-scale distribution, the vast majority of studies unraveling patterns of microbial diversity provides evidence that the community composition and/or structure strongly vary across space and time for all marine microbial taxa. Water depth appears to be one of the major forces structuring bacterial, archaeal, and protist communities in pelagic ecosystems. This is mainly due to major differences in light, oxygen, and resource availability, salinity, and potentially reduced predation and grazing in mesopelagic environments (De Vargas et al. 2015; Monier et al. 2015; Sintes et al. 2015; Sunagawa et al. 2015; Zinger et al. 2011). These communities also display horizontal variations that broadly fit defined ecoregions differing in their sea surface temperature and nutrient availability (Rusch et al. 2007; Sunagawa et al. 2015; Zinger et al. 2011). However, the existence of the classical latitudinal gradient of biodiversity remains unclear for marine microbes, with some studies suggesting a peak of species richness in the tropics and others in temperate water masses (reviewed in Fuhrman

2009; Sunagawa *et al.* 2015). However, microbial planktonic communities exhibit other noticeable macroecological patterns. These include the distance-decay of similarity (i.e., spatially close communities tend to be more similar than distant ones), the taxa-area relationship (i.e., the increase of species richness with increasing area of observation), or the Rapoport's rule (i.e., smaller latitudinal range of species in low latitude areas; Amend *et al.* 2013; Sintes *et al.* 2015; Sunagawa *et al.* 2015; Zinger *et al.* 2014). Pelagic microbial communities also display strong and predictable variations across time, at various temporal scales (i.e., from hours to years, reviewed in Fuhrman *et al.* 2015). All of these findings will undoubtedly enable the construction of models of marine microbial distribution in the near future (Larsen *et al.* 2012) and predict how these important actors of the ocean functioning will respond to global change.

Benthic microbial communities also display some of these patterns. In particular, benthic microorganisms are less subject to physical mixing. They usually display steeper spatial variation due to either stronger environmental heterogeneity or dispersal limitation, as well as strong dependency on the productivity of the upper layers (Bienhold *et al.* 2012, 2016; Ruff *et al.* 2015; Zinger *et al.* 2011, 2014). The decline of microbial diversity toward polar regions does not seem to occur in these environments. On the contrary, it seems that both biomass and diversity of benthic microbial communities in the deep sea increase with increasing latitudes (Danovaro *et al.* 2016). However, studying benthic ecosystems in terms of both abiotic conditions and diversity remains a technological challenge due to the remoteness of these systems, especially those in deep sea. Many groups have received little attention such as marine Fungi (but see Richards *et al.* 2015 and Tisthammer *et al.* 2016), viruses (but see Brum *et al.* 2015 and Marston *et al.* 2013) and nematodes (but see Dell'Anno *et al.* 2015 and Vanreusel *et al.* 2010). These limitations weaken global-scale conclusions on the patterns and main drivers of these communities.

Beyond these basic research aspects, there is now growing interest in the application of eDNA to microbial communities to track and treat pollution in recreational waters, fisheries, or industries (Bourlat *et al.* 2013; Tan *et al.* 2015), for example, by identifying

fecal and food waste pollution sources (Harwood *et al.* 2014), waterborne pathogens, or antibiotic resistance (Kelly *et al.* 2014b). Benthic diatoms are also good bioindicator candidates for coastal and estuarine waters eutrophication (Desrosiers *et al.* 2013), and could be monitored with eDNA.

13.3 Environmental DNA for marine macroorganisms

Environmental DNA-based biodiversity research in marine ecosystems is now rapidly expanding for macroorganisms in both basic and applied research. The first attempts to use this methodology for studying macroorganism diversity were conducted to monitor the expansion of echinoderms in Tasmanian waters (Deagle *et al.* 2003). It is only recently that eDNA started to be used to monitor benthic metazoans (Chariton *et al.* 2010; Fonseca *et al.* 2010; Guardiola *et al.* 2015, 2016; Leray & Knowlton 2015), fish (Kelly *et al.* 2014a; Miya *et al.* 2015; Thomsen *et al.* 2012a; Yamamoto *et al.* 2017), or marine mammals (Foote *et al.* 2012). In ecological assessments, accurate estimation of species richness and correct identification of bioindicators is a prerequisite. Several attempts have been made to reduce the biases introduced by PCR or improve species detection and identification at both the molecular (Clarke *et al.* 2017; Miya *et al.* 2015) and analytical levels (Brown *et al.* 2015; Morgan *et al.* 2013). Despite the fact that biases in the molecular approach may still remain, the molecular inventories of metazoans are usually consistent with classical observations (Aylagas *et al.* 2016; Cowart *et al.* 2015; Lejzerowicz *et al.* 2015; Leray & Knowlton 2015; Thomsen *et al.* 2016), and the main limitation of DNA metabarcoding appears to be the detection of rare species that would require more intense sampling. While eDNA is becoming the tool of choice to monitor marine mesofauna and macrofauna, its use has not yet been reported, to our knowledge, to unravel seagrass and coral diversity.

Non-indigenous, invasive species are one of the major threats to marine ecosystems, and eDNA-based monitoring could provide a promising method to track down their dynamics. Such studies remain, however, very limited for marine ecosystems compared to studies on freshwater ecosystems. Many invasive marine species have a planktonic life stage that render them difficult to detect visually. So far, metabarcoding has been successfully implemented for the early detection of invasive species using species-specific primers, such as for the bivalve *Rangia cuneata* in Europe (Ardura *et al.* 2015). It has also been used in the non-directed detection of invasive species using universal metazoan primers (Brown *et al.* 2015; Hatzenbuhler *et al.* 2017; Zaiko *et al.* 2015). This application remains, however, in its infancy in marine ecosystems, and development of sampling and molecular protocols is needed.

Beyond applications at the community level, eDNA is now beginning to be used for monitoring fish populations (Sigsgaard *et al.* 2016). In this study, the authors sequenced two mitochondrial DNA polymorphic regions to determine the genetic diversity and size of whale shark populations. Error rates and eDNA decay in water samples were assessed in order to correct estimates of population characteristics. The authors reported an exponential decay of eDNA, with very little remaining DNA detected after six days. Using qPCRs on these mitochondrial regions together with markers of the whale shark's most likely prey (fish spawn from mackerel tuna), they were able to identify a positive relationship between the predator and prey eDNA quantities, even when the whale shark was not observed in the field. This reveals the power of this approach for both population genetics and research on trophic interactions.

## CHAPTER 14

Terrestrial ecosystems

While eDNA-based analyses have rapidly gained momentum in the freshwater ecology community, first for single species detection and more recently for diversity surveys, their success has been less immediate among terrestrial ecologists. Soil microbiologists are a notable exception, as they quickly realized that targeting DNA directly in the environment could free them from cultivating microorganisms prior to any community census (Hugenholtz et al. 1998; Pace et al. 1986). Nowadays, eDNA approaches still lie at the heart of large research initiatives, which are mostly led by soil microbial ecologists such as the Earth Microbiome Project (EMP) (http://www.earthmicrobiome.org/; see Section 14.4). These approaches are now applied to a continuously wider range of terrestrial macroorganisms that leave behind DNA traces in soil (Arribas et al. 2016; Fahner et al. 2016; Zinger et al. 2016, 2017), but also in more improbable terrestrial compartments like vegetation (Nichols et al. 2015), or air (Fröhlich-Nowoisky et al. 2012; Kraaijeveld et al. 2015).

14.1 Detectability, persistence, and mobility of eDNA in soil

The lack of initial enthusiasm for eDNA outside the microbial world is possibly connected to the fate and behavior of DNA in the main matrix of terrestrial ecosystems (i.e., soil). Detectability and persistence of eDNA in soil are two complex and still poorly understood phenomena, which can be influenced by many parameters (Barnes & Turner 2016; Levy-Booth et al. 2007; Nielsen et al. 2007). These pertain either to the DNA molecules themselves and their characteristics (e.g., GC content, intracellular or

extracellular state, conformation), to the host species (e.g., biomass, physiology, behavior, and so on), or to the environmental conditions, both abiotic (temperature, pH, UV radiations, nature of substrate) and biotic (microbial and enzymatic activities). We know that eDNA can remain detectable in soil for days to years, and even decades if environmental conditions are not favorable to its degradation. For example, Andersen et al. (2012) analyzed eDNA from soil samples collected in zoos and farms, and they were able to recover amplifiable DNA from camels six years after the species had left the area. They also found that eDNA from the soil surface was a better indicator of the current vertebrate community composition above-ground, while deeper soil strata preserved eDNA longer. Interestingly, host social organization, behavior, and biomass had a considerable influence on how well vertebrate eDNA could be detected. In another study, Yoccoz et al. (2012) monitored crop eDNA in previously cultivated alpine fields and highlighted a positive relationship between the frequency of crop amplicons and the year of crop abandonment. After several decades of non-cultivation, frequencies of crop sequences were, however, very low and comparable to background noise (i.e., a few units per thousand).

The capacity of soil to retain amplifiable DNA molecules for long periods of time is double-edged. On one hand, soil can be seen as a repository of past biodiversity, which can be exploited to reconstruct past ecosystems (see Chapter 15). On a finer time scale, this integrating property of soil is interesting to escape, to a certain extent, the temporal or seasonal variability typically associated with traditional diversity surveys like botanical relevés. On the other hand, the modern and historical signals conveyed by soil can be difficult to disentangle, leading for example to a

Environmental DNA for Biodiversity Research and Monitoring. Pierre Taberlet, Aurélie Bonin, Lucie Zinger, & Eric Coissac,
Oxford University Press (2018). © Pierre Taberlet, Aurélie Bonin, Lucie Zinger, & Eric Coissac 2018.
DOI: 10.1093/oso/9780198767220.001.0001

higher risk of false positive detection for contemporary species. However, one should keep in mind that most of the total eDNA in soil usually corresponds to current material, as proven by the uncovering of seasonal shifts of communities using soil eDNA (Zinger *et al.* 2009b; see also Section 4.1.2).

After its release into soil, DNA is generally adsorbed to a substrate such as clay particles or organic material. This greatly hinders its spatial diffusion, even if long-distance transportation through water has sometimes been demonstrated (Douville *et al.* 2007). In their study of alpine fields, Yoccoz *et al.* (2012) did not detect crop eDNA in the areas with no history of cultivation, even if barley, rye,

and potato DNA could be amplified at low levels in the previously cultivated fields 1 km away. Furthermore, they also revealed major differences in plant diversity patterns for boreal meadows and heath plots situated less than 100 m apart (Fig. 14.1), clearly indicating that eDNA had not leached between the two types of habitats.

The less eDNA moves in soil, the less its signal will be shared between two adjacent sites. This observation is magnified by the fact that the spatial distribution of soil organisms is intrinsically highly structured both horizontally and vertically. Strong small-scale (e.g., ≤ 10 m) horizontal variations have been observed with either traditional observations

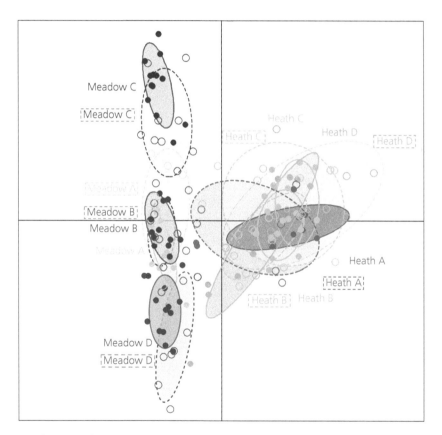

Fig. 14.1 Comparison of two types of boreal plant communities (meadow or heath) inferred using either eDNA metabarcoding or above-ground botanical surveys (from Yoccoz *et al.* 2012). This figure represents a non-symmetrical correspondence analysis based on presence/absence data at the genus level for the eDNA metabarcoding data set (open circles), and species level for the botanical survey data set (full circles). Letters A to D correspond to the four sampled sites. Ellipses show within-community variability in the botanical survey data set (continuous), or the eDNA metabarcoding data set (dotted).

Yoccoz, N. G., Bråthen, K. A., Haile, G. J. *et al.* DNA from soil mirrors plant taxonomic and growth form diversity. Reprinted by permission from *Molecular* Ecology. John Wiley and Sons. Copyright (2012).

(reviewed in Ettema & Wardle 2002) or through DNA metabarcoding (Tedersoo *et al.* 2016; Zinger *et al.* 2017). Similarly, there are clear vertical discrepancies in the community composition of Fungi, Bacteria, and meiofauna between the organic and mineral horizons of soil (Baldrian *et al.* 2012; Yang *et al.* 2014). This has profound implications for the sampling design. These are developed further in Chapter 4, which explains how to determine the appropriate number of sampling units across the study area, and how to arrange them spatially depending on the context and scientific question addressed. Terrestrial ecologists should nonetheless be aware that the sampling effort will always be more intense in environments where eDNA diffuses poorly, as in soil. Fortunately, using composite samples (i.e., a mix of soil cores taken at various locations within the study area, or even in different plots) is an efficient way to increase the spatial representativeness of soil sampling, while limiting experimental costs. This comes with a loss of information on the precise spatial distribution of the different organisms. Therefore, this design should be chosen only if it does not compromise the success of the study.

These considerations suggest that the soil matrix is a poor integrator of the whole local biodiversity, as compared to aquatic ecosystems. Environmental DNA in soil is also less likely to belong to above-ground organisms than to soil organisms. This holds especially true for vertebrates, whose eDNA is relatively difficult to detect in soil (but see Andersen *et al.* 2012). Nevertheless, this problem can be bypassed using sampler organisms, that is, organisms that feed on plants, arthropods, or even on vertebrate skin, carcasses, blood, or feces (Calvignac-Spencer *et al.* 2013a). This aspect of terrestrial biodiversity is developed further in Section 17.2.3.

14.2 Plant community characterization

To this day, conventional above-ground botanical inventories have gathered a considerable corpus of knowledge on plant communities that can now be contrasted with patterns inferred from plant eDNA found in soil. In the Yoccoz *et al.* (2012) study already mentioned in Section 14.1, boreal plant communities were reconstructed by analyzing a short fragment of the P6 loop of the chloroplast *trn*L (UAA) intron (Taberlet *et al.* 2007) amplified from soil eDNA. The results were highly consistent with those from classical botanical surveys (Fig. 14.1), and eDNA even detected species that were not observed in the field, but known to occur in the area. However, based on sequence frequencies, eDNA metabarcoding tended to overestimate forbs and underestimate woody species. The opposite was true for botanical surveys based on above-ground biomass, while proportions were congruent between the two approaches for graminoids (Fig. 14.2). These three functional groups show different root:shoot biomass ratios (Aerts & Chapin 1999). One can thus hypothesize that for plants, soil eDNA is more representative of biomass turnover than of actual biomass, and this highlights the fact that DNA read frequencies can be difficult to relate to taxon abundances without a proper calibration (Yoccoz *et al.* 2012).

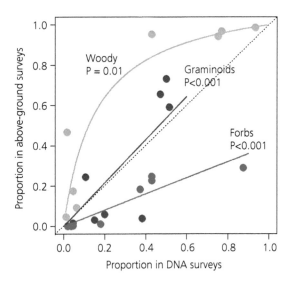

Fig. 14.2 Fitted relationships between proportions in above-ground botanical surveys and DNA surveys for three functional groups of boreal plants (from Yoccoz *et al.* 2012). For each group, the corresponding *p*-value is indicated. The dotted line corresponds to a 1/1 equivalence. For the woody group, proportions were logit-transformed.

Yoccoz, N. G., Bråthen, K. A., Haile, G. J. *et al.* DNA from soil mirrors plant taxonomic and growth form diversity. Reprinted by permission from *Molecular Ecology.* John Wiley and Sons. Copyright (2012).

14.3 Earthworm community characterization

Earthworms are among the most important ecosystem engineers in soil (Lavelle *et al.* 1997). Through their burrowing and casting activities, they have a considerable influence on soil structure and properties, and they play a significant role in nutrient cycling, water retention or infiltration, and soil fertility (Blanchart *et al.* 1999; Edwards 2004; Lavelle *et al.* 1997). As they also represent a substantial proportion of soil animal biomass and are sensitive to factors such as land use and soil contaminants (Chan 2001; Römbke *et al.* 2005), they are considered as good bioindicators of soil health (Pérès *et al.* 2011). Traditional earthworm inventories rely either on passive physical separation of earthworms from soil (e.g., by hand-sorting or soil washing; Edwards & Bohlen 1996), or on behavioral methods where earthworms are forced to the surface through physical or chemical stimulus (Coja *et al.* 2008). However, all these methods are invasive, labor-, and time-consuming. They also require strong taxonomic skills and their results can be skewed by factors such as soil properties, season, earthworm life stage, and species characteristics (Bartlett *et al.* 2006; Lawrence & Bowers 2002).

This problematic step of earthworm physical isolation, extraction, and taxonomic identification can be bypassed by carrying out earthworm community census studies directly on soil samples using eDNA metabarcoding. In a pioneer study, Bienert *et al.* (2012) designed two short metabarcode regions in the mitochondrial 16S rRNA gene, which are specific to earthworms, and tested them on soil collected in the French Alps. The earthworm communities characterized by eDNA were very similar to those recovered by hand-sorting. However, eDNA-based inventories were more efficient at detecting endogeic species (i.e., species living deep in soil) because these tend to flee during manual digging for hand-sorting. Conversely, eDNA metabarcoding samples did not contain a leaf litter layer, so they missed several epigeic species (i.e., the surface dwellers), unlike hand-sorting. After this proof of concept, Pansu *et al.* (2015a) improved the sampling protocol vertically, by including litter in the soil cores they collected. They also increased the horizontal representativeness of the sampling scheme,

by preparing composite samples from soil cores covering the entire surface of the studied area. As this procedure considers the spatial heterogeneity of earthworm species distribution, it allowed for the detection of more species in French alpine soils than hand-sorting. It also uncovered a significant effect of land use on earthworm community composition, a pattern that was not brought to light by classical survey methods. Nevertheless, the incompleteness of reference databases of the mitochondrial 16S rRNA gene remains a major impediment for precise earthworm taxonomic identification, even if the situation is undoubtedly called to improve with the recent and tremendous increase in sequencing throughputs.

14.4 Bacterial community or metagenome characterization

As outlined at the beginning of this chapter, today, soil microbiologists are probably the most receptive audience for the opportunities offered by eDNA approaches. Initiated in 2010, the EMP is a flagrant evidence of this early appeal for eDNA. This large collaborative research effort relies on eDNA sequencing and mass spectrometry "to characterize global microbial taxonomic and functional diversity for the benefit of the planet and humankind" (http://www.earthmicrobiome.org/; Gilbert *et al.* 2014). Despite the inevitable difficulties and delays associated with such a large collective endeavor, the first findings are now available. For example, O'Brien *et al.* (2016) examined bacterial taxonomic diversity at different spatial scales in a switchgrass stand, from the centimeter scale to the habitat scale (>10 m), up to the global scale. They found that despite extreme variability at a very fine scale, patterns start to emerge at the habitat scale, with a higher alpha diversity observed in fertilized plots. About 20% of the molecular taxonomic units (MOTUs) revealed in this study overlap with those from other EMP samples collected around the world, and this figure reaches 40% when considering only EMP grassland soils. This highlights the existence of a core set of cosmopolitan bacterial groups, a feature which has also been described for marine Bacteria (Gibbons *et al.* 2013).

Some of the earlier works on microbial communities are also worth revisiting in the context of this chapter. In a seminal paper published a few years ago, Fierer *et al.* (2012) characterized the soil functional diversity and the bacterial and archeal taxonomic composition for 16 sites spanning a wide range of biomes (cold and hot deserts, a prairie grassland, temperate, tropical and boreal forests, and tundra). For this purpose, they performed shotgun sequencing at an unprecedented depth and 16S rDNA sequencing, in order to carry out

metagenomic analyses as well as 16S rDNA-based taxonomic inventories. Their results show that cold and hot deserts clearly stand apart from the other biomes for both metagenome and bacterial community compositions (Fig. 14.3). Cold deserts harbor the lowest functional and bacterial taxonomic diversities, a pattern that mainly drives the significant correlation between these two variables (Fig. 14.4). Genes involved in osmoregulation and dormancy/sporulation are overrepresented in desert sites, while the opposite is observed for genes associated

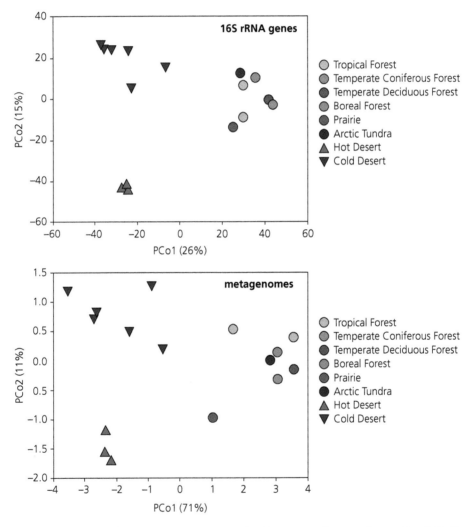

Fig. 14.3 Principal coordinate analysis of bacterial community (upper panel) and metagenome (lower panel) composition of 16 soil samples based on Bray–Curtis distances (from Fierer *et al.* 2012). Bacterial communities and metagenomes were characterized by 16S rDNA metabarcoding and by shotgun sequencing, respectively.

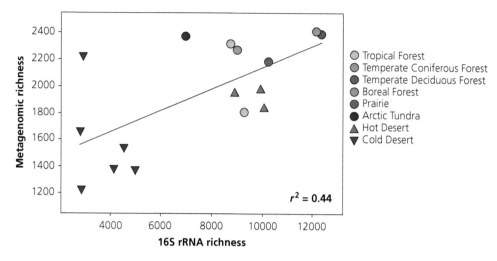

Fig. 14.4 Taxonomic and functional alpha diversities of 16 soils samples collected worldwide (from Fierer *et al.* 2012). Taxonomic richness (x-axis) was assessed by 16S rDNA metabarcoding and corresponds to the number of phylotypes (i.e., MOTUs) in the sample. Functional richness (y-axis) was determined by shotgun sequencing and corresponds to the number of functional gene categories identified for the sample.

with nutrient cycling, degradation of plant organic compounds, or antibiotic resistance. Overall, this suggests that in the desert environment, bacterial community structure is mainly determined by abiotic conditions, instead of microbe–microbe competition.

14.5 Multitaxa diversity surveys

One of the most fascinating opportunities offered by eDNA metabarcoding is the possibility to carry out multitaxa diversity surveys using the same sampling scheme and eDNA extracts. For example, Ramirez *et al.* (2014) collected more than 600 soil cores in Central Park, New York City, to investigate biodiversity and biogeographic patterns in all three domains of life (Bacteria, Archaea, and Eukaryota). Bacterial and archaeal diversities were estimated through the amplification and sequencing of a variable region of the 16S rRNA gene, while a comparable region of the 18S rRNA gene was analyzed to assess eukaryote diversity. For each domain, sequences with 97% or more similarity were clustered together to define phylotypes. The results show that even an urban and managed ecosystem like Central Park can harbor unsuspected levels of below-ground biodiversity for all three

domains of life, most of which had never been described in public databases. Moreover, community composition was highly variable among samples, with less than 20% of phylotypes shared between two random samples. Even more surprising, a set of 52 random samples from Central Park was compared to the same number of soil samples collected in very different biomes around the globe (e.g., Antarctic cold deserts, tropical forests, temperate forests, arctic tundra, and grasslands), and the levels of diversity were similar in magnitude in both sets, with only 6.5% fewer prokaryotic phylotypes and 26% fewer eukaryotic phylotypes in Central Park. However, these results should be considered with caution, as it is difficult to assess from the study how raw data were filtered to discard polymerase chain reaction (PCR) and sequencing artifacts, two parameters which can greatly inflate biodiversity estimates (Reeder & Knight 2009). It is also unclear whether the taxonomic and phylogenetic resolution of the metabarcodes considered in this study (i.e., 90-bp long for Bacteria) was appropriate to allow detecting meaningful biogeographical patterns.

With multitaxa eDNA diversity surveys, it is now possible to go one step further in the comprehension of the factors governing soil community assembly

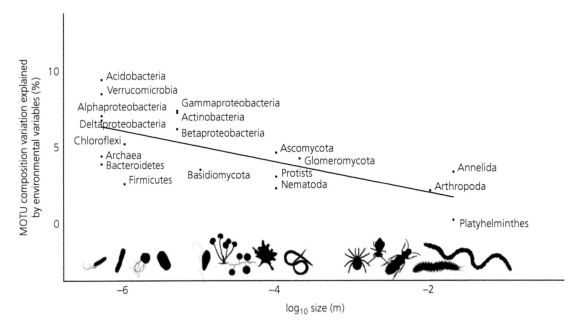

Fig. 14.5 Relationship between body size and the percentage of MOTU composition variation explained by environmental variables. This negative relationship is significant (Pearson's $r = -0.67$, $p = 0.002$), and implies that the distribution of small-bodied organisms is better explained by environmental variables than that of bigger organisms (from Zinger *et al.* 2017).

and diversity. For instance, Zinger *et al.* (2017) examined the fine-grained spatial distribution of soil Bacteria, Archaea, and Eukaryota in a tropical rainforest plot using eDNA metabarcoding. They aimed at determining the underlying community drivers, and the relative importance of niche-based versus neutral processes in shaping these communities. They found that soil community composition was highly variable and poorly explained by contextual data, which suggest an overall random distribution of soil organisms. Nevertheless, they identified some spatial variation that correlated

with spatially structured abiotic conditions such as soil aluminum, topographic variables, or plant characteristics. Most importantly, environmental variables better explained the distribution of small-bodied organisms, while the distribution of larger organisms was more patchy and random (Fig. 14.5), uncovering the differential role of body size on soil community assembly across the tree of life. This study exemplifies how integration of diversity census across multiple taxonomic groups can help test previous hypotheses on complex patterns of diversity and community structure.

## CHAPTER 15

Paleoenvironments

One of the most fascinating facets of eDNA is its potential for reconstructing past environments both in paleoecology and in archaeology. A few seminal papers have been published on this topic, dealing mainly with lake sediment and permafrost analyses. However, many other substrates can be used in the archaeological context, such as midden material.

The use of eDNA for describing paleoenvironments is not straightforward. First, working with ancient DNA as starting material requires following strict methodological rules to avoid contamination (Pääbo *et al.* 2004; Poinar & Cooper 2000; Willerslev & Cooper 2005). The second difficulty is the degradation of DNA over time. In the best preservation conditions represented by the basal sections of deep ice cores, it seems difficult to obtain results from DNA older than 0.8 million years (Willerslev *et al.* 2007). For lake sediments, the oldest samples providing results are between 10–20,000 years old in cold climates (Epp *et al.* 2015; Giguet-Covex *et al.* 2014; Pansu *et al.* 2015b; Pedersen *et al.* 2016) and only a few thousand years old in warmer climates (Boessenkool *et al.* 2014).

15.1 Lake sediments

In the next few years, lake sediments will undoubtedly represent the most commonly used material for characterizing past environments based on eDNA. It contains both eDNA from organisms living within the lake itself, which allows for the assessment of the lentic ecosystem (e.g., Capo *et al.* 2015), and eDNA from organisms living in the catchment, allowing for the reconstruction of the terrestrial environment around the lake (e.g., Pansu *et al.* 2015b). Thus, the different slices of sediment provide opportunities for obtaining DNA-based time series, complementary to the more classical time series based on pollen and/or macrofossils.

15.1.1 Pollen, macrofossils, and DNA metabarcoding

A few studies have compared data from pollen, macrofossils, and eDNA (Parducci *et al.* 2015; Pedersen *et al.* 2013). Usually, studies on pollen detect more taxa than macrofossils and eDNA, but the overlap among these three proxies is not complete (Fig. 15.1). Each approach not only provides some unique taxa undetected by the two others, but also has its own advantages and disadvantages. Pollen data has a bias toward wind-dispersed pollen that can be transported over long distances, leading to a regional view of the plant communities. Macrofossils are more characteristic of the local flora. Environmental DNA is also considered to be of local origin, and can provide data even in the absence of pollen and/or macrofossils. Furthermore, for these three approaches, the taxonomic resolution is not the same according to the taxonomic group considered. For example, the pollen of different species of Poaceae is difficult to identify, while eDNA has the potential of discriminating between most species when using a marker specifically developed for this group (see Poac01 in Appendix 1; Baamrane *et al.* 2012). As a consequence, approaches based on pollen, macrofossils, and eDNA are complementary.

15.1.2 Plants and mammals from Lake Anterne

Lake Anterne is a small alpine lake (0.12 km²) located in the Northern French Alps, at 2063 m above sea level. Its catchment is also relatively small (2.55 km²),

Environmental DNA for Biodiversity Research and Monitoring. Pierre Taberlet, Aurélie Bonin, Lucie Zinger, & Eric Coissac, Oxford University Press (2018). © Pierre Taberlet, Aurélie Bonin, Lucie Zinger, & Eric Coissac 2018.
DOI: 10.1093/oso/9780198767220.001.0001

(A)

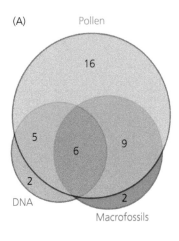

(B)

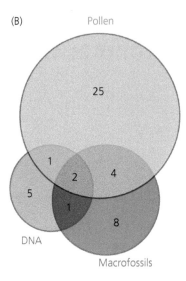

Fig. 15.1 Taxonomic overlap when analyzing peat sediments (A) and lake sediments (B) using three different approaches (pollen, macrofossils, and DNA metabarcoding). (A) Data from Pedersen *et al.* (2013) for Lake Comarum, South Greenland; the taxonomic overlap has been estimated at the family level. (B) Data from Parducci *et al.* (2015) for the Seida site (north-eastern European Russia); the taxonomic overlap has been estimated at the species level (except in a few cases where it was estimated at the genus or family levels).

above the present-day tree line, and it is mainly covered by subalpine grasslands with scattered pockets of heathland. A 20 m long composite sediment core was retrieved from the lake in September 2007 and preserved at 4°C. It corresponds to the last 10,000 years. DNA extractions of 47 core slices were performed more than four years later, using a protocol targeting extracellular DNA (Taberlet *et al.* 2012c). Two extractions were carried out per core slice, and four amplifications were performed per extraction, leading to a total of eight polymerase chain reactions (PCRs) per core slice. The primer pairs used were Sper01 targeting Spermatophyta and Mamm02 targeting Mammalia (see Appendix 1). The sequencing was performed on an Illumina HiSeq 2500 platform (2 × 100 bp paired-end reads).

The most important results concern the reconstruction of plant communities over the last 10,000 years (Pansu *et al.* 2015b) and the farming history (Giguet-Covex *et al.* 2014). Figure 15.2 summarizes these results. After an initial phase corresponding to the post-glacial recolonization (10,000–7,000 BP), the lake was surrounded by a pine forest (6,500–5,000 BP) characterized by the presence of *Pinus* and other herbaceous plant species typical of pine forest undergrowth. At about 2,500 years BP, the landscape radically changed toward open habitats due to pastoralism. Domestic animals such as sheep and cows were reliably detected during a 500-year period, just before 2,000 BP. At the end of this

period, a very strong erosion signal was revealed (Giguet-Covex *et al.* 2011). It was not connected to any climatic event, but was most probably due to overgrazing, which led to the abandonment of intensive pastoralism in this area for about 800 years. Then pastoralism started again at about 1,000 BP, but mainly with cattle. This study clearly demonstrates that eDNA from lake sediments can reliably be used to obtain time series concerning not only plant communities, but also other types of organisms such as domestic animals.

15.1.3 Viability in the ice-free corridor in North America

There was a controversy about how America was colonized by humans coming from eastern Asia. Human presence in America is attested from at least 13,400 years BP (Sanchez *et al.* 2014), but other archaeological evidence suggests earlier dates, from 14,700 years BP or even a few millennia earlier (Dillehay *et al.* 2008, 2015; Gilbert *et al.* 2008). At that time, the northern part of North America was glaciated, making colonization from Asia difficult. The first hypothesis was that humans followed the Pacific coast of North America. The second hypothesis relied on the opening of an ice-free corridor within the large ice sheet (Fig. 15.3). However, the precise dating of this corridor's opening was controversial. This 1,500 km long corridor might not have been suitable for

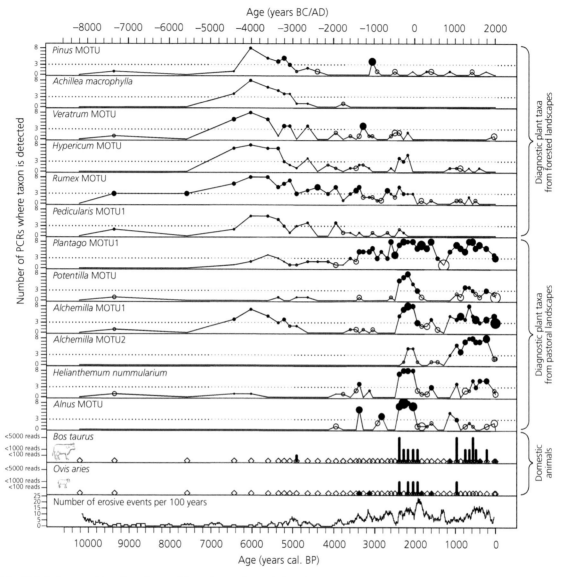

Fig. 15.2 Environmental DNA from plants and domestic animals recorded in Lake Anterne sediments. Data on plants, domestic animals, and erosive events are from Pansu *et al.* (2015b), Giguet-Covex *et al.* (2014), and Giguet-Covex *et al.* (2011). For eDNA results, the x-axis corresponds to the time, the y-axis to a number of positive PCR replicates for the considered MOTU. The circle size is representative of the mean frequency of reads for the MOTU at a given time.

From Pansu, J., Giguet-Covex, C., Ficetola, G. F. *et al.* Reconstructing long-term human impacts on plant communities: an ecological approach based on lake sediment DNA. Reprinted by permission from *Molecular Ecology*. John Wiley and Sons. Copyright (2015).

human colonization before the attested presence of humans south of the ice sheet. Pedersen *et al.* (2016) tested these two hypotheses by analyzing sediment cores from two lakes situated in a bottleneck portion of the corridor. They obtained radiocarbon dates, and analyzed pollen, macrofossils, and eDNA. They found evidence of a steppe vegetation with bison and mammoth starting at 12,600 years BP, of an open forest with moose and elk at 11,500 years BP, and of a boreal forest at 10,000 BP. The corridor was not

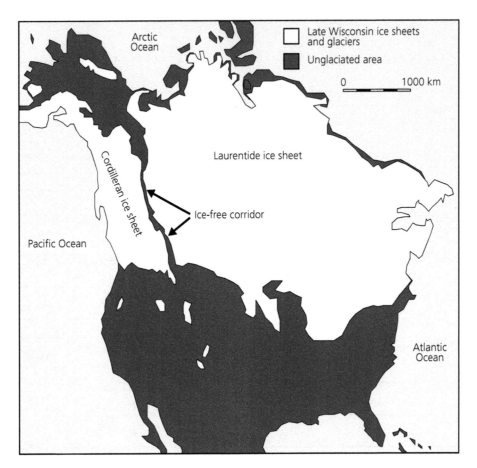

Fig. 15.3 Late Wisconsin (12,000 years BP) extent of ice sheets in North America showing the ice-free corridor between the Cordilleran and the Laurentide ice sheets.

viable for human colonization before 12,600 years BP, making this colonization route very unlikely for the first humans living south of the ice sheet and this confirmed the first hypothesis.

Beside solving the controversy about the colonization of America, the Pedersen *et al.* (2016) study represents the first large-scale trial to analyze eDNA using a shotgun approach. DNA was extracted from a total of 32 sediment cores. After library preparation, the sequencing was carried out on an Illumina HiSeq 2500 platform, using a single-end approach and targeting about 30 million sequence reads per library (i.e., per core sample). The comparison of all these reads with the full nucleotide database from GenBank led to the description of the biological succession within the corridor bottleneck. This shot-

gun-based metabarcoding approach is more difficult and more expensive to perform than the more classical PCR-based metabarcoding. However, it does have several advantages. First, it allows for all types of organisms, from Bacteria to mammals, to be targeted at once. Second, it is possible to obtain a rough estimate of the proportions among the different types of organisms, provided that the reference database is not too biased toward some of them. Finally, the most important advantage is the possibility to authenticate the ancient origin of the DNA molecules by looking at damage on both ends of the fragments obtained. Indeed, ancient DNA molecules are characterized by a higher proportion of damage close to their extremities (Briggs *et al.* 2007; Hofreiter *et al.* 2001; Jónsson *et al.* 2013).

15.2 Permafrost

Contrary to lake sediments, permafrost samples only contain a tiny part of the surrounding biodiversity. Therefore, to properly reconstruct past environments, it is necessary to combine many samples from the considered time period.

15.2.1 Overview of the emergence of permafrost as a source of eDNA

Willerslev *et al.* (2003) demonstrated that it is possible to extract and amplify plant and mammalian eDNA from permafrost samples dating from up to 400,000 years (see Chapter 11). At that time, the analysis of PCR products had to be carried out via cloning and Sanger sequencing, a time-consuming and expensive approach.

With the development of next-generation sequencing, it was tempting to extend the Willerslev *et al.* (2003) study for reconstructing past plant communities at a much larger scale. However, before reaching this objective, it was necessary to overcome some methodological challenges. First, the plant marker used in Willerslev *et al.* (2003) had a low taxonomic resolution. Consequently, a universal plant marker amplifying a short DNA fragment but having a better taxonomic resolution had to be developed (primer pair Sper01 in Appendix 1; Taberlet *et al.* 2007). Second, an extensive reference database for arctic and boreal plants was built (Sønstebø *et al.* 2010). Third, the DNA methodology for assessing plant communities was validated using modern arctic soils (Yoccoz *et al.* 2012). Finally, the pollen, macrofossils, and DNA metabarcoding approaches to reconstruct plant communities were compared (Fig. 15.4). It appeared that they represent three complementary proxies of past vegetation (Jørgensen *et al.* 2012).

15.2.2 Large-scale analysis of permafrost samples for reconstructing past plant communities

Before the emergence of eDNA, analyses of past plant communities in the Arctic have been based mainly on pollen data but also, to a lesser extent, on macrofossils. Willerslev *et al.* (2014) carried out the

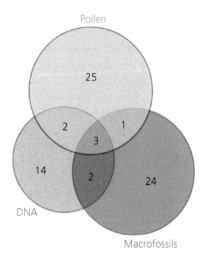

Fig. 15.4 Complementarity of three different approaches (pollen, macrofossils, and DNA metabarcoding) for reconstructing past plant communities based on permafrost samples (data from Jørgensen *et al.* 2012). In the pollen record, 12 tree species originating from boreal areas are probably not of local origin. The taxonomic overlap has been estimated at the lowest common taxonomic level (family, genus, or species) of the three data sets, depending on their respective resolutions.

first large study of permafrost samples to assess the evolution of plant communities in the Arctic during the past 50,000 years. They collected 242 permafrost samples that were precisely radiocarbon dated. DNA was extracted with all necessary precautions when working with ancient DNA. The extracts were then amplified using plant metabarcodes: the universal Sper01 primer pair targeting Spermatophyta, plus three complementary primer pairs, Poac01, Aste01, Cype01, for a better taxonomic identification of three families, Poaceae, Asteraceae, and Cyperaceae, respectively (see Appendix 1 describing these primers).

The results showed that the flora was quite diversified before the last glacial maximum, with 109 plant molecular taxonomic units (MOTUs) detected. Then the diversity strongly decreased during the last glacial maximum, with only 45 MOTUs detected. Finally, with the temperature increase during the postglacial period, 73 MOTUs were observed, with a shift toward graminoids (Poaceae and Cyperaceae) and shrubs. Unexpectedly, forbs (non-graminoid herbaceous plants) dominated during almost all the considered periods, until about 10,000 years ago when the proportion of graminoids strongly increased. Paleobotanists previously thought that graminoids dominated plant communities continuously since 50,000 years ago. This shift from forbs to graminoids coincided with the extinction of the megafauna, but it was not possible to determine the cause and the consequence of both, or whether these two events were independent.

Apart from the analysis of permafrost samples, Willerslev *et al.* (2014) also studied megafaunal diets, based on intestinal or stomach contents and/or coprolites recovered from four woolly mammoths, two woolly rhinoceroses, one bison, and one horse. Except for two mammoth samples, all other samples were dominated by forbs, showing that these were an important component of their diet.

15.3 Archaeological midden material

Midden material from archaeological sites represents a valuable source of information for tracking food habits of ancient human communities. It also provides information about the surrounding biodiversity, using humans as biodiversity samplers. This last approach is comparable to the use of leeches (Schnell *et al.* 2012) or carrion flies (Calvignac-Spencer *et al.* 2013b) as an indirect source of mammalian DNA.

15.3.1 Bulk archaeological fish bones from Madagascar

Taxonomic identification of archaeological fish bones is difficult, particularly from postcranial material. This difficulty can be partially solved by implementing a bulk-bone metabarcoding analysis (Grealy *et al.* 2015) that has been later applied to a coastal archaeological site in southwest Madagascar (Grealy *et al.* 2016). DNA was extracted from several bulk samples of fish bones collected from 100- to 300-year-old midden material, amplified using primers targeting a ~56 bp fragment of mitochondrial 12S gene, and sequenced on the Illumina MiSeq platform. The comparison of the obtained sequences with a reference database allowed for the identification of 23 fish families, including 14 genera and 5 species with a high confidence. Based on these metabarcoding results and on the current habitat preference of the different taxa identified, it was possible to deduce that the human community who caught these fish had a strong reliance on nearshore reefs. The detection of nocturnal predators may be an indication of leaving nets out overnight, or dive and/or night fishing. Finally, it was interesting to note that several taxa identified have not been recorded in modern fisheries data, leading to the conclusion that fish biodiversity decreased, probably due to overfishing.

15.3.2 Midden from Greenland to assess past human diet

To assess past subsistence economies in Greenland, Seersholm *et al.* (2016) collected 34 samples from midden deposits in four different sites, corresponding to four different extinct cultures. The oldest samples date from about 4,000 years BP. After DNA extraction, a shotgun metabarcoding approach was carried out (as in Section 15.1.3), allowing for the identification of several mammals, including cattle, goat, sheep, dog/wolf, harp seal, ringed seal, bowhead whale, narwhal, reindeer, and walrus. The most striking result was the evidence of bowhead whale exploitation 4,000 years ago.

## CHAPTER 16

Host-associated microbiota

The increasing recognition of the crucial role of microorganisms for the fitness and performance of plants and animals has led to the concepts of holobiont (i.e., a host and its microbiota), and hologenome (i.e., the collective genomes of a holobiont) (Zilber-Rosenberg & Rosenberg 2008). While these concepts are still being debated (Moran & Sloan 2015), it clearly appears that the microbiota can be viewed as an extended phenotype of the host, which has great implications for human and ecosystem health. This paradigm shift, together with recent advances in molecular methods, is currently fueling intense research seeking to understand the factors shaping the microbiota composition, its impact on host metabolism, immunity, and behavior, and the homeostasis of these systems at contemporary and evolutionary time scales. The first initiatives launched on this topic targeted model organisms such as humans (Human Microbiome Project Consortium 2012), mice (Xiao *et al.* 2015), *Drosophila* (Broderick & Lemaitre 2012), or *Arabidopsis* (Bai *et al.* 2015). They are now expanding to other organisms of general interest, such as bees (Zheng *et al.* 2017), or rice (Edwards *et al.* 2015). These projects generally use combinations of "omics" tools in order to apprehend taxonomic, phylogenetic, and functional characteristics of the microbiota. We will here provide a brief review of the advances that have been made in this field by using DNA-based approaches. The human model will not be discussed further, and interested readers should refer to dedicated reviews (e.g., Lozupone *et al.* 2012; Robinson *et al.* 2010).

16.1 DNA dynamics

The problem of DNA persistence for these living systems is generally very poorly documented.

One might expect that it will vary, depending on the host species and host surface considered. For example, it is unlikely that the persistence of extracellular DNA is high in the gut environment due to high amounts of digestive endonucleases and an overall low pH (Katoch & Thakur 2012; but see Lennon *et al.* 2017 where stool—not gut—samples were analyzed). For other host organs and surfaces, such as the rhizo/phyllosphere, or the skin/cuticle, extracellular DNA might be more persistent. The large gap of knowledge on this issue and the importance of DNA cycles for the inferences that can be drawn from eDNA call for a better characterization of the DNA production and degradation rates in these particular environments.

16.2 Early molecular-based works

The idea that microorganisms live in association with other organisms of different nature is not new. In the late nineteenth century, de Bary (Oulhen *et al.* 2016) first coined the term "symbiosis" as "the living together of unlike named organisms" involving a macroorganism (host) and microbes, whether they live outside or inside the host cells (exo- or endo-symbionts; Sapp 1994). Until the advent of culture-independent methods, most research in this area focused on two-partner associations, such as in lichens. With the advent of eDNA-based techniques, the extent of microbial diversity in the total environment and the strong differences observed between "free-living" and host-associated communities (Bik *et al.* 2016; Buée *et al.* 2009; Ley *et al.* 2008b; Sullam *et al.* 2012) led to a burst of explorative studies. These aimed at better characterizing microbiotas and at understanding how these particular communities are affected by the

Environmental DNA for Biodiversity Research and Monitoring. Pierre Taberlet, Aurélie Bonin, Lucie Zinger, & Eric Coissac, Oxford University Press (2018). © Pierre Taberlet, Aurélie Bonin, Lucie Zinger, & Eric Coissac 2018.
DOI: 10.1093/oso/9780198767220.001.0001

host condition, genetics, evolutionary history, and environment. These studies were essentially conducted using DNA metabarcoding approaches with the 16S rRNA gene for Bacteria, and 18S rRNA gene or internal transcribed spacer (ITS) for Fungi on various host organs and surfaces. Protist- and Archaea-associated communities have comparatively received far less attention.

A large body of literature deals with the plant rhizosphere microbiota due to its importance for plant nutrient uptake, and because it harbors symbionts that have been studied for a long time through non-eDNA-based approaches (e.g., nitrogen-fixing Bacteria, ecto-, or endo-mycorrhizal Fungi). Both plant host-related parameters (e.g., host developmental stage, species, genotype, and functional traits) and contextual conditions (e.g., soil physicochemical characteristics, season, elevation) have been repeatedly identified as the main determinants of root-associated microbial communities (see Buée *et al.* 2009 for an extended review; Bálint *et al.* 2013; Roy *et al.* 2013). Similar conclusions were reported for the phyllosphere microbiota (Cordier *et al.* 2012; Jumpponen & Jones 2009; Kembel *et al.* 2014), although the microbiotas inhabiting these organs highly differ from those of roots in their structure and composition (Hacquard 2016).

The gut microbiota has received considerable attention in animals as well. Early, yet recent, comparative works have particularly focused on host diet, taxonomy, and geographic location. Host taxonomy and diet are repeatedly found to determine the gut bacterial microbiota across metazoans, such as in mammals (Bergmann *et al.* 2015; Bik *et al.* 2016; Ley *et al.* 2008a), fish (Sullam *et al.* 2012), birds (Kohl 2012), ectotherms (Kohl *et al.* 2017), sponges and corals (Lee *et al.* 2012; Webster & Taylor 2012), and more especially insects (Anderson *et al.* 2012; Jones *et al.* 2013; Yun *et al.* 2014; Douglas 2015; Martinson *et al.* 2017). As for plants, animal microbiotas highly differ from organ to organ (Sullam *et al.* 2012), but vary roughly according to the same factors (Anderson *et al.* 2012; Ley *et al.* 2008b). This heterogeneity might compromise the results when dissection or tissue collection is difficult to carry out (such as in insects) due to cross-contamination problems. In particular, insects harbor an often dense and rich microbiota on their cuticles that might modify the overall picture of the intrabody microbiota (Douglas 2015). It is hence common practice to sterilize the surface of insect bodies prior to DNA extraction. Nevertheless, it has been recently suggested that sterile versus non-sterile DNA extracts of insects yield similar microbial communities (Hammer *et al.* 2015). More complex symbiotic systems are now being explored. For example, the microbiotas of a symbiosis involving three non-bacterial partners, an ant, a plant, and a Hemiptera, are broadly different across the three partners, but several microorganisms are present in all of them (Pringle & Moreau 2017).

16.3 Post-holobiont works

The explorative studies mentioned in Section 16.2 have revealed the incredible diversity of microbes each organism can harbor, leading to the emergence of the holobiont concept, and putting microbial communities and their hosts in a more eco-evolutionary perspective. Obtaining a more mechanistic understanding of how the host-microbiota homeostasis is constructed and maintained is now possible by combining several tools. These include quantitative studies, experimental approaches, the "omics" toolbox, and new statistical approaches able to handle large multivariate data (see e.g., Lozupone *et al.* 2012 for a review).

These integrative approaches are now starting to refine our understanding of the functional properties a host can gain from its microbiota. For example, herbivory is associated with the ability of particular members of herbivore gut microbiota to degrade complex plant polymers such as lignin or cellulose. Metagenomic and metatranscriptomic analysis of the microbiota of wood- or dung-feeding termites support this hypothesis, as contrasted taxonomic and functional profiles are observed depending on the host feeding characteristics (He *et al.* 2013). Using metagenomics, research on the microbiota of fungus gardens cultivated by leaf-cutter ants (Tribe Attini) has shown that they harbor Bacteria with high plant biomass-degrading capacity (Suen *et al.* 2010). The bacterial carbohydrate-active enzymatic (CAZy) profile predicted from the fungus garden metagenomes is found to be very similar to those of bovine rumen or termite gut. This suggests that some members of the microbiota confer particular functionalities to their hosts, irrespective of their phylogenetic origin. The fungus garden (and its microbiota) could hence be viewed as an external herbivorous digestive system, which would allow leaf-cutting ants to construct large colonies and have considerable impact on the surrounding ecosystem.

The potential mechanisms shaping the host microbiota can now be better identified with DNA-based approaches. In plants, for example, the root microbiota is assumed to be acquired from the pool of Bacteria present in the soil. By characterizing the root microbiota of different plants with DNA metabarcoding of the 16S rRNA gene in both controlled conditions and *in natura*, recent studies confirmed this hypothesis and showed that the microbiota in different rhizocompartments display a clear spatial gradient from the bulk soil to the interior of root tissues (Bai *et al.* 2015; Bulgarelli *et al.* 2015; Edwards *et al.* 2015; Peiffer *et al.* 2013). These beta diversity patterns are due to the recruitment of particular microbial taxa during the root colonization, which seems to be controlled by the host genotype (through molecular signaling), microbial interactions, soil's physico-chemical characteristics, and site location. These hypotheses are supported by metagenomic data, where microbial genes encoding for proteins involved in microbial–microbial and host–microbial interactions, and in the inhibition of plant immunity were all found to be under positive selection (Bulgarelli *et al.* 2015).

In animals, the microbiota also varies depending on host taxonomy, but the exact mechanisms underlying this pattern remain difficult to disentangle due to the confounding effects of host evolutionary history, life-history traits, diet, habitat, and geographic location. There is therefore an increasing number of experimental studies on either model or non-model species where conditions are controlled to rule out potential confounding factors. For example, Brooks *et al.* (2016) compared the gut microbiota of four species of wasps, nine species of flies, eight species of mosquitoes, and six species of deer mice. To exclude non-host-related confounding effects that can influence the microbiota, they reared individuals in

similar environmental conditions and controlled for the age, developmental stage, endosymbiont status, and sex of sampled individuals. They observed a correlation between the phylogenetic distance between hosts and the compositional distance between their respective microbiotas, a pattern called "phylosymbiosis." This pattern can emerge from coevolution (i.e., by vertical transmission) and/or cospeciation, as reported for some members of host-associated microbiotas (Ley *et al.* 2008b). However, these processes are unlikely to occur at the scale of the whole microbiota. It has therefore been proposed that phylosymbiosis could mainly result from the recruitment of microbial taxa from the contemporary environmental species pool that are similar between closely related hosts. This could be due to shared ecological traits either between hosts (e.g., immunity, gut secretions) and/or between environmental Bacteria themselves. In addition, Brooks *et al.* (2016) conducted transplant experiments where the fecal microbiota of a donor species was introduced into the diet of a recipient species. They found that the survival and performance of the host were reduced when transplanting the gut microbiota between host species, and that this reduction was more important when the donor was more phylogenetically distant from the recipient species. This result suggests that phylosymbiosis represents a functional association. As phylosymbiosis is difficult to detect in wild individuals due to numerous confounding factors, the authors recommended first conducting laboratory experiments in order to identify potential phylosymbiotic members of the microbiota, and then determining their distribution patterns in natural communities. More recently, a phylogenetically informed approach was used to tease apart the contribution of host evolutionary history and major dietary shifts in the phylosymbiosis of the gut microbiota, as characterized with the 16S rRNA gene, and mammals (Groussin *et al.* 2017). The authors found these factors to have acted at different evolutionary scales, with diet shift and host phylogeny being related to ancient and recent bacterial lineages, respectively. They conclude that cospeciation between a few bacterial lineages and their mammalian hosts has an important role in shaping the gut microbiota.

While all these approaches remain based on observational studies or greenhouse experiments manipulating soil physico-chemical conditions and host genotypes, a preliminary attempt has been made toward cultivating and sequencing all members of *Arabidopsis thaliana*'s microbiota (Bai *et al.* 2015). Not only do these isolates give access to the whole evolutionary history and functional potential of *Arabidopsis*' microbiota, but they also provide the unique opportunity to manipulate each member of the microbiota itself, which will certainly considerably deepen our understanding of host-microbiota interactions.

## CHAPTER 17

Diet analysis

If we exclude the early 16S rDNA-based surveys of bacterial communities from the nineties (Giovannoni *et al.* 1990), diet analyses appear among the first eDNA studies (Deagle *et al.* 2005; Harper *et al.* 2006; Hofreiter *et al.* 2000; Höss *et al.* 1992; Jarman *et al.* 2002;). Several reasons can explain this early success. These include the fact that sampling is generally straightforward, eDNA is usually relatively concentrated and easy to extract in feces or stomach contents, and taxa richness of a diet is limited compared to that of a whole ecosystem. Moreover, data analysis can be conservative as only common diet items usually have a real ecological relevance. Historically, animal diets have been characterized with a wide array of approaches, ranging from direct observations of foraging behavior (Bjugstad *et al.* 1970); sorting and identification of food fragments in feces or bolus using macro or microhistology (Holechek *et al.* 1982; McInnis *et al.* 1983; Moreby 1988; Putman 1984); use of prey-specific antibodies (Boreham & Ohiagu 1978; Symondson 2002) or temperature/denaturing gradient gel electrophoresis (TGGE/DGGE; Deagle *et al.* 2005; Harper *et al.* 2006); analysis of natural alkanes of plant cuticles which differ across plant taxa (Dove & Mayes 1996); near-infrared spectroscopy (NIRS) to estimate nutritional contents in animal feeds (e.g., total nitrogen, moisture, fiber, starch, and so on; Foley *et al.* 1998; Kaneko & Lawler 2006); and tracking of stable isotopic markers (Guest *et al.* 2008; Hata & Umezawa 2011; Ponsard & Arditi 2000). Therefore, for many species, there was already a substantial corpus of knowledge on which to build when eDNA-based diet analyses first emerged, although eDNA brought to light unexpected results in many cases (Kowalczyk *et al.* 2011; Rayé *et al.* 2011).

17.1 Some seminal diet studies

17.1.1 Proof of concept—analyzing herbivore diet using next-generation sequencing

Although eDNA was exploited as early as the nineties to characterize the herbivorous component of diets (Höss *et al.* 1992), Valentini *et al.* (2009) were the first to transpose this type of study to the next-generation sequencing era. They were also the first to adopt a marker exclusively developed for DNA metabarcoding (i.e., for which the amplification bias is expected to be limited among species). More specifically, Valentini *et al.* (2009) examined a fragment of the P6 loop of the chloroplast *trn*L (UAA) intron (Taberlet *et al.* 2007). This fragment was amplified from fecal eDNA obtained from a wide range of herbivorous species (marmot, brown bear, capercaillie, grasshopper, slug, and snail) and pyrosequenced using the 454 platform from Roche. The results showed that this method is fast and robust, and that it can be implemented easily for virtually any herbivorous species, as long as fecal or gut samples can be collected. In addition, it has the advantage of being standardizable between and within species, which opens new avenues in ecology. However, it is challenging to get the relative abundances of the different plant species consumed from the raw number of sequences obtained (but see also Section 17.2.2). The lack of taxonomic resolution associated with the generalist *trn*L eDNA marker can also be an impediment for distinguishing between closely related plant species.

Environmental DNA for Biodiversity Research and Monitoring. Pierre Taberlet, Aurélie Bonin, Lucie Zinger, & Eric Coissac, Oxford University Press (2018). © Pierre Taberlet, Aurélie Bonin, Lucie Zinger, & Eric Coissac 2018. DOI: 10.1093/oso/9780198767220.001.0001

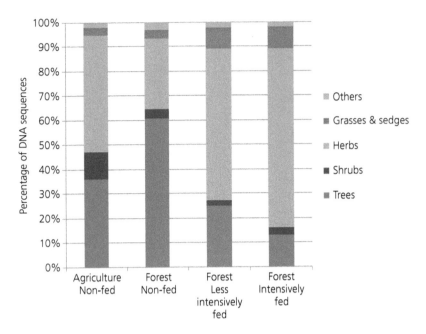

Fig. 17.1 Percentages of grass/sedge, herb, shrub, and tree sequences in the diet of four bison populations differentially supplemented with hay during winter (from Kowalczyk et al. 2011).
Reprinted by permission from *Forest Ecology and Management*, 261/4. Elsevier. Copyright (2011). Rafał Kowalczyk, Pierre Taberlet, Eric Coissac, Alice Valentini, Christian Miquel, Tomasz Kamiński, and Jan M. Wójcik, Influence of management practices on large herbivore diet—Case of European bison in Białowieża Primeval Forest (Poland).

17.1.2 Assessing the efficiency of conservation actions in Białowieża forest

After demonstrating the feasibility and robustness of eDNA-based herbivorous diet analysis, the next step was to answer a real biological question using this approach. This is what Kowalczyk et al. (2011) did in a pioneer study that aimed to evaluate the effect of conservation policies in a protected area. Białowieża forest (Poland) is well-known for being the last European primeval forest and for sheltering herds of European bison (*Bison bonasus*), but it is also an important logging center. As large herbivores, bisons have a substantial influence on their habitat (i.e., on the forest), so maintaining the balance between the forest, the logging activity, and the presence of bisons requires the implementation of active conservation measures. In Białowieża, during winter, bisons feed mainly on trees, so their winter diet has been supplemented with hay since the eighteenth century. The primary objective is to reduce their impact on the forest in general and on the exploited parcels in particular, and thus to limit conflict with inhabitants. However, the overall efficiency of these feeding campaigns on the diet of the Białowieża bisons remained to be proven. Therefore, feces from four populations differentially supplemented with hay were collected and analyzed by DNA metabarcoding with the same plant marker as in Valentini et al. (2009). The study clearly established that bisons fed daily or weekly with hay reduced their consumption of woody species, and that their impact on the forest is thus limited (Fig. 17.1). Several factors contributed to the success of this study. First, the diet composition had to be determined coarsely in terms of woody species versus hay components, which does not imply a precise taxonomic identification at the plant species level. Second, the managers were also interested in the relative diet shift between woody species and hay, a piece of information that DNA metabarcoding can provide even if absolute abundances of plant species in a diet cannot be determined.

17.1.3 Characterizing carnivore diet, or how to disentangle predator and prey eDNA

Environmental DNA extracted from feces or gut content contains DNA coming from the consumed items, but also DNA from the consumer, which is generally more abundant and better preserved as it comes from undigested cells (Deagle *et al.* 2006). This is not a problem when consumer and consumed items do not belong to the same clade, as a marker specific to the consumed clade can generally be designed. In this case, the strategy adopted for herbivorous diets can be easily broadened to carnivorous, hematophagous, or scavenger diets. However, what happens when the predator is closely related to its prey? This arises, for example, when analyzing the vertebrate component of the diet of a mammal predator, when focusing on insect species feeding on other insects (e.g., dragonflies), or when studying hematophagous bats consuming the blood of endothermic animals. In such a situation, a marker targeting the predator/prey clade will preferentially amplify the predator's DNA (Deagle *et al.* 2005; Shehzad *et al.* 2012b). Several solutions can be implemented to prevent the final polymerase chain reaction (PCR) product from being overwhelmed by predator sequences. On the one hand, the predator's sequences can be selectively digested using an endonuclease prior to, during, or after PCR amplification (Blankenship & Yayanos 2005; Dunshea 2009; Green & Minz 2005). However, this approach generally shows poor efficiency and also requires prior knowledge on the expected prey and their metabarcode sequences. On the other hand, the amplification of predator sequences can be significantly reduced (Vestheim *et al.* 2011). It is sometimes possible to design prey-specific primers which do not anneal to predator DNA (King *et al.* 2010; Vestheim *et al.* 2005), but such a design is all the more complicated as the prey are taxonomically diverse and closely related to the predator. Some attempts have also successfully exploited the properties of some DNA analogs such as peptide nucleic acids (PNAs), which hybridize to their complementary DNA sequence to prevent PCR amplification of undesirable sequences (Chow *et al.* 2011). A popular and cheaper variant of this approach relies on the use of a blocking oligonucleotide with a C3 carbon spacer on the 3'-end hampering PCR extension, and with a concentration 10 times higher than that of each amplification primer (Vestheim & Jarman 2008). This blocking oligonucleotide must overlap with one of the amplification primers by at least six nucleotides, and must be complementary of the predator sequence, but as different as possible from the prey (see Section 6.7 for more details).

Shehzad *et al.* (2012b) have implemented this blocking oligonucleotide method to investigate the diet of the leopard cat (*Prionailurus bengalensis*) in Pakistan. Without blocking oligonucleotide, 91.6% of the total number of sequences amplified with vertebrate primers (Vert01 in Appendix 1) belonged to the leopard cat, making it impossible to reliably characterize its diet using eDNA (Fig. 17.2). With a blocking oligonucleotide, this proportion fell to 2.2%, and seven previously undetected preys were identified. In total, 18 different prey taxa were found in the 38 feces examined for this predator species.

17.1.4 Analyzing an omnivorous diet, or integrating several diets in a single one

The difficulty of establishing the composition of an omnivorous diet rests in the broad range of foods consumed by the animal, which originate from diverse taxonomic clades that cannot be analyzed by a single marker. Consequently, this requires adopting a mixed strategy relying on a set of different markers. De Barba *et al.* (2014) present a detailed methodology to deal with such complexity when studying the brown bear (*Ursus arctos*) diet. In this work, the brown bear diet was divided into three main components (herbivorous, carnivorous, and insectivorous). Each of these components was assessed using a different marker: the P6 loop *trn*L marker to detect vascular plants (Sper01 in Appendix 1; Taberlet *et al.* 2007), a 12S rDNA marker dedicated to vertebrates (Vert01 in Appendix 1; Riaz *et al.* 2011), and a very short 16S rDNA marker targeting mollusk, arthropod, and vertebrate DNA (close to Moll01 and Arth01 in Appendix 1; De Barba *et al.* 2014). Working with multiple markers has several consequences. This approach multiplies the number of amplifications and the amount of DNA extract becomes limiting. Furthermore, it requires

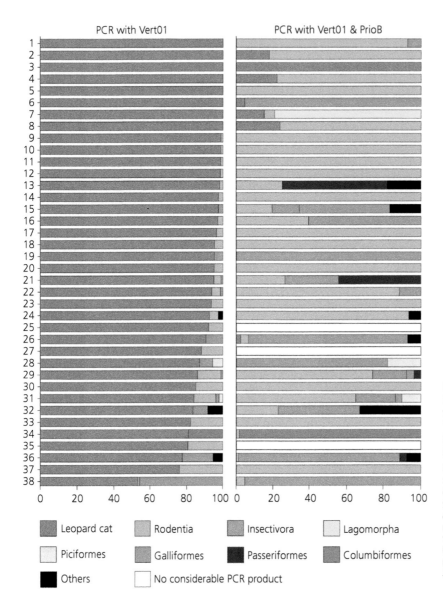

Fig. 17.2 Sequencing results for 38 leopard cat feces amplified with Vert01 primers (Appendix 1) without (left) and with (right) blocking oligonucleotide. The horizontal bar represents the percentage of predator or prey sequences identified in the sample (from Shehzad et al. 2012b). Shehzad, W., Riaz, T., Nawaz, M. A. et al. Carnivore diet analysis based on next-generation sequencing: application to the leopard cat (Prionailurus bengalensis) in Pakistan. Reprinted by permission from *Molecular Ecology*. John Wiley and Sons. Copyright (2012).

the integration of data from several independent metabarcodes for a single sample. Finally, and most importantly, it has to deal with a difficulty specific to omnivorous diets: how is it possible to distinguish the signal from the noise when the DNA template from a diet component can be highly diluted? De Barba *et al.* (2014) proposed two solutions to tackle these problems. First, they multiplexed the three primer pairs in a single PCR reaction to limit the number of amplifications, but this requires the adjustment of the relative amplification yield of each multiplexed marker. The second solution was to include many negative and positive controls in the experiment. Negative controls consisted of PCR reactions performed without DNA templates. Positive controls corresponded to PCR amplifications of mock communities obtained by mixing DNA extracted separately from plants, mammals, and insects. Later, during the data analysis step, these controls were used to fine-tune the filtering parameters, in order to completely clean the negative controls and to adjust the positive control composition to that of the actual mock communities.

17.2 Methodological and experimental specificities of eDNA diet analyses

17.2.1 eDNA sources

When establishing the experimental design of a diet study, it is important to consider what the source of eDNA will be (e.g., feces or stomach contents). Even if the choice is limited for practical reasons, one has to remain aware of the potential biases associated with each source, which are further developed in the following paragraphs.

17.2.1.1 Feces

The most convenient eDNA source for diet studies is often feces (or regurgitated pellets when available). Feces can be exploited to characterize the diet of large herbivores (Kowalczyk *et al.* 2011; Rayé *et al.* 2011; Valentini *et al.* 2009) or carnivores (Shehzad *et al.* 2012a, 2012b, 2015; Xiong *et al.* 2017), but also that of small animals like insects (Ibanez *et al.* 2013). Feces sampling is generally non-invasive, and direct observation is usually not necessary, which facilitates the study of elusive animals. However, several limitations are associated with feces as an eDNA source for diet analyses. Stool DNA has gone through the entire digestive tract, hence it is highly degraded and the size of the examined metabarcode must be as short as possible (Fig. 17.3). If the animals were not observed directly while producing the stools, it is highly recommended to identify the species which actually provided the feces from the fecal DNA. Indeed, experience has shown that even skilled samplers can confuse a species' feces for another's in the field. Moreover, the temporal relationship between ingested food and the resulting DNA detected in the stool is not a simple time shift. The physiology of animal digestion can induce a bolus homogenization over several meals, which hampers the establishment of a precise diet time series from DNA stool (Landmann *et al.* unpublished data).

DNA can be extracted from feces using a standard tissue DNA extraction kit (Shehzad *et al.* 2012b), but quicker and cheaper methods based on phosphate buffer (Taberlet *et al.* 2012c) can be easily adapted to this purpose (see Section 5.4). As with soil samples, the phosphate buffer method allows for the processing of large numbers of samples efficiently and at a reasonable cost. If DNA cannot be extracted right after sampling, fecal samples can be preserved in silica gel, or by using a freeze dryer before proceeding to DNA extraction. Silica gel is very useful as it is easily applicable in the field, even in remote places.

17.2.1.2 Gut content

When feces cannot be collected easily for individual animals, or when identification of the species producing the stool is not straightforward, an alternative is to extract eDNA from the gut (or rumen or stomach) content. This strategy has been adopted to infer diets not only for invertebrates (Blankenship & Yayanos 2005; Eitzinger & Traugott 2011), but also for mammals (Soininen *et al.* 2009, 2013). Gut sampling has, of course, the major drawback of being invasive (i.e., of requiring animal sacrifice or at least heavy experimental handling to collect the sample). Nevertheless, gut eDNA is expected to be of better quality than fecal eDNA (Fig. 17.3), as the digestion process has not been completed entirely.

17.2.1.3 Whole body

When studying the diet of small animals, DNA can be extracted directly from the whole body (Jurado-Rivera *et al.* 2009). Sometimes only parts of the animal, for example the head and/or the thorax, are ground for the DNA extraction (Kitson *et al.* 2013; Navarro *et al.* 2010). These parts correspond to the beginning of the digestive tract, so they are expected to contain better quality prey/food DNA. When extracted from the whole body, the DNA extract is obviously enriched in predator (or consumer) versus prey (or consumed) species' DNA, and it is therefore more difficult to infer a reliable diet from such samples. But an experiment based on the predator beetle *Pterostichus melanarius* fed with a single type of prey (mealworm, i.e., the larva of *Tenebrio molitor*) shows that the detectability of the prey DNA in whole body DNA extracts is similar to that measured from gut DNA extracts (see also Fig. 17.3).

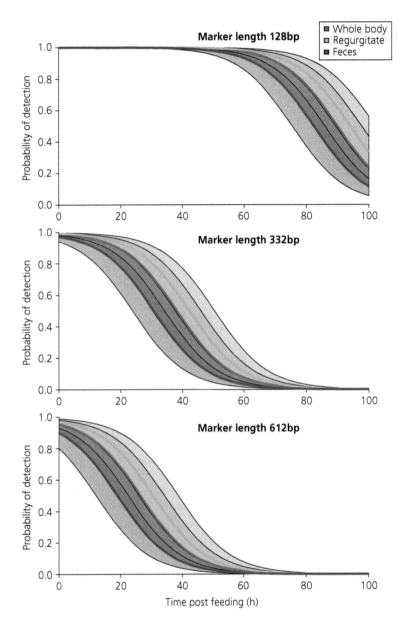

Fig. 17.3 Detection of prey DNA after feeding and starvation of the predatory carabid beetle *Pterostichus melanarius.* Three DNA sources (regurgitates, whole bodies, and feces) and three DNA marker sizes (128 bp, 332 bp, and 612 bp; from the top to the bottom of the figure, respectively) are considered. Bold solid lines indicate the logistic regressions of the observed detection probability and the shaded area the 95% confidence interval envelopes of the fit (from Kamenova *et al.* 2017).

17.2.2 Quantitative aspects

When studying trophic interactions, having a good estimate of the quantity of consumed food items is often crucial, so whatever the method used to analyze the diet, any potential bias in the quantification has to be measured. eDNA-based diet assessment is no exception to this rule, and two distinct questions have to be addressed: (i) is the DNA quantity extracted from the sample related to the biomass of ingested food? And (ii) is the method able to properly quantify the DNA content of the extract?

17.2.2.1 Relationship between the amount of ingested food and DNA quantity in the sample

The first question is actually not restricted to diet analysis and can be generalized as follows: how does an eDNA sample reflect the studied environment?

In the context of diet analysis, the differential digestibility of ingested DNA is often questioned (Deagle *et al.* 2010), but little is actually known on this topic. A study on copepods (Durbin *et al.* 2012) shows little influence of the prey species on the speed of DNA degradation, but this conclusion cannot be generalized to other diets. For herbivorous diets, plant DNA detection can differ from the results based on macroscopic observations of the rumen content, leading to under or overestimation of the plant amount depending on the plant functional group (Nichols *et al.* 2016). Similar biases exist for all the available methods: for example, microhistology, which relies on the observation of plant cuticle remains in stools, overestimates grass content at the expense of forbs (Anthony & Smith 1974). The second problem is the temporal uncertainty introduced by the digestion process. Figure 17.3 (data from Kamenova *et al.* 2017) illustrates that prey DNA can be detected in the DNA extract several days after ingestion. Along the same lines, an experiment on elephant, a hindgut fermenter herbivore, shows that food remains are present in the stools from 12 hours after ingestion up to several days later (Landmann *et al.* unpublished data). These two studies demonstrate that the DNA content extracted from a sample actually reflects the mixture of several meals. Still, diets of large carnivores (e.g., Shehzad *et al.* 2012a, 2015) are often characterized by the detection of a single prey by scat, which is more compatible with a sequential digestion process.

17.2.2.2 Quantifying DNA with PCR and next-generation sequencing

Assaying each DNA marker variant in a DNA mixture can be achieved using quantitative PCR (e.g., Bowles *et al.* 2011) or by combining traditional PCR with next-generation sequencing in a DNA metabarcoding experiment (e.g., Kowalczyk *et al.* 2011; Rayé *et al.* 2011; Thomas *et al.* 2014). Quantitative PCR (qPCR) is the method of choice to detect a particular DNA molecule from a mixture (see Chapter 9), but specific primers and probes have to be developed and the qPCR has to be calibrated for each of the consumed species to be investigated.

DNA metabarcoding is presumably less precise than qPCR, but it has several advantages of its own.

It does not require *a priori* knowledge of the diet composition, and it allows for larger experiments, as there is less processing per sample, especially if the diet is complex. However, the DNA metabarcoding experiment has to be well set up at each step. The choice of metabarcode is a key factor in avoiding PCR biases. As all the metabarcode variants (i.e., corresponding to the different consumed species in presence) will be amplified concurrently in the same PCR tube, the priming sites have to be highly conserved to give the same amplification probability to each variant (see Chapter 2). The PCR protocol also influences the quantitative results. Some DNA polymerases give better quantitative results than others (see Section 6.2). An annealing temperature that is too high will increase the PCR selectivity, but decrease the amplification efficiency for species with primer mismatches (see Section 6.3). A long elongation time will favor the conservation of the initial DNA proportions (see Section 6.3). Finally, sequencing can also introduce biases. Some marker variants are systematically sequenced with an average quality lower than others, due to some intrinsic characteristics of their sequence. As a result, they can be easily rejected during the initial quality control, or filtered out during data analysis based on a quality threshold, skewing the final species proportions.

Despite all these potential biases, diet relative quantification with DNA metabarcoding is not worse than with other methods, and could perhaps be considered as better. Two strategies can be proposed to quantify food items in an eDNA-based diet. The first, and easiest, way is to consider relative frequencies of sequencing reads as a proxy of the relative abundances of the food items (e.g., Kowalczyk *et al.* 2011; Rayé *et al.* 2011; Thomas *et al.* 2014). The second method involves counting the occurrence of each species in a set of samples. The more often an item is observed, the larger fraction of the animal diet it will contribute to. Therefore, the occurrence frequencies provide quantitative information on the diet at the population level. Interestingly, for herbivorous diets where each meal consists of a large number of plants, both quantification methods give similar results (unpublished data; Fig. 17.4).

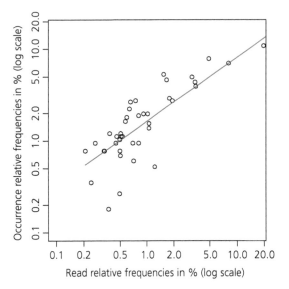

Fig. 17.4 Relationship between relative frequencies of plant sequencing read counts and plant sample occurrences. The occurrences are estimated from 145 *Bison bonasus* feces analyzed by DNA metabarcoding using the Sper01 primers (Appendix 1), amplifying the *trnL* P6 loop intron. Frequencies are expressed in percentage and plotted according to a log scale. Each dot corresponds to a plant species.

17.2.2.3 Empirical correction of abundances

Through controlled feeding experiments, it has been empirically confirmed that read abundances reflect food item proportions. Notably, this has been carried out for sheep (Willerslev *et al.* 2014) but the correspondence is not perfect. An interesting procedure was proposed by Thomas *et al.* (2014) to reduce the quantification error in a seal diet study. It involves preparing mixtures of the flesh of two items of prey in the same amount, with one item present in all the mixtures. Tissue DNA is extracted directly from each mixture, and then the metabarcode is amplified from the obtained extracts and sequenced. As there is an item of prey common to all mixtures, the observed read count ratios allow for the estimation of a correction factor for each pair of prey items, which can be used to correct read counts for real diet assessment. This strategy, applied to a controlled seal diet, halves the average error on food item proportions, but the efficiency of this correction decreases with diet complexity.

17.2.3 Diet as a sample of the existing biodiversity

In recent years, researchers have become interested in some animal diets not *per se*, but for the information these diets can provide on the prey or food biodiversity the animal encounters in its environment. Some invertebrates are even called "vertebrate samplers" in the literature (Calvignac-Spencer *et al.* 2013a), because vertebrate DNA can theoretically be easily recovered from their meals and used to explore vertebrate biodiversity (Bohmann *et al.* 2014; Calvignac-Spencer *et al.* 2013b; Schnell *et al.* 2012) or to target species distribution (Kent 2009; Schubert *et al.* 2015). Up to now, only hematophagous insects (Gariepy *et al.* 2012; Kent 2009), leeches (Schnell *et al.* 2012), and carrion flies (Calvignac-Spencer *et al.* 2013b; Schubert *et al.* 2015) have successfully joined the ranks of these vertebrate samplers, but the list of parasitic, coprophagous, or saprophagous invertebrates that could potentially be exploited for this purpose is much longer (Calvignac-Spencer *et al.* 2013a). Interestingly, cryptic and even rare species can be accessed through this indirect sampling approach (Calvignac-Spencer *et al.* 2013b; Schnell *et al.* 2012).

A number of characteristics will influence the suitability of a particular sampler for vertebrate biodiversity assessment, including its distribution range, diet breadth, ease of sampling, body size, digestion efficiency, mobility in the environment, and so on (see a review in Calvignac-Spencer *et al.* 2013a). Perhaps the most important is host preference, both at the individual and species level, as it can greatly bias the resulting representation of biodiversity (see a review in Calvignac-Spencer *et al.* 2013a; Schnell *et al.* 2015b). Consequently, non-detection of a particular vertebrate species in the diet cannot be reliably interpreted as its absence in the studied environment (Schnell *et al.* 2015b).

This sampling approach based on ubiquitous invertebrates has been mostly used to investigate the vertebrate component of biodiversity. This is all the more interesting as the vertebrate biomass, and thus the amount of eDNA they release on the environment, is generally negligible compared to that of other groups. Therefore, vertebrates are fairly difficult to reach using eDNA sampling methods. Other taxonomic groups like plants or insects can be targeted through indirect sampling. For example, Vesterinen *et al.* (2016) used DNA metabarcoding to show that there was a large overlap between the diet of the insectivorous bat *Myotis daubentonii* and the community of insects available as prey in its environment. Similarly, the droppings of frugivorous birds were analyzed to evaluate the diversity of potentially dispersed plants and ultimately assess the efficiency of restoration efforts in a protected area in Northern Italy (Galimberti *et al.* 2016). Even if not strictly speaking a diet item, honey has also proven to be a good indicator of both plant and insect diversity (Prosser & Hebert 2017; Richardson *et al.* 2015; Valentini *et al.* 2010). In the near future, exciting ecological applications, such as the study of phenology at the level of the ecosystem, can be contemplated via the detection of pollen in honey, or fruit in the diet of frugivores.

17.2.4 Problematic diets

Despite many advantages, one must remain aware that DNA-based diet analysis has intrinsic limitations that will always be challenging, or even impossible to overcome. For example, for animals feeding on Fungi or Bacteria (e.g., nematodes; Moens & Vincx 1997), DNA metabarcoding cannot distinguish between what was really consumed and what comes from an environmental contamination (e.g., in the digestive tract or after defecation). Cannibalism is another difficult case, as predator and prey sequences cannot be disentangled, so it is generally overlooked in carnivorous diet studies (Ray *et al.* 2016; Shehzad *et al.* 2012b).

CHAPTER 18

Analysis of bulk samples

As indicated in Chapter 1, environmental samples can provide DNAs with very heterogeneous qualities, from high quality DNA similar to extracts from tissues, to highly degraded and low-concentration DNA similar to extracts from ancient material. In this chapter, we focus on the analysis of high quality eDNA from bulk samples.

18.1 What is a bulk sample?

A bulk sample is a particular environmental sample that mainly contains organisms from the target taxonomic group under study. Typically, the mix of different insect species captured with a Malaise trap or other insect traps is a bulk sample (e.g., Ji et al. 2013). DNA extracted from a bulk sample can sometimes be called "community DNA" (Creer et al. 2016). The mix of nematode species extracted from soil can also be considered as a bulk sample (e.g., Porazinska et al. 2010). However, the situation is often more complex and there is no clear separation between a classical environmental sample and a bulk sample. For example, if a marine sediment is size-fractionated, some fractions will be enriched in meiofauna and can be considered as bulk samples (Chariton et al. 2010; Fonseca et al. 2010). Filtered or size-fractionated water samples containing whole organisms such as unicellular eukaryotes can also be considered as bulk samples (e.g., De Vargas et al. 2015). Even a typical insect bulk sample will produce DNA from insects, but also from many other eukaryotes associated with their diet and from diverse Bacteria.

A difference that is consistently noted between bulk samples and other environmental samples is that the DNA extracted from a bulk sample

appropriately preserved is of high quality and quantity. This has several consequences for the molecular protocols that can be implemented.

18.2 Case studies

18.2.1 Bulk insect samples for biodiversity monitoring

From arthropod bulk samples collected with Malaise traps, Yu et al. (2012) isolated 547 individuals, extracted their DNA, and sequenced the standard cytochrome c oxidase subunit I (COI) barcode using a Sanger protocol. Seven mock communities were built by mixing between 121 and 292 DNA extracts corresponding to different barcode sequences. Yu et al. (2012) proposed a new metabarcoding protocol for the analysis of these mock communities based on degenerated COI primers (Table 18.1; primers Fol-degen-for and Fol-degen-rev). Polymerase chain reaction (PCR) products were sequenced on the 454 sequencing platform from Roche. The results obtained showed that overall 74% of the initial species were recovered, with a particularly low percentage for Hymenoptera (54%), and that read numbers did not correlate with abundance.

Based on this protocol, Ji et al. (2013) implemented a larger study to compare DNA metabarcoding with standard traditional biodiversity surveys. They collected samples in three different biomes (subtropical forest in China, temperate woodland in the United Kingdom, and tropical rainforest in Malaysia) using different insect traps (light, pitfall, and Malaise traps), resulting in more than 50,000 insect specimens. A huge effort was made to identify these specimens using traditional approaches,

Environmental DNA for Biodiversity Research and Monitoring. Pierre Taberlet, Aurélie Bonin, Lucie Zinger, & Eric Coissac,
Oxford University Press (2018). © Pierre Taberlet, Aurélie Bonin, Lucie Zinger, & Eric Coissac 2018.
DOI: 10.1093/oso/9780198767220.001.0001

and the results were compared to those of DNA metabarcoding. The two data sets showed statistically correlated alpha- and beta-diversities, and produced similar conclusions concerning conservation applications, either for restoration ecology or for systematic conservation planning.

18.2.2 Nematode diversity in tropical rainforest

To evaluate the diversity of nematodes in a tropical rainforest in Costa Rica, Porazinska *et al.* (2010) collected soil, litter, and canopy samples (four samples per habitat). After isolating the nematodes from the different types of habitats using classical methods, they extracted DNA from these bulk samples, and amplified a ~400 bp fragment of the 18S rRNA gene. The sequencing was carried out on the 454 platform. A total of 214 putative nematode species were found, without a single species being present in all samples, and with 42% of the species observed only within a single sample. This led to an estimated total number of nematode species between 390 and 497 for the entire data set. Clearly, the diversity of nematode communities is quite high in tropical rainforest, much higher than in temperate forest. It was also interesting to observe that the nematode diversity is higher in litter than in soil or canopy.

18.2.3 Marine metazoan diversity in benthic ecosystems

From nine marine sediment samples (each composed of three cores) collected at low tide, Fonseca *et al.* (2010) isolated the meiofaunal size fraction, extracted DNA, and amplified the extract with universal eukaryote primers targeting a ~370 bp of 18S rRNA gene. In addition to the sediment samples, a mock community composed of 41 nematodes was also analyzed and used to calibrate the data analysis. The sequencing on the 454 platform led to the identification of a few hundred taxa, dominated by nematodes, followed by platyhelminthes, arthropods, and mollusks. Remarkable differences in diversity occurred at microgeographical scales.

Another marine metabarcoding study was carried out by Leray & Knowlton (2015). In this case, the targets were eukaryotic organisms colonizing multilayered settlement surfaces (autonomous reef monitoring structures) over a six-month period in different oyster reefs in temperate and subtropical locations. Mobile organisms larger than 2 mm were barcoded using a standard COI approach based on Sanger sequencing. Other organisms, either sessile or of small size, were pooled and analyzed by metabarcoding with a COI marker (primers mlCOIintF and jgHCO amplifying a 313 bp fragment; Table 18.1). The sequencing of the 54 bulk samples was carried out on an Ion Torrent platform from Thermo Fisher Scientific. The results showed almost no overlap between the temperate and the subtropical locations. This pilot experiment indicated that DNA metabarcoding has a good potential for ecological studies and environmental monitoring using such autonomous reef monitoring structures. However, the experimental design did not include any known community that would have allowed the estimation of the proportion of undetected organisms.

18.3 Metabarcoding markers for bulk samples

As seen in the previous section, different types of markers have been used to analyze bulk samples. A first approach involves using the standard animal barcode COI, or a shorter version of this barcode (Table 18.1). The COI gene is the most common provider of metabarcodes for insects or arthropods, and more generally for metazoans (e.g., Bista *et al.* 2017; Clarke *et al.* 2017; Elbrecht & Leese 2017; Hajibabaei *et al.* 2011; Ji *et al.* 2013; Leray & Knowlton 2015; Ransome *et al.* 2017). Another scientific community, working on nematodes, meiofauna, and protists prefers using a fragment of the 18S rDNA gene (e.g., Chariton *et al.* 2010; Fonseca *et al.* 2010; Visco *et al.* 2015).

What are the differences between these two alternative approaches? The main advantage of using the standard animal barcode COI is the associated large reference database managed by the International Barcode of Life consortium (BOLD; http://www.boldsystems.org). Due to the high level of variability of the COI gene, the taxonomic identification is quite precise, usually to the species level. However, the COI gene has a significant

Table 18.1 Commonly used COI primers for analyzing bulk samples. The following primer pairs can be used (length in brackets, excluding primers): LCO1490/HCO2198 (658 bp), dgLCO-1490/dgHCO-2198 (658 bp), Fol-degen-for/Fol-degen-rev (658 bp), mlCOlintF/jgHCO (313 bp), dgLCO-1490/mlCOlintR (319 bp), BF1/BR1 (217 bp), BF1/BR2 (316 bp), BF2/BR1 (322 bp), BF2/BR2 (421 bp). Primers mlCOlintF and mlCOlintR have been developed for marine metazoans. Primers BF1, BF2, BR1, and BR2 have been developed for freshwater macroinvertebrates. The positions of these primers are given in Figure 18.1

Code	Sequence 5'–3'	Reference
LCO1490	GGTCAACAAATCATAAAGATATTGG	(Folmer *et al.* 1994)
HCO2198	TAAACTTCAGGGTGACCAAAAAATCA	(Folmer *et al.* 1994)
dgLCO-1490	GGTCAACAAATCATAAAGAYATYGG	(Meyer 2003)
dgHCO-2198	TAAACTTCAGGGTGACCAAARAAYCA	(Meyer 2003)
Fol-degen-for	TCNACNAAYCAYAARRAYATYGG	(Yu *et al.* 2012)
Fol-degen-rev	TANACYTCNGGRTGNCCRAARAAYCA	(Yu *et al.* 2012)
mlCOlintF	GGWACWGGWTGAACWGTWTAYCCYCC	(Leray *et al.* 2013)
mlCOlintR	GGRGGRTASACSGTTCASCCSGTSCC	(Leray *et al.* 2013)
jgHCO	TAIACYTCIGGRTGICCRAARAAYCA	(Leray & Knowlton 2015)
BF1	ACWGGWTGRACWGTNTAYCC	(Elbrecht & Leese 2017)
BF2	GCHCCHGAYATRGCHTTYCC	(Elbrecht & Leese 2017)
BR1	ARYATDGTRATDGCHCCDGC	(Elbrecht & Leese 2017)
BR2	TCDGGRTGNCCRAARAAYCA	(Elbrecht & Leese 2017)

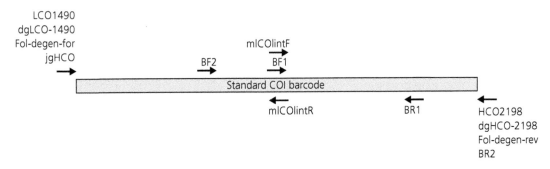

Fig. 18.1 Position of commonly used COI primers for the analysis of bulk samples. See Table 18.1 for the sequences and references.

drawback. It is a protein coding gene, and all third positions of the codons are variable. Consequently, even degenerated primers will not amplify all the different species in the same way, and 20–30% of the species present in DNA extracts can be missed (e.g., Yu *et al.* 2012). Inversely, it is possible to design universal primers on 18S rRNA gene (i.e., primers targeting sequences that are much less variable that any protein coding gene). For example, the primers designed by Hardy *et al.* (2010) are true universal eukaryotic primers, with very low

levels of variation in their target sequences (see Euka01 in Appendix 1). However, metabarcodes based on universal primers such as these also have some drawbacks. First, their taxonomic resolution is not as good as COI and is highly variable among taxonomic groups, being quite low for vertebrates and plants, but relatively good for many other metazoan groups such as nematodes and arthropods. Second, in some cases, the universal 18S primers amplify a fragment that is variable in length, leading to a less efficient amplification of some taxonomic groups. For instance, the length of the amplified fragment of universal eukaryotic primers Euka01 or Euka02 (Appendix 1) is much larger for amphipods/isopods, meaning that these groups will not be reliably detected. Another universal eukaryotic primer pair solves this difficulty by being much more homogeneous with regards to the size of the amplicon (Euka03 in Appendix 1). Unfortunately, they poorly amplify amphipods/isopods. If it is important to not miss these crustacean groups, additional markers can be used, such as Peka01 and Peka02 (Appendix 1).

According to the characteristics of COI and 18S metabarcoding markers, different strategies can be implemented. If missing 20–30% of the species or if the proportions among the different taxonomic groups are not important, and if taxonomic identification at the species level is required, then a COI metabarcoding marker can be chosen. If it is important to amplify the very large majority of the different taxonomic groups to get an estimate of the proportions in the DNA extract, and if a taxonomic identification at the genus or the family level is sufficient, a universal 18S metabarcoding marker can be chosen. One can also envisage a mixed strategy by conjointly using these two types of markers. In this case, the COI marker will provide better taxonomic identification for a subset of the data, and the 18S marker a better overview of the global diversity.

18.4 Alternative strategies

The drawbacks of metabarcoding COI markers stimulated further technical developments to limit the impact of taxonomic biases introduced by PCR. Through mitochondrial enrichment via differential centrifugation, followed by genomic DNA extraction and deep sequencing on an Illumina platform (HiSeq 2000), Zhou et al. (2013) demonstrated that the taxonomic bias can be strongly reduced, and that a good estimate of the initial biomass can be obtained. For this purpose, they built a known insect community composed of 73 specimens. After morphological identification, they sequenced the standard barcode of all specimens and found 37 different molecular operational taxonomic units (MOTUs). After shotgun sequencing of this genomic DNA enriched in mitochondria, they applied two bioinformatic pipelines. The first one relied on a COI reference library. By aligning the reads with the reference library, 36 MOTUs out of 37 (97.3%) could be detected. The missing species was represented by a single specimen with a body length smaller than 2 mm. The second pipeline did not use any reference library. The reads were *de novo* assembled into different scaffolds. In this case, 34 MOTUs out of 37 (91.9%) could be recovered, plus one that was not identified when building the known community.

Liu et al. (2016) furthered this approach by designing capture probes for mitochondrial DNA. They tested their system against a 49-insect species mock community and observed a 100-fold enrichment of mitochondrial DNA. They also showed that the enrichment by capture preserved the relative abundance among species. Finally, another interesting experiment based on capture/shotgun sequencing was developed by Shokralla et al. (2016) and is detailed in Section 19.2.2.

## CHAPTER 19

The future of eDNA metabarcoding

What are the trends for eDNA metabarcoding going to be in a few years? It is always a difficult exercise to try to predict the future. However, a few recent papers provide some interesting cues. Currently, DNA metabarcoding faces several important technical difficulties. The main challenges concern the primer bias and polymerase chain reaction (PCR) errors, and more generally the need for standardization of experiments at all levels (Chapters 2–8) to allow for cross-experiment comparisons. Figure 19.1 illustrates the three alternative strategies for obtaining metabarcoding data, including the now-traditional PCR-based approach, as well as two different alternatives relying on shotgun sequencing. We will first discuss the relative merits of these three strategies, taking into account the technical difficulties and the cost while keeping in mind that sequencing is becoming cheaper and cheaper, and that simple protocols will always be more popular than complex ones. We will then discuss how reference databases will evolve in the near future and, finally, a few open and exciting questions will be presented.

19.1 PCR-based approaches

As largely discussed in this book, the main problems with the PCR-based approach are associated with primer bias and PCR errors. It is generally possible to design metabarcoding primers with limited taxonomic bias (Chapter 2) that do not necessarily target the same regions as standard barcodes (cytochrome c oxidase subunit I for animals, or *rbc*L or *mat*K for plants), but this makes it necessary to build custom-made reference databases (Chapter 3). Even with these imperfections, PCR-based metabarcoding can remain a valuable tool.

19.1.1 Single-marker approach

There are situations where the PCR-based approach will still be valuable in the future. It will first concern studies where the number of samples to be analyzed is of prime importance to solve the scientific question. It is relatively easy to set up PCRs for hundreds or even thousands of samples (Chapter 6), but it would be difficult to do the same with approaches involving either the direct shotgun sequencing of the DNA extract, or the capture-based shotgun approach (Fig. 19.1). The PCR-based approach will also be valuable when a complete reference database has already been built for a metabarcode that exhibits limited primer biases. Many diet studies, or large-scale soil studies of macroorganisms, belong to this category where the number of samples is important, and where a comprehensive reference database can be easily obtained. It is thus likely that the PCR-based approach will continue to play a key role in eDNA analysis.

19.1.2 Multiplex approach

In some situations, the DNA extracts are precious, and cannot easily be resampled. The analysis of such extracts using the single-marker approach presented here has a strong drawback: during the amplification of a particular marker, all other DNA fragments of the extract that are not amplified by the primers are simply lost. One potential solution to extract as much information as possible from the same extract involves multiplexing many markers within the same PCR (see e.g., De Barba *et al.* 2014; and Section 6.9). For example, when analyzing precious DNA extracts from lake sediments, it might

Environmental DNA for Biodiversity Research and Monitoring. Pierre Taberlet, Aurélie Bonin, Lucie Zinger, & Eric Coissac, Oxford University Press (2018). © Pierre Taberlet, Aurélie Bonin, Lucie Zinger, & Eric Coissac 2018. DOI: 10.1093/oso/9780198767220.001.0001

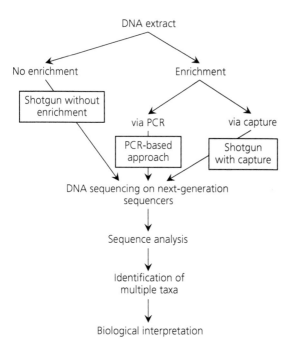

DNA extract

No enrichment   Enrichment

Shotgun without
enrichment

via PCR   via capture

PCR-based   Shotgun
approach   with capture

DNA sequencing on next-generation
sequencers

Sequence analysis

Identification of
multiple taxa

Biological interpretation

Fig. 19.1 The three different alternative strategies for implementing a DNA metabarcoding study.

be suitable to multiplex a eukaryote marker (e.g., Euka02; Appendix 1) together with plant, insect, and mammal markers (Sper01, Inse01, Mamm02, respectively; Appendix 1). In this example, the melting temperature is different for the different primer pairs. To adjust the system, it is possible to play with both the melting temperature and with the final concentrations in the PCR mix. One or two nucleotides can be removed on the 5′-end of the primers with a higher melting temperature, and the concentration of the primer pairs with a lower melting temperature can be increased. Such a PCR-based multiplex approach could represent a cheap and relatively easy to implement alternative to the enrichment by capture (see Section 19.2.2).

19.2 Shotgun-based metabarcoding

As the sequencing costs have strongly decreased in the last few years, it is now possible to envisage large-scale shotgun sequencing experiments for analyzing eDNA, with or without an initial capture step.

19.2.1 Without enrichment by capture

One of the main advantages of the shotgun without enrichment by capture is that it represents a fully open approach, without any *a priori* knowledge about the target taxa. A few papers implementing a shotgun approach on modern samples have already been published.

The first examples involved targeting mitochondrial DNA of bulk arthropod samples, either after an enrichment step by differential centrifugation (Zhou *et al.* 2013; see Section 18.4 for more details), or without enrichment (Andújar *et al.* 2015). In both studies, shotgun sequencing allowed the identification of members of the entire community via their mitochondrial DNA. This strategy has been called "metagenome skimming" by Linard *et al.* (2015) and Papadopoulo *et al.* (2015).

Srivathsan *et al.* (2015, 2016) were the first to implement shotgun-based metabarcoding studies without enrichment to target non-bacterial organisms in diets. In their two papers, they analyzed two and six fecal samples of two leaf-eater monkeys (*Pygathrix nemaeus* and *Presbytis femoralis*) to assess their diet. After extracting DNA from the feces, they sequenced the extracts on Illumina platforms, producing between ~67 and ~108 million paired-end reads per sample. Globally, this shotgun approach produced slightly better results than the more classical PCR-based metabarcoding amplifying the *trn*L P6 loop (Sper01 primers, Appendix 1); however, the sequencing cost per sample is at least 1,000 times higher. In this context, it is clear that the shotgun approach without enrichment does not represent the future of diet analysis.

Two other papers have implemented a shotgun metabarcoding approach without enrichment for ancient samples. The first aimed to assess the viability of a past ice-free corridor in North America by analyzing lake sediments (Pedersen *et al.* 2016; see Section 15.1.3 for more details), and the second to characterize past human diet using midden from Greenland (Seersholm *et al.* 2016; see Section 15.3.2 for more details). In both cases, the shotgun approach solved the biological question at hand. However, the identification of the different taxa was in some cases based on a very low number of informative reads.

When working with ancient samples, another crucial advantage of the shotgun approach is the possibility to authenticate ancient DNA molecules by looking at the deamination of cytosine (C) bases. This phenomenon leads to uracil residues that are read by DNA polymerases as thymine (T) bases. The deaminations occur in higher frequencies close to the ends of DNA fragments, resulting in higher frequencies of C to T substitutions at the extremities of ancient fragments (Briggs *et al.* 2007; Hofreiter *et al.* 2001; Jónsson *et al.* 2013).

19.2.2 With enrichment by capture

Shotgun sequencing with enrichment by capture (hereafter capture/shotgun) is a very attractive strategy that we will illustrate by presenting two papers. The first paper deals with the analysis of ancient cave sediments, looking at mammalian mitochondrial DNA in general, and more specifically at possible remains of Neandertal and Denisovan DNA (Slon *et al.* 2017). The authors collected 85 cave sediment samples from seven archaeological sites with known Hominidae occupation. After DNA extraction from about 100 mg of sediment, aliquots of the extracts were used to build single-stranded DNA libraries (Gansauge & Meyer 2013; Gansauge *et al.* 2017). These libraries were then subjected to two capture enrichments using mammalian and human mitochondrial probes. Over the whole data set, a total of 12 mammalian families were found, including Bovidae, Canidae, Cervidae, Cricetidae, Elephantidae, Equidae, Felidae, Hominidae, Hyaenidae, Mustelidae, Rhinocerotidae, and Ursidae. The human probe gave positive results validated with deamination on fragment extremities for 15 sediment samples from four sites. In each case, between 10 and 165 Hominidae sequence reads were recovered. Additional libraries were prepared from 10 of these 15 sites, and after enrichment with the human probe, the presence of Neanderthal and Denisovan DNA was confirmed in nine samples out of ten. This paper demonstrates the extreme power of capture/shotgun for analyzing rare and degraded DNA.

The second paper (Shokralla *et al.* 2016) compared the classical PCR-based metabarcoding on the standard animal barcode (COI) with a capture/ shotgun approach targeting the same marker, with the objective of avoiding the well-known qualitative and quantitative PCR bias linked to the COI marker. Four different insect bulk samples, one Malaise trap content, and three benthic macroinvertebrate samples, were analyzed. Globally, it appeared that capture/shotgun provides highly consistent results and is more powerful than PCR for detecting taxa. For the Malaise trap sample containing 1,066 morphologically identifiable individuals, 53% and 70% more genera and families, respectively, were detected by capture/shotgun. However, concerning the quantitative aspect, the COI primer bias was not solved with this approach. This is likely due to the design of the capture probes of differing efficiency toward some local insect orders. Nevertheless, the design of more balanced capture probes should be able to strongly decrease the quantitative bias. The capture/shotgun approach is attractive, as it has the potential of allowing an efficient and reliable use of the standard barcoding markers, without the bias due to primers designed on protein-encoding genes.

19.3 Toward more standardization

Meta-analyses have an important impact on scientific progress, as they allow the results obtained independently in several primary studies to be generalized. The Barcoding of Life initiative represents a significant effort toward reaching this goal, by imposing standard barcodes for taxonomic identification of plants and animals. However, the standardization of metabarcoding experiments is more difficult to implement, as the experimental design should be adjusted according to the scientific question and system studied. Furthermore, standardization has the intrinsic risk of blocking or slowing down further developments. This was the case for DNA barcoding developed during the Sanger sequencing era, and which later had difficulties coping with next-generation sequencing.

19.3.1 For sound comparisons across studies

Comparisons among different metabarcoding experiments are difficult to achieve, for several reasons. First, the sampling strategy can be extremely different among studies. For example, it is difficult to compare studies on soil biodiversity when the starting material is 100 mg or 15 g of soil. Second, the possibilities for processing the samples at the bench are numerous. It is well known that different extraction protocols or different metabarcodes can lead to diverging results. Different DNA polymerases and amplification parameters also generate different results (Chapter 6). Finally, sequence analysis can have a strong impact on the final results. Thus, there is a strong need for standardization in eDNA studies (Nilsson *et al.* 2011), not only in the experimental approach, but also in recording the associated metadata (Tedersoo *et al.* 2015).

A first attempt toward this goal has already been made concerning the metabarcoding of microorganisms. The Earth Microbiome Project is a "massively collaborative effort to characterize microbial life" on earth (http://www.earthmicrobiome.org; see Section 14.4), initiated in 2010 (Gilbert *et al.* 2010, 2014). In order to fulfill the scientific objectives, standardized protocols were proposed, allowing reliable comparisons among the different studies performed at the worldwide level to be made. These standardized protocols encompass (i) a guide for the metadata, including an ontology for the description of habitats; (ii) a DNA extraction protocol starting with 0.1–0.25 g of soil; (iii) a detailed protocol for amplifying and sequencing the V4 region of 16S rRNA gene of Bacteria and Archaea (using primer pair Bact04, Appendix 1); (iv) a detailed protocol for amplifying and sequencing the V9 region of 18S rRNA gene of Eukaryota (using primer pair Euka04, Appendix 1); (v) a detailed protocol for amplifying and sequencing the ITS1 region of Fungi (primers ITS1f and ITS2); (vi) instructions for running the sequence analysis; and (vii) shipping instructions.

It was not too difficult to find a consensus within the scientific community for proposing a standard approach for microorganisms, although it is not systematically adopted. However, the case of macroorganisms is more complex. There is a dilemma between the use of the standard barcodes and the need for providing a comprehensive and balanced view of the target group of organisms. Consequently, any standardization is much more difficult to establish, although capture/shotgun with appropriate probes might be a solution for

obtaining sound results on standard barcodes for both plants and animals.

For any target group, it might be advisable to systematically include the same complex eDNA template as a positive control in all experiments. Even if the experimental protocol is not exactly the same in different studies, it becomes possible to carry out meta-analyses if this complex positive control provides the same results. Using internal DNA standards at the extraction step, or quantifying species-specific relative correction factors are other alternatives that would facilitate cross-experiment comparisons (Smets *et al.* 2016; Thomas *et al.* 2016; see Section 19.5.3).

While innovations in molecular biology and bioinformatics will continue to improve our description of an eDNA extract, our lack of understanding on how DNA from different organisms is released, and how it is recycled in different environmental compartments will always limit the use of this data for comparative analyses and for inferring ecological processes. Some systematic laboratory experiments can be carried out to monitor the DNA released from the target organisms, and its eDNA degradation in the studied environmental matrix (e.g., Dejean *et al.* 2011; Sigsgaard *et al.* 2016; Thomsen *et al.* 2012a).

19.3.2 For environmental monitoring

Using eDNA for environmental monitoring is quite attractive (e.g., Baird & Hajibabaei 2012; Holdaway *et al.* 2017; Lim *et al.* 2016; Ransome *et al.* 2017; Thomsen *et al.* 2012b; Thomsen & Willerslev 2015; Valentini *et al.* 2016), especially in aquatic ecosystems. However, the constraints for implementing standardized metabarcoding protocols for biomonitoring are strong. The experiments must be fully replicable in time, in space, and across laboratories. At the present, no clear solution emerges for plants and animals, mainly because most of the efforts are dedicated toward the amplification of standard barcodes. The advantage of targeting standard barcodes is that the current reference databases can be used. Nonetheless, many alternative primer pairs have been proposed for amplifying standard barcodes, with all of them introducing various biases in the analyses, and there is no consensus about which one to use. As indicated, the solution might be to develop a capture/shotgun approach using standardized probes.

Recently, Apothéloz-Perret-Gentil *et al.* (2017) developed a taxonomy-free metabarcoding approach for estimating a diatom index in rivers. It does not rely on any taxonomic reference database, the primer pair used is highly conserved among diatoms (Zimmermann *et al.* 2011), and its implementation is relatively simple. Thus, this high-throughput approach has the potential of becoming a popular biomonitoring tool for rivers in the near future.

19.4 Next-generation reference databases

In Section 3.3.2, we proposed the use of genome skimming (Dodsworth 2015; Straub *et al.* 2012) for building reference databases. Genome skimming has been suggested as an extension of the current standard barcoding system (Coissac *et al.* 2016), as it offers many advantages. First, it enables the assembly of organelle DNA and nuclear ribosomal DNA tandem repeats (e.g., with Organelle Assembler; http://metabarcoding.org/asm), which gathers all potential metabarcodes in a single reference database. Second, it is very easy to implement at the bench, as it requires only DNA extraction and sequencing library preparation. Third, the cost of genome skimming is only marginally higher than traditional barcoding, especially for plants for which several standard barcodes must be sequenced. Fourth, it works even with highly degraded specimens, including type specimens from museums or herbariums (Besnard *et al.* 2014). Fifth, genome skimming permits a much better implementation of shotgun-based metabarcoding experiments by providing genome-wide reference databases. Finally, it leads to the design and evaluation of highly efficient metabarcodes.

Considering all these advantages, it is likely that most of the reference databases will rely on genome skimming in the near future. It is already being implemented for plants in two large projects involving all alpine plants (PhyloAlps project; http://phyloalps.org) and all Norwegian plants (PhyloNorway project). The same approach will undoubtedly emerge for animals.

19.5 Open questions

19.5.1 What will be the impact of new sequencing technologies on eDNA analysis?

The Nanopore-based sequencing by Oxford Nanopore Technologies is developing rapidly (Jain *et al.* 2015, 2016; Ramgren *et al.* 2015; Steinbock & Radenovic 2015). Among other Nanopore platforms, the MinION (https://nanoporetech.com/products/minion) is a portable device that can be plugged into a laptop. It generates 5–10 Gb of several kb-long reads in a single run, and can be used in the field. Its main limitation is currently the sequence quality, which is much lower than that on Illumina sequencers. Nevertheless, the company is investing considerable effort in improving data quality, leaving open the possibility of using the MinION for DNA metabarcoding. Due to its flexibility and moderate cost, this device will certainly motivate the implementation of small-scale experiments. Its rapidity to produce results is also an important criterion for biomonitoring. All these properties will certainly lead to a democratization of eDNA metabarcoding in the near future.

19.5.2 Will some specific repositories be developed for DNA metabarcoding?

There is currently no optimal solution for storing DNA metabarcoding data. The only standard public repository currently available for this purpose is the Short Read Archive system (SRA; https://www.ncbi.nlm.nih.gov/sra), which is shared between the National Center for Biotechnology Information (NCBI), the European Molecular Biology Laboratory (EMBL), and the DNA Data Bank of Japan (DDBJ). This system unfortunately presents several drawbacks in a DNA metabarcoding context. First, sequences from each sample must be uploaded as a separate file. This is not very convenient, and it has the additional disadvantage of leading to a loss of information. Indeed, raw data should be filtered for dispatching individual reads into different files, so for example, reads that cannot be assigned to any sample are lost, even if they can provide clues about potential sequencing problems. Similarly, this precludes context-dependent filtering of spurious sequences. Moreover, the system's data uploading or browsing aspects are not user-friendly and lack detailed explanation, although recent improvements have been made. For now, the SRA system is incompatible with both current large-scale metabarcoding experiments and meta-analyses requirements. The Dryad Digital Repository (https:// datadryad.org) is a convenient alternative that enables a full DNA metabarcoding experiment and its associated metadata to be stored. However, data are not deposited in a standard format and it might be difficult to browse currently available data for a given taxon or ecosystem without being aware of the papers associated with these data.

All of these considerations call for the development of a proper repository that will enable archiving, sharing, and analyzing metabarcoding data to address broad-scale ecological questions (Holdaway *et al.* 2017). This requires setting up several standards for raw/preprocessed sequencing data (Chapter 8), fully processed data (MOTU taxonomic information and abundances, and so on), information on sample origin and characteristics, as well as for the molecular and sequencing protocols used for the experiment. While standards for sequence data are now relatively well established (e.g., FASTQ files), further work needs to be done for the other types of data. Microbial ecologists have recently proposed several options that are still only being partially adopted, but that could constitute the way forward. The BIOM format (McDonald *et al.* 2012) stores several sparse-formatted tables containing characteristics on MOTUs and samples in one single file. It is now recognized as the standard for the Earth Microbiome Project, and as a candidate standard for the Genomic Standards Consortium. The minimum information necessary, along with sequences, has also been already defined (Yilmaz *et al.* 2011b; gensc.org/projects/ mixs-gsc-project) and includes both the biological (e.g., sample biome, location) and experimental (e.g., PCR primers and conditions, sequencing methods) contexts. Internal controls (e.g., negative/positive DNA extraction controls, PCR controls) remain, however, largely overlooked in these considerations, which may hinder data reliability. This calls for the development of both annotation and handling methods for experimental controls

(see Chapter 8 for a few proposals). Finally, better connections with other biodiversity repositories such as the Global Biodiversity Information Facility (GBIF; https://www.gbif.org) are currently under consideration, which would greatly help to integrate eDNA-based surveys within global biodiversity assessments and existing monitoring programs.

19.5.3 Will metabarcoding provide quantitative results?

Throughout this book, we have discussed several times the problem of MOTU abundances and their meaning. The observed MOTU abundance does not necessarily reflect the original number of copies of the corresponding metabarcode in the sample due to potential biases at the sampling, extraction, amplification, sequencing, and bioinformatic steps. This means that the MOTU abundance might not be a good proxy for the relative number of individuals or for biomass.

We have presented several ways to reduce such biases (Chapters 2, 5–7). Estimating the total number of metabarcode copies in the sample (e.g., via qPCR) could further help to correct observed raw MOTU abundances. However, this method is costly and difficult to implement in large-scale experiments. Alternatively, an internal control can be added to the samples or to the eDNA extracts (Section 6.4.5; Smets *et al.* 2016; Ushio *et al.* 2017). This control should contain one metabarcode (or several) for which the initial copy number is known within each sample or PCR. Most preferably, the metabarcode should be absent from the ecosystem under study in order to avoid any interference with the non-control data set. Internal controls can potentially provide good estimates of the absolute number of metabarcode copies in the samples.

Translating the metabarcode copy number to biomass or individual abundance represents another challenge, as the number of metabarcode copies varies across cells, tissues, individuals, species, and probably time. In this context, only very rough estimates of biomass or individual abundance can be expected, even if several studies found that MOTU abundances tend to correlate with species' relative biomass or abundance (e.g., Evans *et al.* 2016;

Yoccoz *et al.* 2012). Several methods have been proposed to correct read abundances (Kembel *et al.* 2012; Thomas *et al.* 2016), but they all require an *a priori* knowledge of the relationship between the number of reads and the biomass or number of individuals for all concerned species. One recent study used intraspecific metabarcoding on a hypersimplified seawater system to study the population genetics of whale sharks, by focusing on the mitochondrial DNA control region (Sigsgaard *et al.* 2016). Based on the nucleotidic diversity of the retrieved haplotype frequencies, they could provide estimates of the female population size in the studied area. These results are lower but within the confidence interval of those obtained from classical approaches.

To summarize, it is clear that DNA metabarcoding can potentially provide relative quantitative information. The absolute quantification of biomass or abundance in the field via PCR-based DNA metabarcoding would require additional experiments that are too heavy to be implemented at large scales or in complex ecosystems. Furthermore, the obtained estimates would certainly be too vague to be reliably integrated into traditional ecological models.

19.5.4 Will metabarcoding be fully integrated into ecological models and theories?

As underlined in Chapter 8, most theoretical models in ecology rely on the concept of species and individuals. Over this last decade, a tremendous effort has been made toward reducing the biases of metabarcoding and making them fit these two concepts. However, even unbiased metabarcoding data will remain intrinsically different from our traditional classification of living forms (see Section 19.5.3 and Chapter 8). This may partly explain why the adoption of metabarcoding by ecologists remains limited. As long as this problem is not solved, biodiversity estimates and ecological processes inferred

from current theoretical models and metabarcoding data will still be highly questionable. On the other hand, unbiased metabarcoding still provides unusual, yet valid information on biodiversity, but comparatively little effort has been made in adapting current models to this new paradigm. At a time when predictive ecology crucially lacks informative data (Urban *et al.* 2016), we believe that reinforcing collaborations between the metabarcoding and the modeling communities will enable new powerful models able to deal with the particularity of metabarcoding data to be developed. One possibility in that direction lies in semi-mechanistic community-level modeling (Mokany & Ferrier 2011). This approach should be able to address uncertainties in the spatial distributions and attributes of taxa. It uses spatial diversity distribution and its correlation with the environment to form the initial conditions for a mechanistic model generating change in biodiversity across space and time.

19.5.5 How do we train students and managers to effectively integrate this tool into academic and operational ecological research and monitoring?

Metabarcoding has the potential to provide an unlimited resource of biodiversity records, but its democratization remains limited for several reasons. First, implementing a DNA metabarcoding study requires multiple and disparate skills. Second, the pertinent information is dispersed in the scientific literature, which mainly involves pilot experiments and reviews, in a context lacking standards. Finally, only large-scale experiments can be carried out due to the substantial number of reads produced by current sequencing platforms. We hope that this book will encourage the teaching of eDNA analysis in universities, and the ecologist to embrace this innovative approach. There is no doubt that eDNA will become a key element in the toolbox for ecological research and biomonitoring in the near future.

Examples of primer pairs available for DNA metabarcoding

(with emphasis on short metabarcodes, see Chapter 2 for explanations)

Bact01

Target taxonomic group: Bacteria
       NCBI taxid: 2
Forward primer: GGATTAGATACCCTGGTAGT
       Reference: Fliegerova *et al.* (2014)
Reverse primer: CACGACACGAGCTGACG
       Reference: Fliegerova *et al.* (2014)
Recommended annealing temperature: 54°C
Target gene: 16S rDNA (V5-V6)
Coverage for the target group: NA
Min. length: 30 bp   **Mean length:** 258 bp   **Max. length:** 399 bp
Taxonomic resolution in the target group:

Species	Genus	Family	Order
24.6% (138,929)	60.2% (2,527)	57.0% (370)	62.7% (166)

Comments: According to its logo, the reverse primer could be degenerated as follows: CACRACACGAGCTGACG. However, as the ambiguous nucleotide is close to the 5′-end, the non-degenerated primer will work properly.

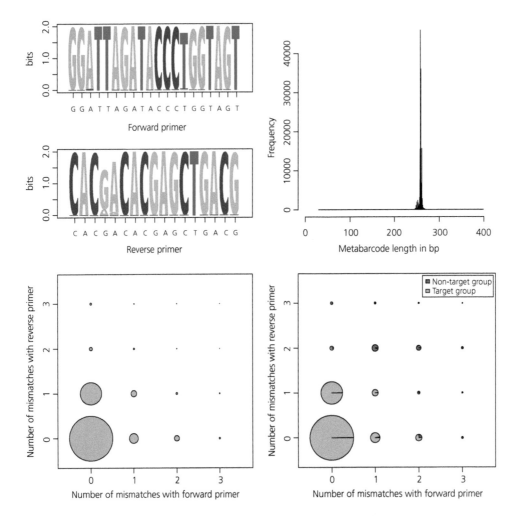

Bact02

Target taxonomic group: Bacteria
      NCBI taxid: 2
Forward primer: GCCAGCMGCCGCGGTAA
      Reference: this book
Reverse primer: GGACTACCMGGGTATCTAA
      Reference: this book
Recommended annealing temperature: 53°C
Target gene: 16S rDNA (V4)
Coverage for the target group: NA
Min. length: 45 bp   **Mean length:** 254 bp   **Max. length:** 748 bp
Taxonomic resolution in the target group:

Species	Genus	Family	Order
19.6% (161,038)	55.7% (2,513)	55.1% (370)	60.2% (166)

Comments: Amplifies both Bacteria and Archaea, excluding eukaryotes. This primer pair has been designed by ecoPrimers (based on a reference database containing 1,079 bacterial and 110 archaeal complete genome sequences). This primer pair is close to Bact03 recommended by the Earth Microbiome Project (www.earthmicrobiome.org/emp-standard-protocols/16s/).

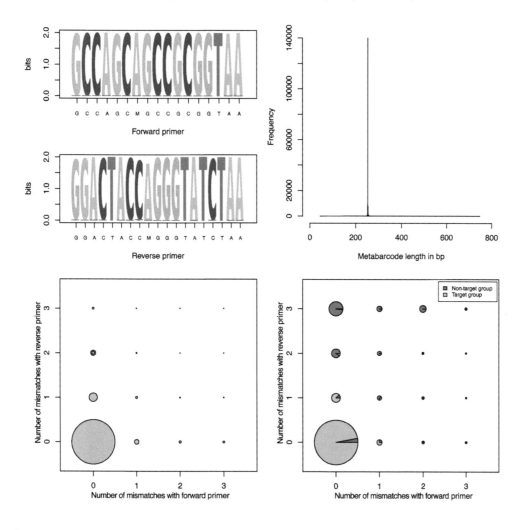

Bact02

Target taxonomic group: Archaea
   NCBI taxid: 2157
Forward primer: GCCAGCMGCCGCGGTAA
   Reference: this book
Reverse primer: GGACTACCMGGGTATCTAA
   Reference: this book
Recommended annealing temperature: 53°C
Target gene: 16S rDNA (V4)
Coverage for the target group: NA
Min. length: 217 bp **Mean length:** 255 bp **Max. length:** 334 bp
Taxonomic resolution in the target group:

Species	Genus	Family	Order
36.1% (3,486)	63.6% (132)	65.7% (35)	61.9% (21)

Comments: Amplifies both Bacteria and Archaea. This primer pair is close to Bact03 recommended by the Earth Microbiome Project (www.earthmicrobiome.org/emp-standard-protocols/16s/).

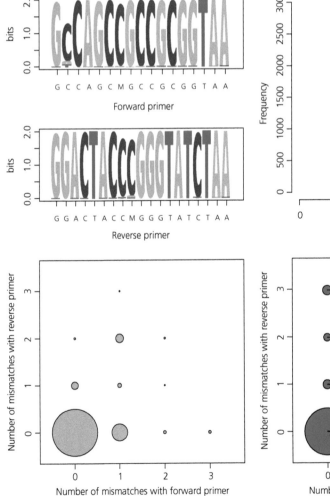

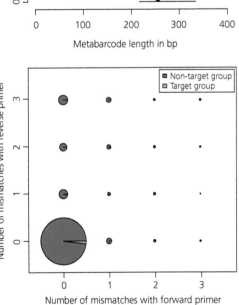

Bact03

Target taxonomic group: Bacteria
        NCBI taxid: 2
Forward primer: GTGYCAGCMGCCGCGGTAA
        Reference: Baker *et al.* (2003), Quince *et al.* (2011)
Reverse primer: GGACTACNVGGGTWTCTAAT
        Reference: Apprill *et al.* (2015)
Recommended annealing temperature: 52°C
Target gene: 16S rDNA (V4)
Coverage for the target group: NA
Min. length: 44 bp   **Mean length:** 253 bp   **Max. length:** 747 bp
Taxonomic resolution in the target group:

Species	Genus	Family	Order
19.6% (158,290)	55.8% (2,507)	55.1% (370)	60.2% (166)

Comments: According to the logos, it does not seem necessary to degenerate both primers when targeting only Bacteria (but see Apprill *et al.* 2015). Recommended by the Earth Microbiome Project (http://www.earthmicrobiome.org/emp-standard-protocols/16s/).

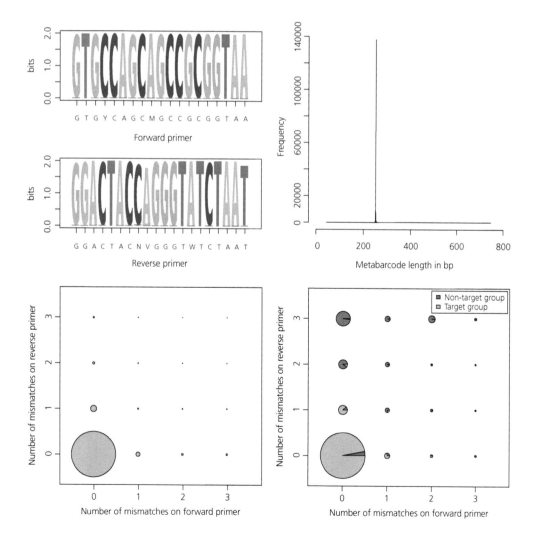

Bact03

Target taxonomic group: Archaea
    NCBI taxid: 2157
Forward primer: GTGYCAGCMGCCGCGGTAA
    Reference: Baker *et al.* (2003), Quince *et al.* (2011)
Reverse primer: GGACTACNVGGGTWTCTAAT
    Reference: Apprill *et al.* (2015)
Recommended annealing temperature: 52°C
Target gene: 16S rDNA
Coverage for the target group: NA
Min. length: 216 bp   **Mean length:** 254 bp   **Max. length:** 333 bp
Taxonomic resolution in the target group:

Species	Genus	Family	Order
36.1% (3,473)	63.6% (132)	65.7% (35)	61.9% (21)

Comments: It does not seem necessary to degenerate both primers when targeting only Archaea. Recommended by the Earth Microbiome Project (http://www.earthmicrobiome.org/emp-standard-protocols/16s/).

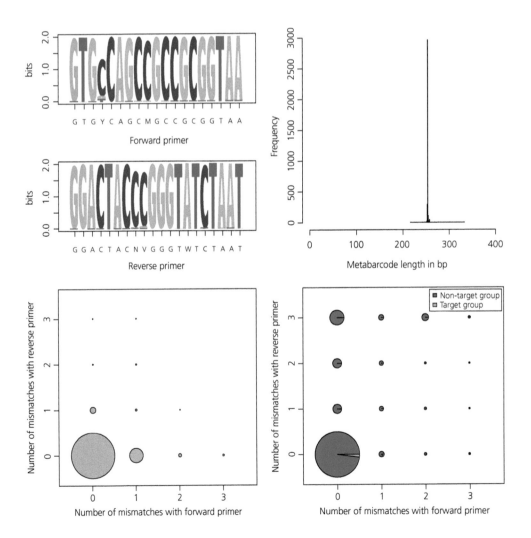

Bact04

Target taxonomic group: Bacteria
      NCBI taxid: 2
Forward primer: GTGYCAGCMGCCGCGGTAA
      Reference: Baker *et al.* (2003), Quince *et al.* (2011)
Reverse primer: CCGYCAATTYMTTTRAGTTT
      Reference: Quince *et al.* (2011)
Recommended annealing temperature: 46°C
Target gene: 16S rDNA (V4 + part of V5)
Coverage for the target group: NA
Min. length: 54 bp   **Mean length:** 373 bp   **Max. length:** 763 bp
Taxonomic resolution in the target group:
Species   Genus   Family   Order
27.7% (143,763)   62.6% (2,541)   58.4% (370)   62.7% (166)
Comments: This primer pair also amplifies a relatively large number of eukaryotes. Recommended by the Earth Microbiome Project (http://www.earthmicrobiome.org/emp-standard-protocols/16s/).

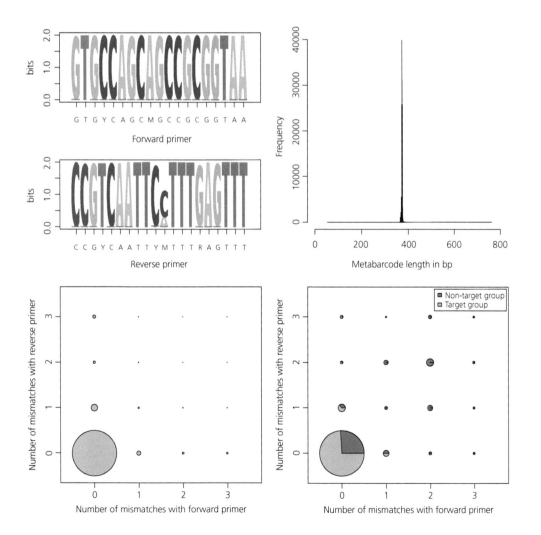

Bact04

Target taxonomic group: Archaea
      NCBI taxid: 2157
Forward primer: GTGYCAGCMGCCGCGGTAA
      Reference: Baker *et al.* (2003), Quince *et al.* (2011)
Reverse primer: CCGYCAATTYMTTTRAGTTT
      Reference: Quince *et al.* (2011)
Recommended annealing temperature: 46°C
Target gene: 16S rDNA (V4 + part of V5)
Coverage for the target group: NA
Min. length: 170 bp   **Mean length:** 378 bp   **Max. length:** 553 bp
Taxonomic resolution in the target group:

Species	Genus	Family	Order
45.0% (3,133)	67.9% (131)	71.4% (35)	61.9% (21)

Comments: This primer pair also amplifies a relatively large number of eukaryotes. Recommended by the Earth Microbiome Project (http://www.earthmicrobiome. org/emp-standard-protocols/16s/).

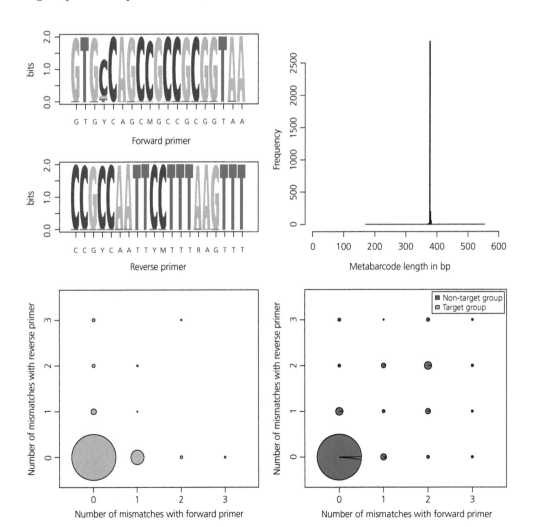

Cyan01

Target taxonomic group: Cyanobacteria
     NCBI taxid: 1117
Forward primer: CTYAAGCCGACATTCTCAC
     Reference: this book
Reverse primer: GACAACYAGGAGGTTTGC
     Reference: this book
Recommended annealing temperature: 51°C
Target gene: 23S rDNA
Coverage for the target group: NA
Min. length: 172 bp   **Mean length:** 180 bp   **Max. length:** 242 bp
Taxonomic resolution in the target group:

Species	Genus	Family	Order
80.6% (36)	100.0% (13)	100.0% (6)	100.0% (2)

Comments: Also amplifies some other Bacteria and eukaryotes.

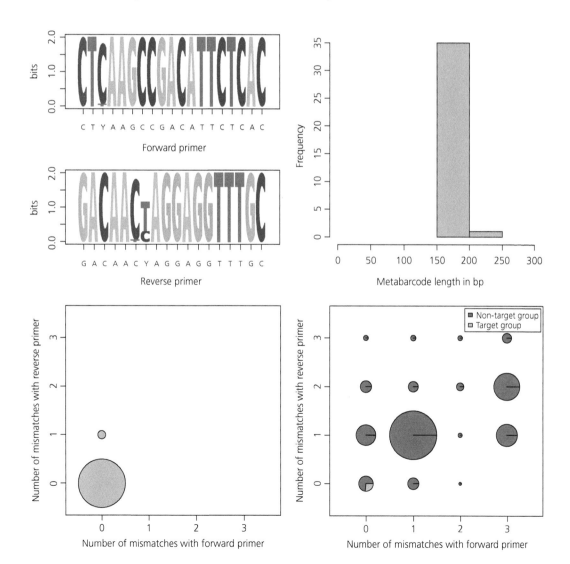

Arch01

Target taxonomic group: Archaea
       NCBI taxid: 2157
Forward primer: CCTGCTCCTTGCACACAC
       Reference: this book
Reverse primer: CCTACGGCTACCTTGTTAC
       Reference: this book
Recommended annealing temperature: 55°C
Target gene: 16S rDNA (V9)
Coverage for the target group: NA
Min. length: 32 bp   **Mean length:** 85 bp   **Max. length:** 166 bp
Taxonomic resolution in the target group:

Species	Genus	Family	Order
31.3% (825)	73.1% (119)	90.6% (32)	85.7% (21)

Comments: Highly specific of Archaea.

Forward primer

Reverse primer

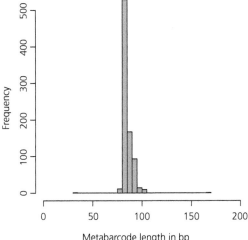

Metabarcode length in bp

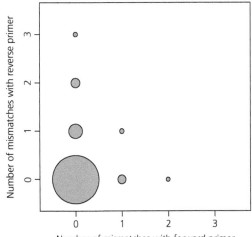

Number of mismatches with forward primer

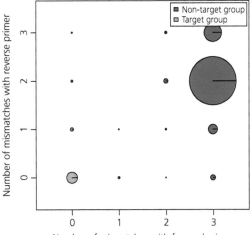

Number of mismatches with forward primer

Euka01

Target taxonomic group: Eukaryota
       NCBI taxid: 2759
Forward primer: TGGTGCATGGCCGTTCTTAGT
       Reference: Hardy *et al.* (2010)
Reverse primer: CATCTAAGGGCATCACAGACC
       Reference: Hardy *et al.* (2010)
Recommended annealing temperature: 59°C
Target gene: 18S rDNA (V7)
Coverage for the target group: NA
Min. length: 48 bp   **Mean length:** 161 bp   **Max. length:** 777 bp
Taxonomic resolution in the target group:

Species	Genus	Family	Order
48.2% (49,080)	60.3% (20,041)	69.1% (5,338)	68.2% (977)

Comments: Highly specific of eukaryotes. Relatively long for some insects and for amphipods and isopods.

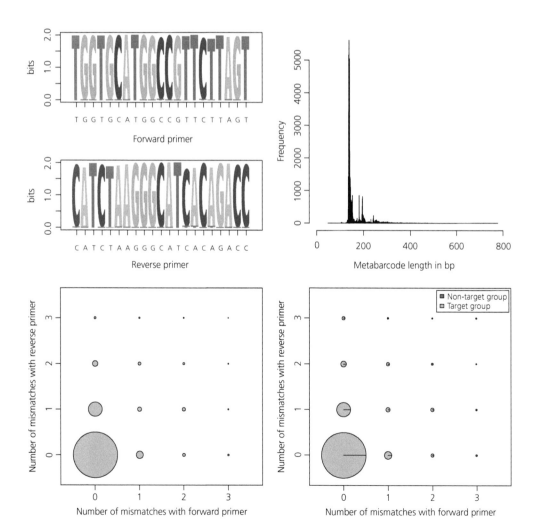

Euka02

Target taxonomic group: Eukaryota
      NCBI taxid: 2759
Forward primer: TTTGTCTGSTTAATTSCG
      Reference: Guardiola *et al.* (2015)
Reverse primer: CACAGACCTGTTATTGC
      Reference: Guardiola *et al.* (2015)
Recommended annealing temperature: 45°C
Target gene: 18S rDNA (V7)
Coverage for the target group: NA
Min. length: 36 bp   **Mean length:** 123 bp   **Max. length:** 892 bp
Taxonomic resolution in the target group:

Species	Genus	Family	Order
47.0% (49,108)	59.5% (20,045)	68.3% (5,320)	67.1% (969)

Comments: Highly specific of eukaryotes. Amplifies the same region than Euka01, but is about 40 bp shorter for almost the same taxonomic resolution. Relatively long for some insects and for amphipods and isopods. If the targets are only Fungi, the forward primer should be TTTGTCTGCTTAATTGCG. If the targets are not Fungi, the forward primer should be TTTGTCTGGTTAATTCCG.

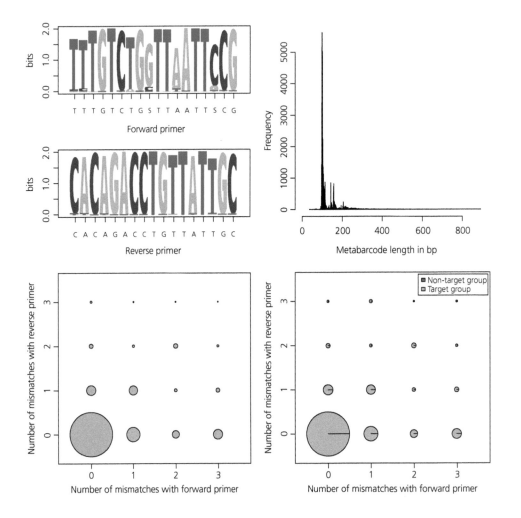

Euka03

Target taxonomic group: Eukaryota
   NCBI taxid: 2759
Forward primer: CCCTTTGTACACACCGCC
   Reference: this book
Reverse primer: CTTCYGCAGGTTCACCTAC
   Reference: this book
Recommended annealing temperature: 55°C
Target gene: 18S nuclear ribosomal DNA (V9)
Coverage for the target group: NA
Min. length: 49 bp   **Mean length:** 133 bp   **Max. length:** 264 bp
Taxonomic resolution in the target group:

Species	Genus	Family	Order
58.1% (11,815)	74.3% (5,160)	84.0% (2,343)	88.6% (673)

Comments: Highly specific of eukaryotes. Does not amplify properly amphipods and isopods (see Peca02). Amplifies the same region than the primers Euka04 described in Amaral-Zettler *et al.* (2009), but Euka03 primers are optimized to amplify a wider range of eukaryotes. The main advantage of this Euka03 when compared to Euka01 or Euka02 is the more homogeneous size of the metabarcode and a higher taxonomic resolution, but the V9 (Euka03) region has been less sequenced than the V7 region (Euka01 and Euka02), leading to a less comprehensive reference database.

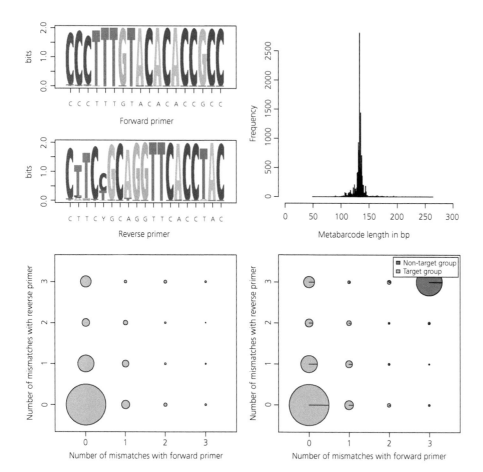

Euka04

Target taxonomic group: Eukaryota
    NCBI taxid: 2759
Forward primer: GTACACACCGCCCGTC
    Reference: http://www.earthmicrobiome.org/emp-standard-protocols/18s/
Reverse primer: TGATCCTTCTGCAGGTTCACCTAC
    Reference: http://www.earthmicrobiome.org/emp-standard-protocols/18s/
Recommended annealing temperature: 53°C
Target gene: 18S rDNA (V9)
Coverage for the target group: NA
Min. length: 45 bp   **Mean length:** 128 bp   **Max. length:** 260 bp
Taxonomic resolution in the target group:

Species	Genus	Family	Order
56.9% (9,982)	75.2% (4,409)	84.9% (2,068)	90.7% (623)

Comments: Highly specific of eukaryotes. Close to the primers designed by Amaral-Zettler *et al.* (2009). The reverse primer is not optimal.

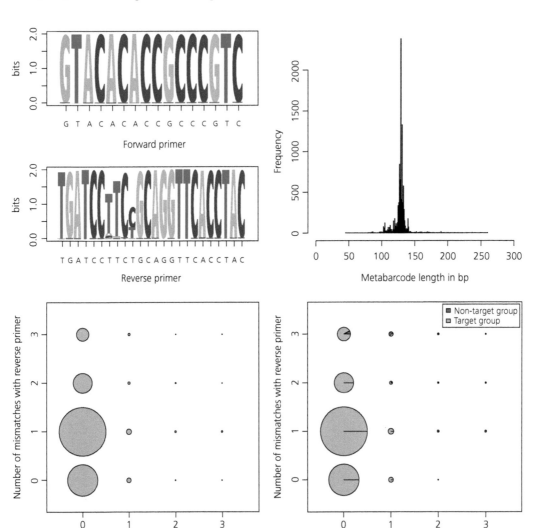

Fung01

Target taxonomic group: Fungi
> **NCBI taxid:** 4751

Forward primer: GGAAGTAAAAGTCGTAACAAGG
> **Reference:** White *et al.* (1990)

Reverse primer: CAAGAGATCCCGTTGYTGAAAGTT
> **Reference:** this book

Recommended annealing temperature: 56°C

Target gene: ITS1 nuclear rDNA

Coverage for the target group: NA

Min. length: 68 bp   **Mean length:** 226 bp   **Max. length:** 919 bp

Taxonomic resolution in the target group:

Species	Genus	Family	Order
72.7% (10,618)	90.2% (1,493)	87.7% (381)	85.5% (124)

Comments: Does not properly amplify Glomeromycota due to a mismatch on the last nucleotide of the reverse primer (see Fung02). Forward primer not specific of Fungi. Reverse primer specific of Fungi. A small proportion of Fungi have a very long fragment for this marker.

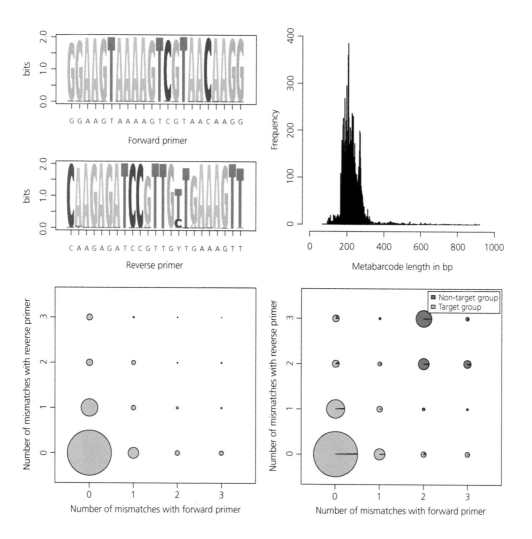

Fung02

Target taxonomic group: Fungi
       NCBI taxid: 4751
Forward primer: GGAAGTAAAAGTCGTAACAAGG
       Reference: White *et al.* (1990)
Reverse primer: CAAGAGATCCGTTGYTGAAAGTK
       Reference: this book
Recommended annealing temperature: 56°C
Target gene: ITS1 nuclear rDNA
Coverage for the target group: NA
Min. length: 68 bp   **Mean length:** 225 bp   **Max. length:** 919 bp
Taxonomic resolution in the target group:

Species	Genus	Family	Order
72.5% (10,846)	90.2% (1,498)	87.7% (381)	85.5% (124)

Comments: Reverse primer from Epp *et al.* (2012) modified for properly amplifying Glomeromycota. Another solution to amplify Glomeromycota would be to use CCAAGAGATCCGTTGYTGAAAGT as reverse primer specific of Fungi. A small proportion of Fungi have a very long fragment for this marker.

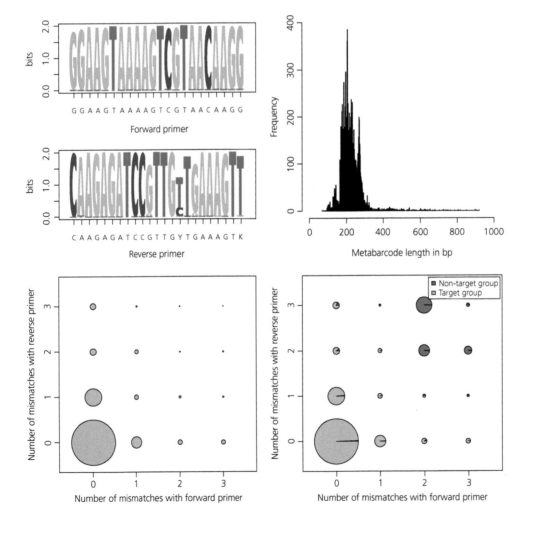

Baci01

Target taxonomic group: Bacillariophyta (Diatoms)
        NCBI taxid: 2836
Forward primer: ATTCCAGCTCCAATAGCGTA
        Reference: this book
Reverse primer: CTTTRAACRCKCTRATTTYTTCAC
        Reference: this book
Recommended annealing temperature: 53°C
Target gene: 18S nuclear rDNA (V4)
Coverage for the target group: NA
Min. length: 108 bp   **Mean length:** 150 bp   **Max. length:** 225 bp
Taxonomic resolution in the target group:

Species	Genus	Family	Order
66.0% (1,064)	76.1% (201)	82.1% (67)	71.9% (32)

Comments: Also amplifies other eukaryotes. The forward primer is close to the primer D514for from Zimmermann *et al.* (2011).

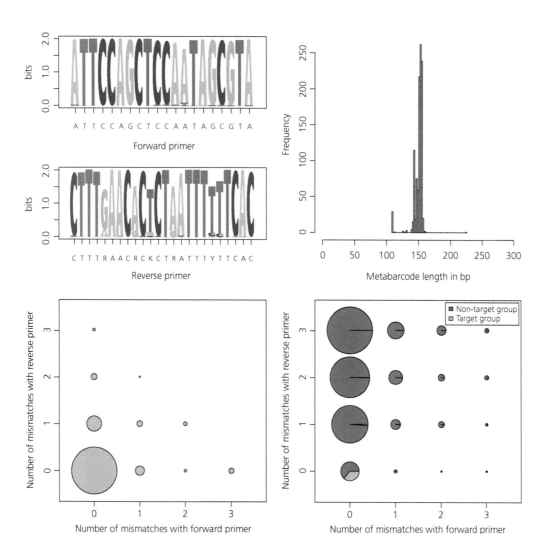

Sper01

Target taxonomic group: Spermatophyta (seed plants)
        NCBI taxid: 58024
Forward primer: GGGCAATCCTGAGCCAA
        Reference: Taberlet *et al.* (2007)
Reverse primer: CCATTGAGTCTCTGCACCTATC
        Reference: Taberlet *et al.* (2007)
Recommended annealing temperature: 52°C
Target gene: P6 loop of the *trn*L intron, chloroplast DNA
Coverage for the target group: 98.8% (739 species amplified *in silico* out of 748)
Min. length: 10 bp   **Mean length:** 48 bp   **Max. length:** 220 bp
Taxonomic resolution in the target group:

Species	Genus	Family	Order
21.5% (48,494)	36.9% (8,417)	77.4% (371)	89.6% (67)

Comments: Widely used for analyzing degraded template. Correspond to the g/h primers from Taberlet *et al.* (2007).

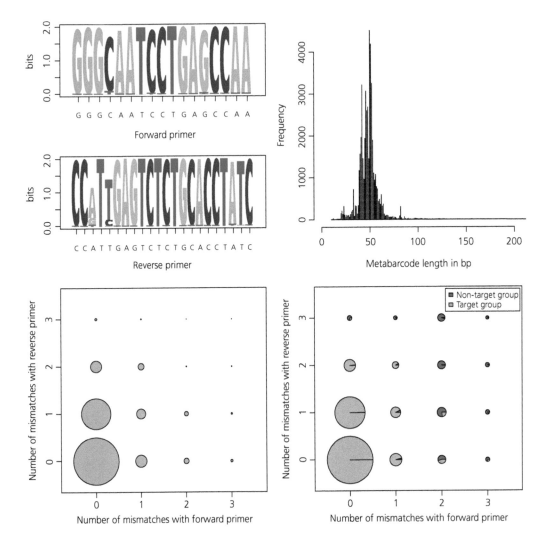

Bryo01

Target taxonomic group: Bryophyta
   NCBI taxid: 3208
Forward primer: GATTCAGGGAAACTTAGGTTG
   Reference: Epp *et al.* (2012)
Reverse primer: CCATYGAGTCTCTGCACC
   Reference: this book
Recommended annealing temperature: 54°C
Target gene: Part of the *trn*L intron, chloroplast DNA
Coverage for the target group: 83.33% (5 species amplified *in silico* out of 6)
Min. length: 41 bp   **Mean length:** 53 bp   **Max. length:** 87 bp
Taxonomic resolution in the target group:

Species	Genus	Family	Order
28.8% (2,151)	38.7% (608)	50.0% (102)	57.1% (28)

Comments: The reverse primer has been slightly modified (with one degenerated nucleotide) when compared to the original primer from Epp *et al.* (2012).

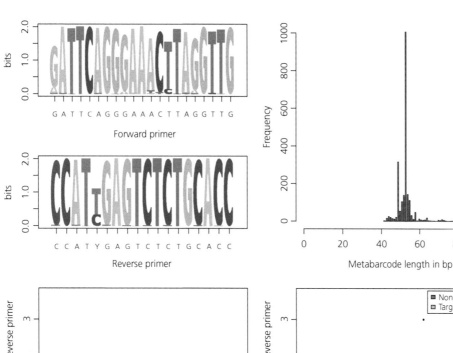

Forward primer

Reverse primer

Metabarcode length in bp

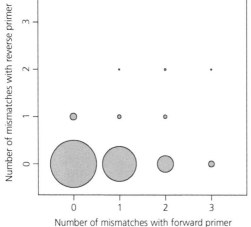

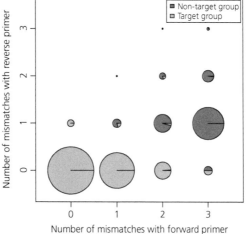

Trac01

Target taxonomic group: Tracheophyta (vascular plants)
    NCBI taxid: 58023
Forward primer: GATATCCRTTGCCGAGAGTC
    Reference: this book
Reverse primer: GAAGGAGAAGTCGTAACAAGG
    Reference: Fuertes Aguilar *et al.* (1999)
Recommended annealing temperature: 56°C
Target gene: ITS1 of nuclear rDNA
Coverage for the target group: NA
Min. length: 28 bp   **Mean length:** 267 bp   **Max. length:** 484 bp
Taxonomic resolution in the target group:

Species	Genus	Family	Order
83.9% (14,442)	96.7% (2,534)	99.0% (210)	96.2% (52)

Comments: This primer pair has been designed to avoid Fungi amplification. The forward primer is highly specific to vascular plants and has been modified from Baamrane *et al.* (2012). The reverse primer is close to ITS-5 (White *et al.* 1990). Not empirically tested.

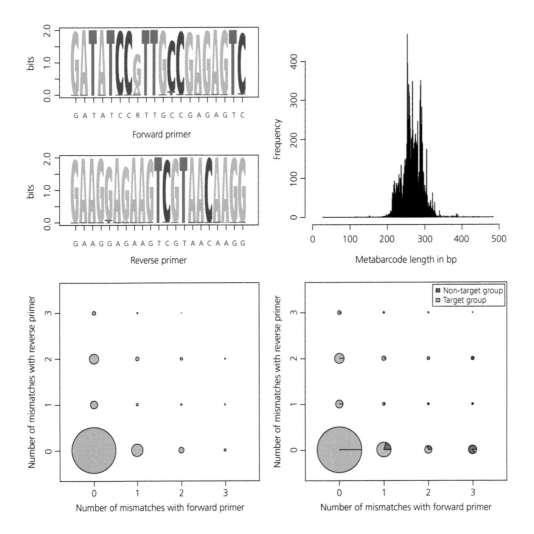

Poac01

Target taxonomic group: Poaceae (grass family)
NCBI taxid: 4479
Forward primer: GATATCCGTTGCCGAGAGTC
Reference: Baamrane *et al.* (2012)
Reverse primer: CSDAHGGCGTCAAGGAACAC
Reference: this book
Recommended annealing temperature: 56°C
Target gene: ITS1 of nuclear rDNA
Coverage for the target group: NA
Min. length: 53 bp   **Mean length:** 72 bp   **Max. length:** 88 bp
Taxonomic resolution in the target group:

Species	Genus	Family	Order
52.8% (3,482)	76.1% (494)	100.0% (1)	100.0% (1)

Comments: Reverse primer modified from Baamrane *et al.* (2012). Not empirically tested.

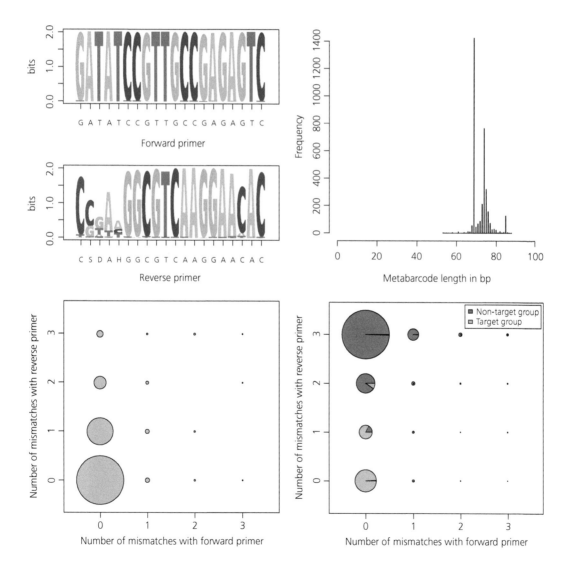

Aste01

Target taxonomic group: Asteraceae (daisy family)
NCBI taxid: 4210
Forward primer: GATATCCGTTGCCGAGAGTC
Reference: Baamrane *et al.* (2012)
Reverse primer: CGGCACRRNAYGTGCCAAGG
Reference: this book
Recommended annealing temperature: 56°C
Target gene: ITS1 of nuclear rDNA
Coverage for the target group: NA
Min. length: 52 bp   **Mean length:** 83 bp   **Max. length:** 240 bp
Taxonomic resolution in the target group:

Species	Genus	Family	Order
53.2% (7,308)	81.1% (1,177)	100.0% (1)	100.0% (1)

Comments: Reverse primer modified from Baamrane *et al.* (2012). Not empirically tested.

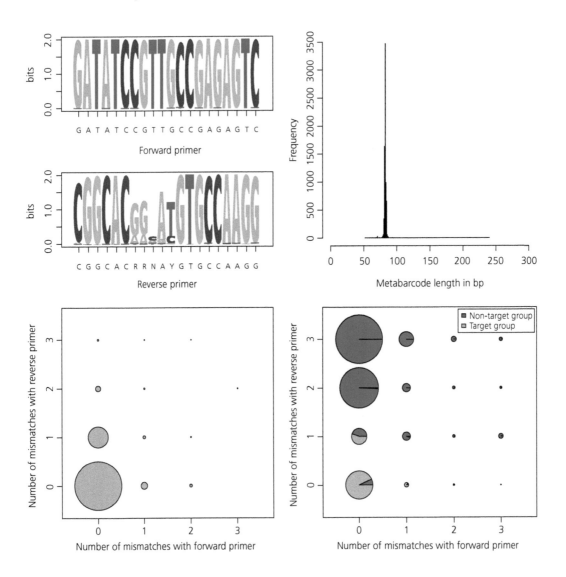

Cype01

Target taxonomic group: Cyperaceae (sedge family)
NCBI taxid: 4609
Forward primer: GATATCCGTTGCCGRGAGTC
     Reference: this book
Reverse primer: GGGWTGWCGCCAAGGAAYA
     Reference: this book
Recommended annealing temperature: 56°C
Target gene: ITS1 of nuclear rDNA
Coverage for the target group: NA
Min. length: 44 bp   **Mean length:** 68 bp   **Max. length:** 79 bp
Taxonomic resolution in the target group:

Species	Genus	Family	Order
62.9% (1,258)	84.6% (52)	100.0% (1)	100.0% (1)

Comments: Forward primer modified from Baamrane *et al.* (2012). Reverse primer modified from Willerslev *et al.* (2014). Not empirically tested.

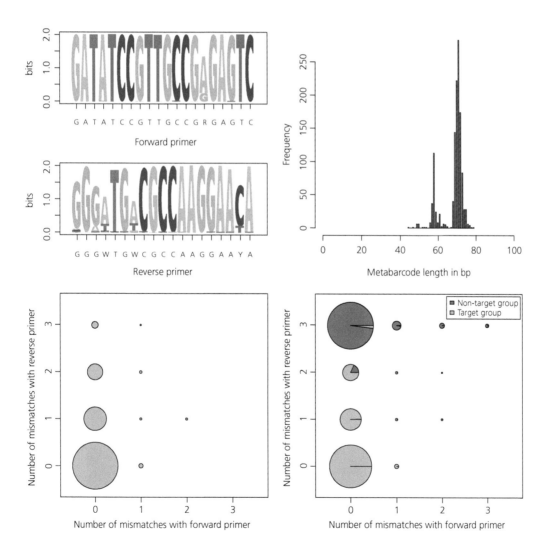

Apia01

Target taxonomic group: Apiaceae (carrot family)
 NCBI taxid: 4037
Forward primer: GATATCCGTTGYCGAGAGTC
 Reference: this book
Reverse primer: GGCGCKGAMTGCGCCAAGGA
 Reference: this book
Recommended annealing temperature: 56°C
Target gene: ITS1 of nuclear rDNA
Coverage for the target group: NA
Min. length: 47 bp   **Mean length:** 65 bp   **Max. length:** 79 bp
Taxonomic resolution in the target group:

Species	Genus	Family	Order
58.5% (1,556)	80.9% (277)	100.0% (1)	100.0% (1)

Comments: Forward primer modified from Baamrane *et al.* (2012). Also amplifies some other families, including mainly Asteraceae, Fabaceae, and Poaceae. Not empirically tested.

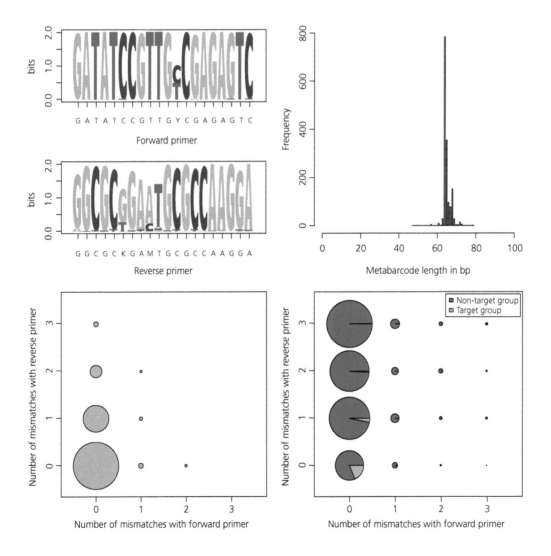

Rosa01

Target taxonomic group: Rosaceae (rose family)
      NCBI taxid: 3745
Forward primer: GATATCCGTTGCCGAGAGTC
      Reference: Baamrane *et al.* (2012)
Reverse primer: CGGCGYGWRTTGCGCCAAGGAA
      Reference: this book
Recommended annealing temperature: 56°C
Target gene: ITS1 of nuclear rDNA
Coverage for the target group: NA
Min. length: 69 bp   **Mean length:** 80 bp   **Max. length:** 92 bp
Taxonomic resolution in the target group:

Species	Genus	Family	Order
44.0% (1,285)	83.1% (89)	100.0% (1)	100.0% (1)

Comments: Also amplifies some other families, including mainly Fabaceae and Rubiaceae. Not empirically tested.

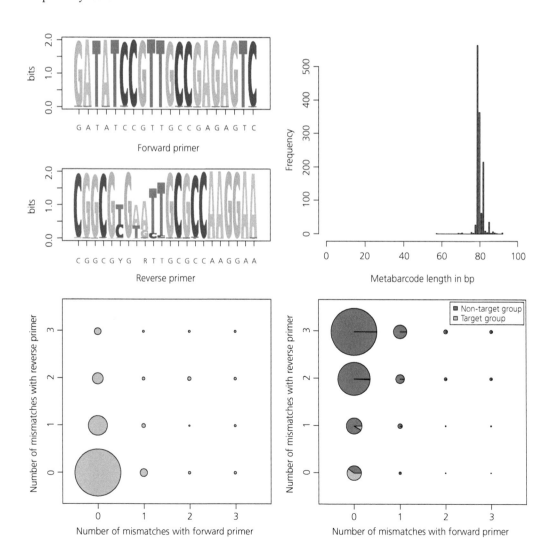

Sapo01

Target taxonomic group: Sapotaceae (sapodilla family)
       NCBI taxid: 3737
Forward primer: GATATCCGTTGCCGAGAGTC
       Reference: Baamrane *et al.* (2012)
Reverse primer: CGACGCGARTYGCGTCAAGGA
       Reference: this book
Recommended annealing temperature: 56°C
Target gene: ITS1 of nuclear rDNA
Coverage for the target group: NA
Min. length: 66 bp   **Mean length:** 86 bp   **Max. length:** 107 bp
Taxonomic resolution in the target group:

Species	Genus	Family	Order
68.4% (512)	87.7% (57)	100.0% (1)	100.0% (1)

Comments: Relatively good taxonomic resolution in a family where the primers Sper01 are inefficient. Not empirically tested.

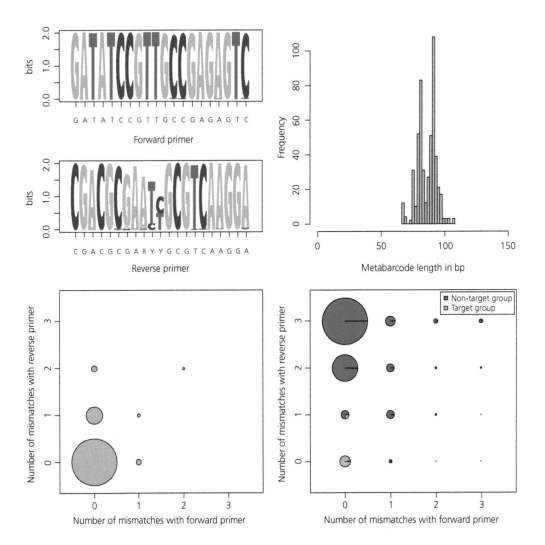

Meta01

Target taxonomic group: Metazoa (metazoans)
      NCBI taxid: 33208
Forward primer: YYCRMCTGTTTAYCAAAAACA
      Reference: this book
Reverse primer: CCGGTYTGAACTCARATC
      Reference: this book
Recommended annealing temperature: 49°C
Target gene: 16S mitochondrial rDNA
Coverage for the target group: 92.65% (5,067 species amplified *in silico* out of 5,469)
Min. length: 18 bp   **Mean length:** 548 bp   **Max. length:** 1,469 bp
Taxonomic resolution in the target group:

Species	Genus	Family	Order
95.8% (8,676)	98.8% (4,839)	99.9% (1,658)	100.0% (350)

Comments: Close to primers designed by Palumbi (Palumbi *et al.* 2002), but optimized according to data available in release 126 of EMBL. This metabarcode is too long for working with degraded DNA, but can be useful for building a reference database containing some of the 16S metabarcodes described here.

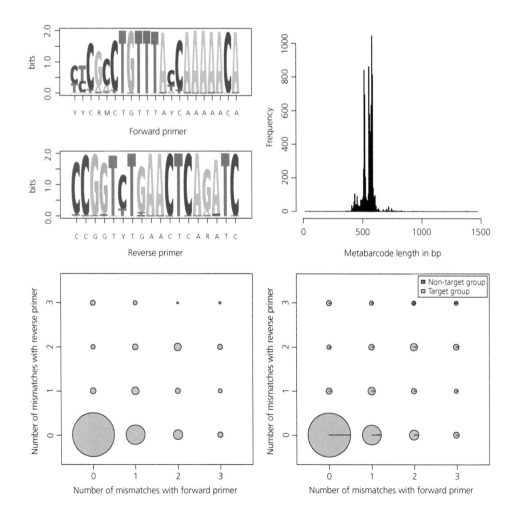

Echi01

Target taxonomic group: Echinodermata (echinoderms)
    NCBI taxid: 7586
Forward primer: ARTGATTATGCTACCTTYGCAC
    Reference: this book
Reverse primer: GMCTGTTTAYCAAAAACATMGC
    Reference: this book
Recommended annealing temperature: 54°C
Target gene: 16S mitochondrial rDNA
Coverage for the target group: 95.35% (41 species amplified *in silico* out of 43)
Min. length: 64 bp   **Mean length:** 72 bp   **Max. length:** 79 bp
Taxonomic resolution in the target group:

Species	Genus	Family	Order
83.2% (113)	96.4% (55)	100.0% (33)	100.0% (18)

Comments: Good taxonomic resolution in echinoderms. Also amplifies a lot of other eukaryotes. Not empirically tested.

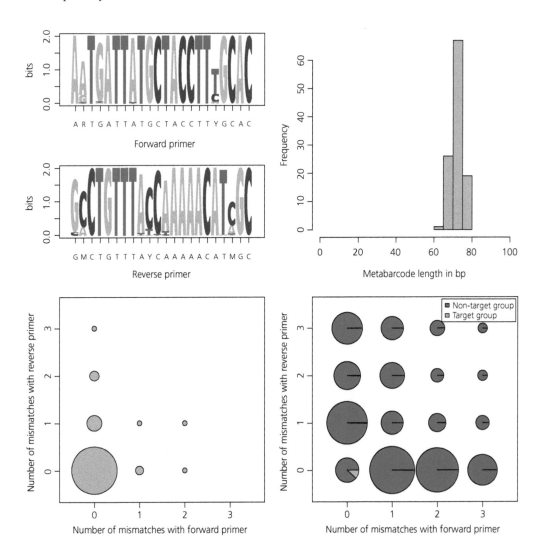

Echi02

Target taxonomic group: Echinodermata (echinoderms)
  NCBI taxid: 7586
Forward primer: TCCAACATCGAGGTCGCAA
  Reference: this book
Reverse primer: ACGAGAAGACCCTRTYGAG
  Reference: this book
Recommended annealing temperature: 53°C
Target gene: 16S mitochondrial rDNA
Coverage for the target group: 95.35% (41 species amplified *in silico* out of 43)
Min. length: 182 bp   **Mean length:** 247 bp   **Max. length:** 298 bp
Taxonomic resolution in the target group:

Species	Genus	Family	Order
93.9% (736)	95.8% (332)	92.8% (97)	83.9% (31)

Comments: Excellent taxonomic resolution in echinoderms. Also amplifies a lot of other eukaryotes. Not empirically tested.

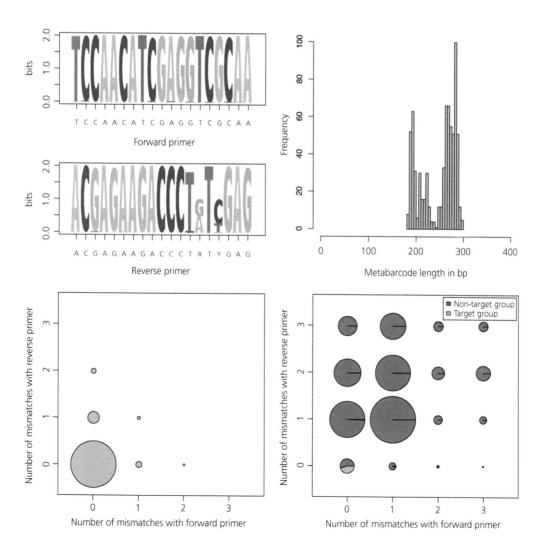

Moll01

Target taxonomic group: Mollusca (mollusks)
      NCBI taxid: 6447
Forward primer: CCAACATCGAGGTCRYAA
      Reference: De Barba *et al.* (2014)
Reverse primer: ARYTACYNYAGGGATAACAG
      Reference: this book
Recommended annealing temperature: 48°C
Target gene: 16S mitochondrial rDNA
Coverage for the target group: 97.90% (233 species amplified *in silico* out of 238)
Min. length: 14 bp   **Mean length:** 30 bp   **Max. length:** 41 bp
Taxonomic resolution in the target group:

Species	Genus	Family	Order
42.8% (9,107)	55.3% (2,191)	61.7% (439)	72.7% (33)

Comments: Reverse primer modified from De Barba *et al.* (2014). Same forward primer as Arth01, reverse primer slightly different. Also amplifies a lot of other eukaryotes close to Moll01 and Arth01, including vertebrates. Mammalian blocking primer: agggataacag-CGCAATCCTATT-C3. Does not amplify birds. Not empirically tested.

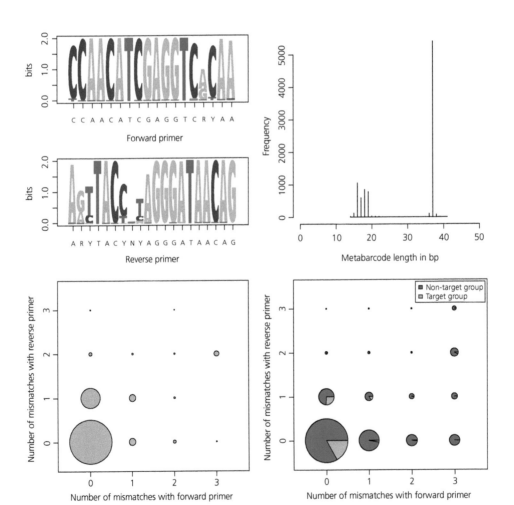

Gast01

Target taxonomic group: Gastropoda (gastropods)
          NCBI taxid: 6448
Forward primer: CCGGTCTGAACTCAGATCA
          Reference: this book
Reverse primer: TTTGTGACCTCGATGTTGGA
          Reference: this book
Recommended annealing temperature: 54°C
Target gene: 16S mitochondrial rDNA
Coverage for the target group: 95.7% (89 species amplified *in silico* out of 93)
Min. length: 60 bp   **Mean length:** 63 bp   **Max. length:** 70 bp
Taxonomic resolution in the target group:

Species	Genus	Family	Order
58.3% (1,280)	70.8% (456)	88.3% (128)	100.0% (3)

Comments: Also amplifies a lot of other eukaryotes. Not empirically tested.

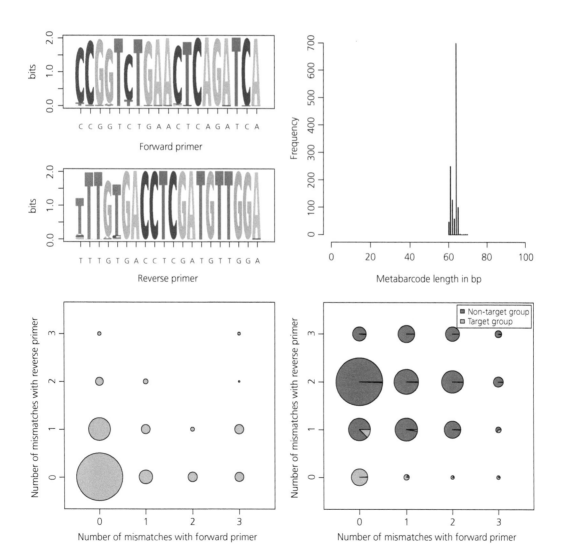

Poly01

Target taxonomic group: Polychaeta (polychaetes)
      NCBI taxid: 6341
Forward primer: CCGGTYTGAACTCAGMTCA
      Reference: this book
Reverse primer: TGGCACCTCGATGTTGGCT
      Reference: this book
Recommended annealing temperature: 55°C
Target gene: 16S mitochondrial rDNA
Coverage for the target group: 100% (27 species amplified *in silico* out of 27)
Min. length: 42 bp   **Mean length:** 63 bp   **Max. length:** 70 bp
Taxonomic resolution in the target group:

Species	Genus	Family	Order
89.0% (145)	94.7% (94)	100% (33)	100.0% (11)

Comments: Excellent taxonomic resolution for a short metabarcode.

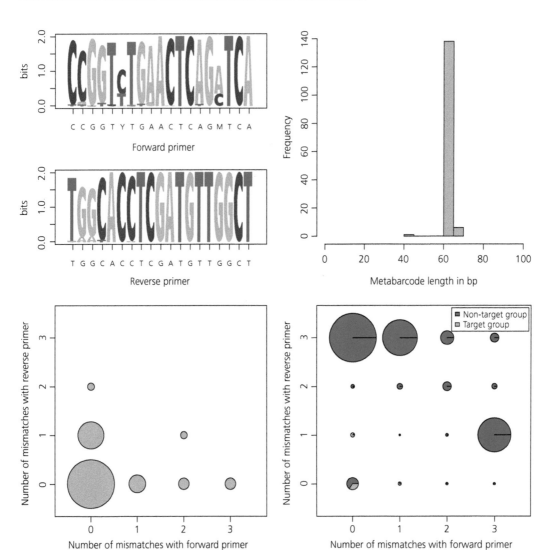

Olig01

Target taxonomic group: Oligochaeta (oligochaetes)
        NCBI taxid: 6382
Forward primer: CAAGAAGACCCTATAGAGCTT
        Reference: Bienert *et al.* (2012)
Reverse primer: CCTGTTATCCCTAAGGTARCT
        Reference: this book
Recommended annealing temperature: 55°C
Target gene: 16S mitochondrial rDNA
Coverage for the target group: 100% (9 species amplified *in silico* out of 9)
Min. length: 93 bp   **Mean length:** 122 bp   **Max. length:** 196 bp
Taxonomic resolution in the target group:

Species   Genus   Family   Order
89.3% (832)   95.7% (188)   100.0% (22)   100.0% (1)

Comments: Excellent taxonomic resolution and coverage. Reverse primer modified from
Bienert *et al.* (2012). Not empirically tested.

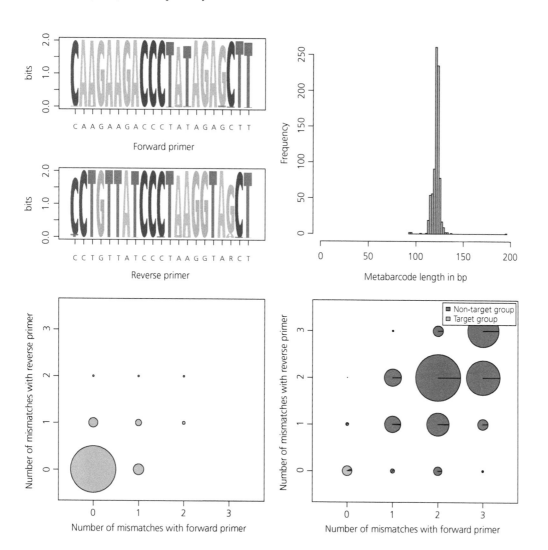

Lumb01

Target taxonomic group: Lumbricina (earthworms)
          NCBI taxid: 6391
Forward primer: CAAGAAGACCCTATAGAGCTT
          Reference: Bienert *et al.* (2012)
Reverse primer: GGTCGCCCCAACCGAAT
          Reference: Bienert *et al.* (2012)
Recommended annealing temperature: 55°C
Target gene: 16S mitochondrial rDNA
Coverage for the target group: 100% (8 species amplified *in silico* out of 8)
Min. length: 26 bp   **Mean length:** 31 bp   **Max. length:** 36 bp
Taxonomic resolution in the target group:

Species	Genus	Family	Order
79.5% (518)	92.4% (105)	85.7% (14)	100.0% (1)

Comments: Excellent resolution for a very short metabarcode.

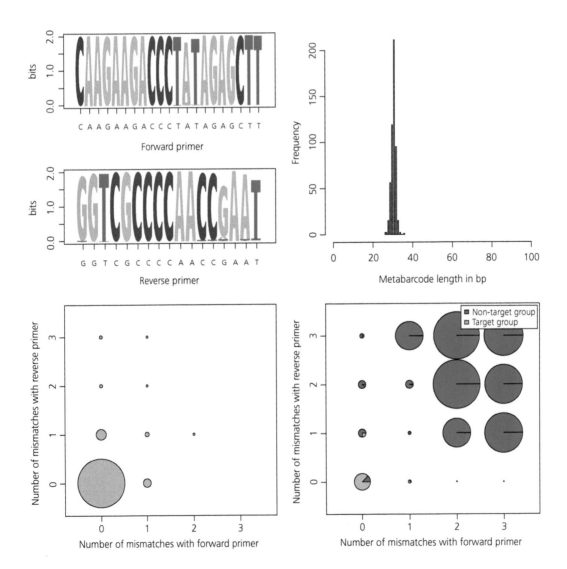

Lumb02

Target taxonomic group: Lumbricina (earthworms)
      NCBI taxid: 6391
Forward primer: ATTCGGTTGGGGCGACC
      Reference: Bienert *et al.* (2012)
Reverse primer: CTGTTATCCCTAAGGTAGCTT
      Reference: Bienert *et al.* (2012)
Recommended annealing temperature: 55°C
Target gene: 16S mitochondrial rDNA
Coverage for the target group: 100% (8 species amplified *in silico* out of 8)
Min. length: 67 bp   **Mean length:** 74 bp   **Max. length:** 81 bp
Taxonomic resolution in the target group:

Species	Genus	Family	Order
85.3% (523)	92.9% (99)	100.0% (13)	100.0% (1)

Comments: Excellent resolution for a short metabarcode. If working in South America, we recommend degenerating the forward primer (ATTCGGTTGGGGCGACY) as a few species have a mismatch on the last nucleotide.

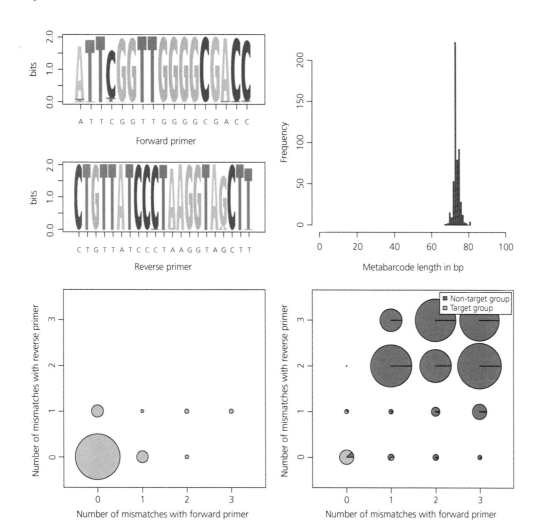

Ench01

Target taxonomic group: Enchytraeidae (potworms family)
NCBI taxid: 6388
Forward primer: GCTGCACTTTGACTTGAC
Reference: Epp *et al.* (2012)
Reverse primer: AGCCTGTGTACTGCTGTC
Reference: Epp *et al.* (2012)
Recommended annealing temperature: 51°C
Target gene: 12S mitochondrial rDNA
Coverage for the target group: NA
Min. length: 33 bp   **Mean length:** 48 bp   **Max. length:** 54 bp
Taxonomic resolution in the target group:

Species	Genus	Family	Order
98.2% (111)	100.0% (18)	100.0% (1)	100.0% (1)

Comments: Excellent taxonomic resolution for a short metabarcode.

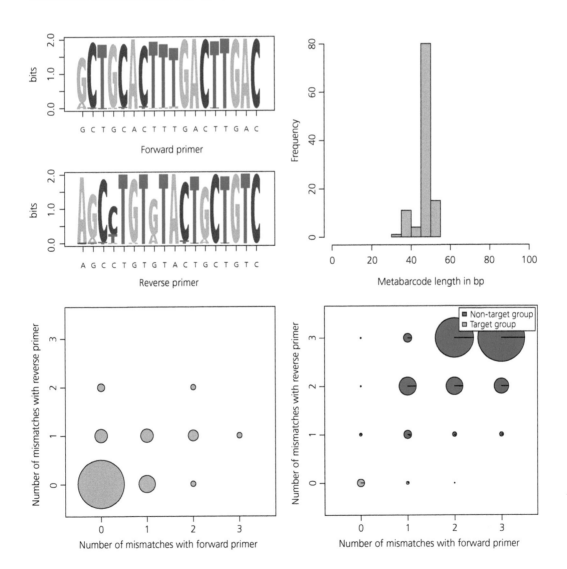

Arth01

Target taxonomic group: Arthropoda (arthropods)
   NCBI taxid: 6656
Forward primer: CCAACATCGAGGTCRYAA
   Reference: De Barba *et al.* (2014)
Reverse primer: ARTTACYNTAGGGATAACAG
   Reference: De Barba *et al.* (2014)
Recommended annealing temperature: 48°C
Target gene: 16S mitochondrial rDNA
Coverage for the target group: 98.39% (1,097 species amplified *in silico* out of 1,115)
Min. length: 18 bp **Mean length:** 37 bp **Max. length:** 97 bp
Taxonomic resolution in the target group:

Species	Genus	Family	Order
41.7% (28,017)	54.0% (9,633)	60.2% (1,279)	64.0% (89)

Comments: Same forward primer as Moll01, reverse primer slightly different, modified from De Barba *et al.* (2014). Mammalian blocking primer: ctagggataacagCGCAATCCTATT-C3. Does not amplify birds. Not empirically tested.

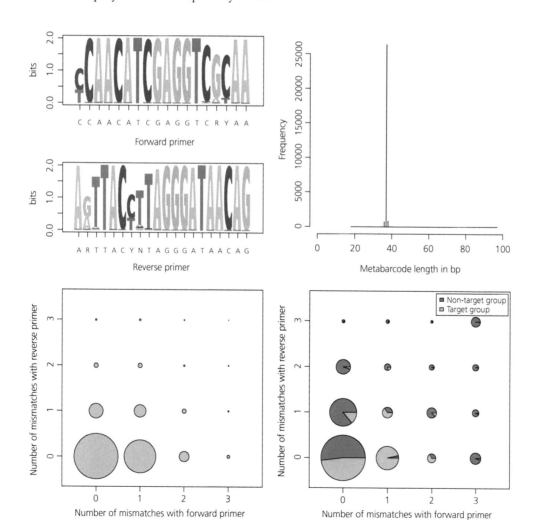

Arth02

Target taxonomic group: Arthropoda (arthropods)
       NCBI taxid: 6656
Forward primer: GATAGAAACCRACCTGGYT
       Reference: this book
Reverse primer: AARTTACYTTAGGGATAACAG
       Reference: this book
Recommended annealing temperature: 49°C
Target gene: 16S mitochondrial rDNA
Coverage for the target group: 95.87% (1,069 species amplified *in silico* out of 1,115)
Min. length: 76 bp   **Mean length:** 142 bp   **Max. length:** 168 bp
Taxonomic resolution in the target group:

Species	Genus	Family	Order
68.6% (6,294)	89.6% (2,400)	97.5% (599)	100.0% (80)

Comments: Taxonomic resolution lower than the Inse01 primer pair, but targets all arthropods with a relatively good coverage. Not empirically tested.

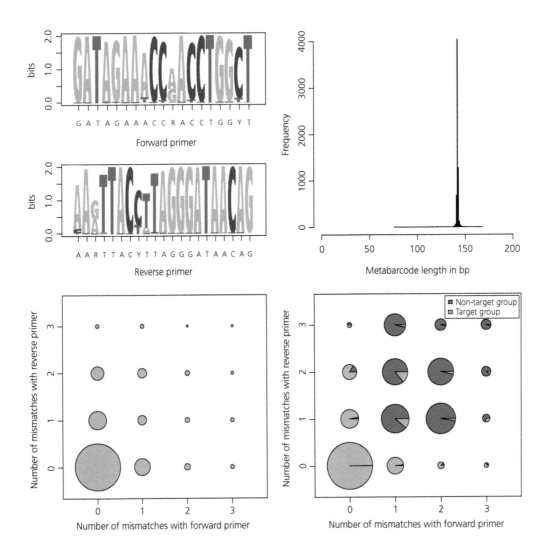

Inse01

Target taxonomic group: Insecta (true insects)
      NCBI taxid: 50557
Forward primer: RGACGAGAAGACCCTATARA
      Reference: this book
Reverse primer: ACGCTGTTATCCCTAARGTA
      Reference: this book
Recommended annealing temperature: 52°C
Target gene: 16S mitochondrial rDNA
Coverage for the target group: 98.82% (751 species amplified *in silico* out of 760)
Min. length: 75 bp   **Mean length:** 155 bp   **Max. length:** 265 bp
Taxonomic resolution in the target group:

Species	Genus	Family	Order
87.8% (21,019)	96.8% (7,605)	95.4% (756)	79.3% (29)

Comments: Close to the primer pair designed by Clarke *et al.* (2014).

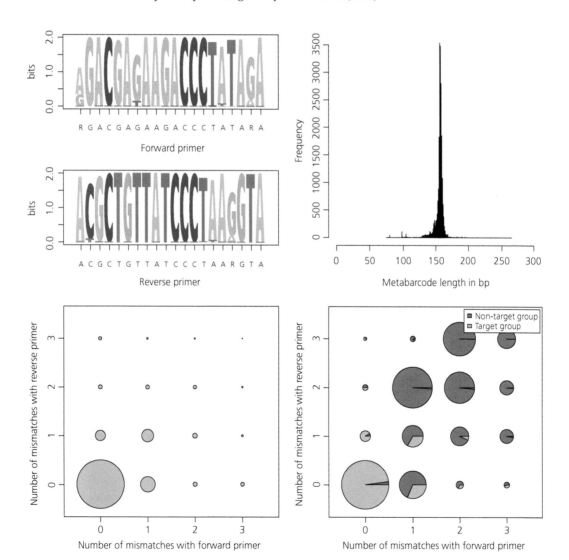

Isop01

Target taxonomic group: Isoptera (termites)
    NCBI taxid: 7499
Forward primer: ATTTCAGGTCAAGGTGCAGCTT
    Reference: this book
Reverse primer: ATTACAACCAAATCCAATTTCA
    Reference: this book
Recommended annealing temperature: 50°C
Target gene: 12S mitochondrial rDNA
Coverage for the target group: 100% (27 species amplified *in silico* out of 27)
Min. length: 63 bp   **Mean length:** 66 bp   **Max. length:** 71 bp
Taxonomic resolution in the target group:

Species	Genus	Family	Order
76.0% (359)	88.2% (152)	100.0% (6)	100.0% (1)

Comments: Good coverage and good taxonomic resolution for a short metabarcode.

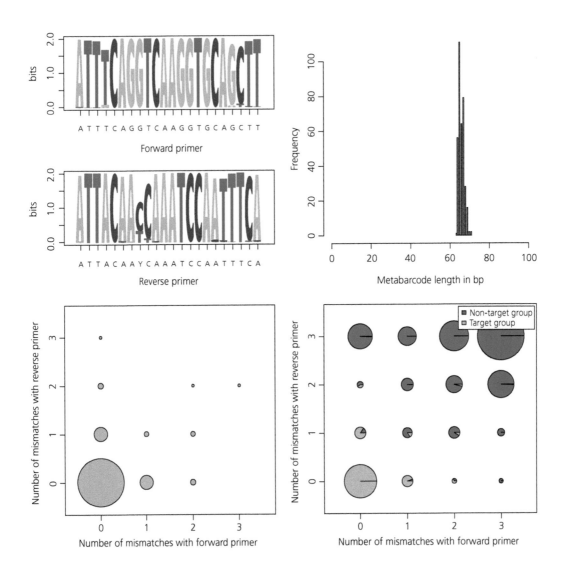

Cole01

Target taxonomic group: Coleoptera (beetles)
      NCBI taxid: 7041
Forward primer: TGCWAAGGTAGCATAATMATTAG
      Reference: this book
Reverse primer: TCTATAGGGTCTTCTCGTC
      Reference: this book
Recommended annealing temperature: 53°C
Target gene: 16S mitochondrial rDNA
Coverage for the target group: 98.33% (59 species amplified *in silico* out of 60).
Min. length: 92 bp   **Mean length:** 107 bp   **Max. length:** 115 bp
Taxonomic resolution in the target group:

Species	Genus	Family	Order
72.6% (8,555)	93.6% (2,699)	94.4% (142)	100.0% (1)

Comments: Both primers modified from Epp *et al.* (2012). Also amplifies some other metazoa, including primates, decapods, frogs, and toads.

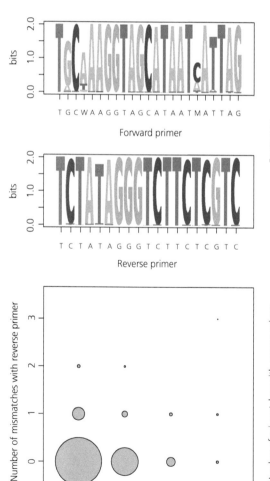

Forward primer

Reverse primer

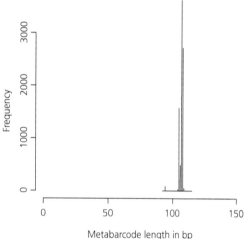

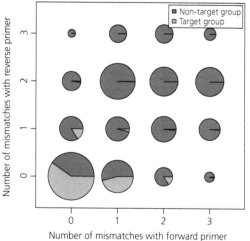

Culi01

Target taxonomic group: Culicidae (mosquitos)
      NCBI taxid: 7157
Forward primer: ACGCTGTTATCCCTAAGGTAACTTA
      Reference: Schneider *et al.* (2016)
Reverse primer: GACGAGAAGACCCTATAGATCTTTAT
      Reference: Schneider *et al.* (2016)
Recommended annealing temperature: 60°C
Target gene: 16S mitochondrial rDNA
Coverage for the target group: 96.77% (30 species amplified *in silico* out of 31)
Min. length: 144 bp   **Mean length:** 145 bp   **Max. length:** 147 bp
Taxonomic resolution in the target group:

Species	Genus	Family	Order
86.0% (57)	100% (5)	100.0% (1)	100.0% (1)

Comments: Also amplifies other insects, including mainly Coleoptera.

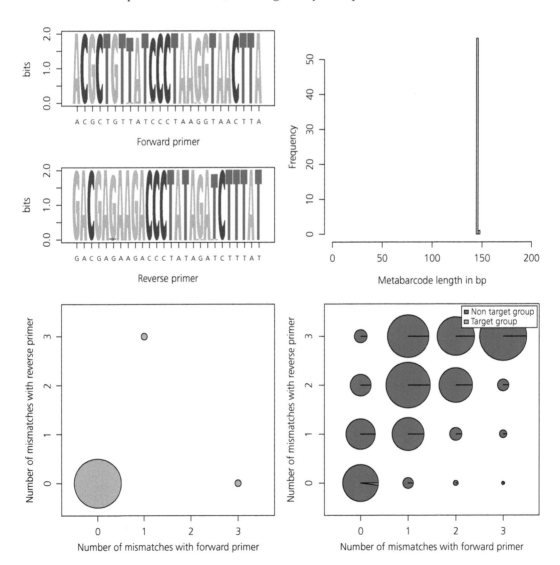

Coll01

Target taxonomic group: Collembola (springtails)
        NCBI taxid: 30001
Forward primer: ACGCTGTTATCCCTWAGG
        Reference: Janssen *et al.* (2017)
Reverse primer: GACGATAAGACCCTWTAGA
        Reference: Janssen *et al.* (2017)
Recommended annealing temperature: 51°C
Target gene: 16S mitochondrial rDNA
Coverage for the target group: 100% (10 species amplified *in silico* out of 10)
Min. length: 76 bp   **Mean length:** 132 bp   **Max. length:** 192 bp
Taxonomic resolution in the target group:

Species	Genus	Family	Order
80.5% (123)	87.2% (47)	75.0% (12)	NA

Comments: Also amplifies Coleoptera and Decapoda. Excellent taxonomic resolution, with even intraspecific variation.

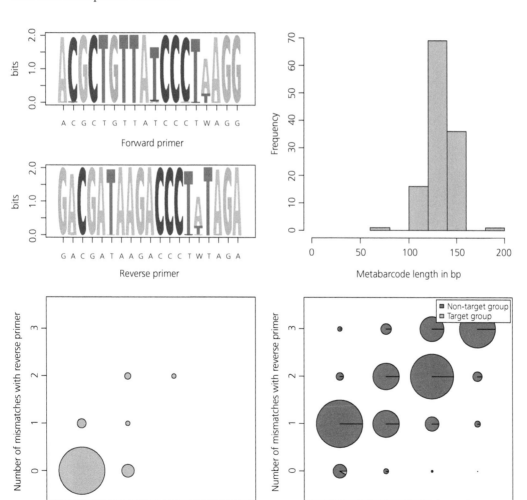

Acar01

Target taxonomic group: Acari (mites and ticks)
  NCBI taxid: 6933
Forward primer: TACTCTAGGGATAACAG
  Reference: this book
Reverse primer: TAATCCAACATCGAGGT
  Reference: this book
Recommended annealing temperature: 45°C
Target gene: 16S mitochondrial rDNA
Coverage for the target group: 96.92% (63 species amplified *in silico* out of 65)
Min. length: 41 bp **Mean length:** 42 bp **Max. length:** 44 bp
Taxonomic resolution in the target group:

Species	Genus	Family	Order
29.6% (389)	75.4% (57)	80.0% (25)	60.0% (5)

Comments: Very short metabarcode, with a limited taxonomic resolution. Not empirically tested.

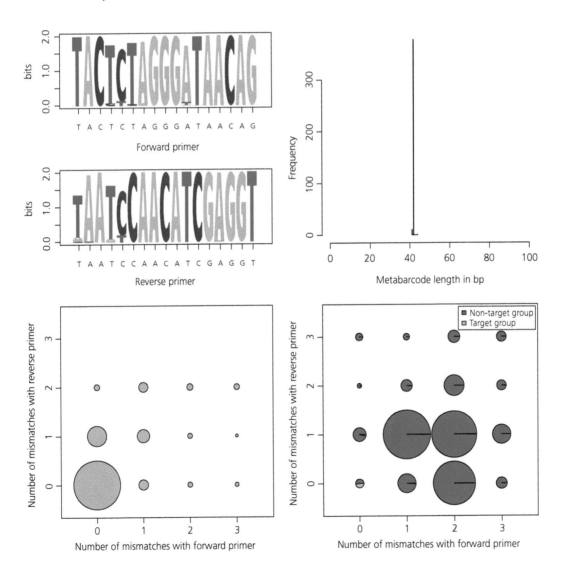

Aran01

Target taxonomic group: Aranae (spiders)
      NCBI taxid: 6893
Forward primer: TTRYGACCTCGATGTTGAATT
      Reference: this book
Reverse primer: CGGTYTGAACTCARATCATGT
      Reference: this book
Recommended annealing temperature: 51°C
Target gene: 16S mitochondrial rDNA
Coverage for the target group: 100% (28 species amplified *in silico* out of 28)
Min. length: 56 bp   **Mean length:** 58 bp   **Max. length:** 160 bp
Taxonomic resolution in the target group:

Species	Genus	Family	Order
75.7% (1,287)	93.0% (440)	96.0% (50)	100.0% (1)

Comments: Good taxonomic resolution for a short metabarcode. Also amplifies some other arthropods. Not empirically tested.

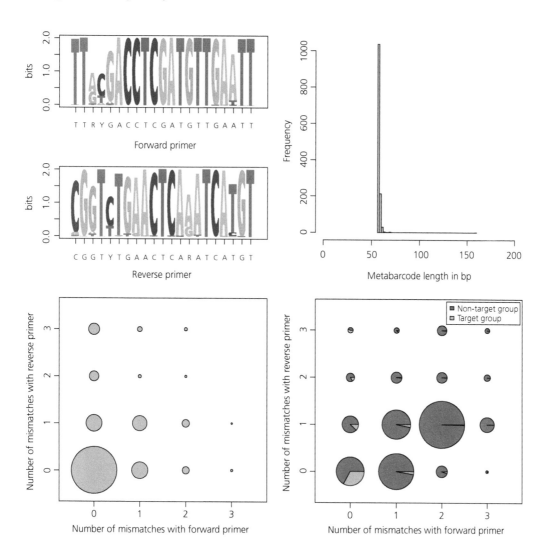

Opil01

Target taxonomic group: Opiliones (harvestmen)
        NCBI taxid: 43271
Forward primer: YTYAACTGTTTATCAAAAACAT
        Reference: this book
Reverse primer: GCTACCTTAGCACAGTCA
        Reference: this book
Recommended annealing temperature: 47°C
Target gene: 16S mitochondrial rDNA
Coverage for the target group: 100% (2 species amplified *in silico* out of 2)
Min. length: 53 bp   **Mean length:** 66 bp   **Max. length:** 72 bp
Taxonomic resolution in the target group:

Species	Genus	Family	Order
79.2% (48)	81.8% (11)	33.3% (3)	100.0% (1)

Comments: Good taxonomic resolution for a short metabarcode. Not empirically tested.

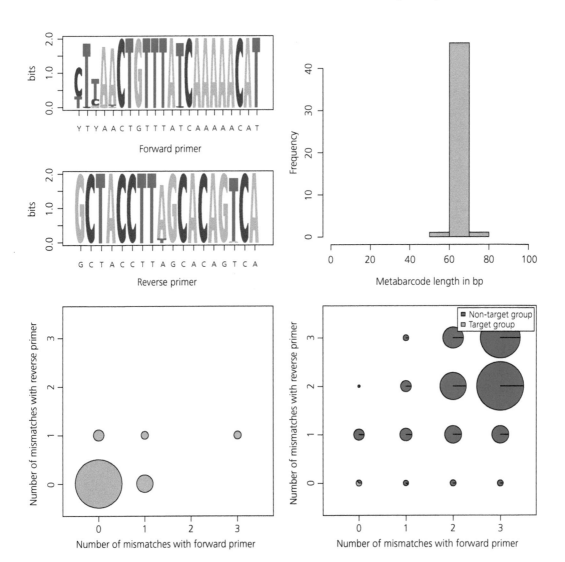

Cope01

Target taxonomic group: Copepoda (copepods)
   NCBI taxid: 6830
Forward primer: GVTMCYYTAGGGATAACAGC
   Reference: this book
Reverse primer: YCGRTYTTAACTCARATCATGTA
   Reference: this book
Recommended annealing temperature: 51°C
Target gene: 16S mitochondrial rDNA
Coverage for the target group: 88.89% (8 species amplified *in silico* out of 9)
Min. length: 110 bp **Mean length:** 112 bp **Max. length:** 114 bp
Taxonomic resolution in the target group:

Species	Genus	Family	Order
100.0% (58)	100.0% (25)	100.0% (14)	100.0% (4)

Comments: Also amplifies many other groups, including mainly mammals and other arthropods. A human blocking oligonucleotide (aacagcGCAATCCTATTCTAGAGTCCA-C3) can be used if human contamination is a problem. Not empirically tested.

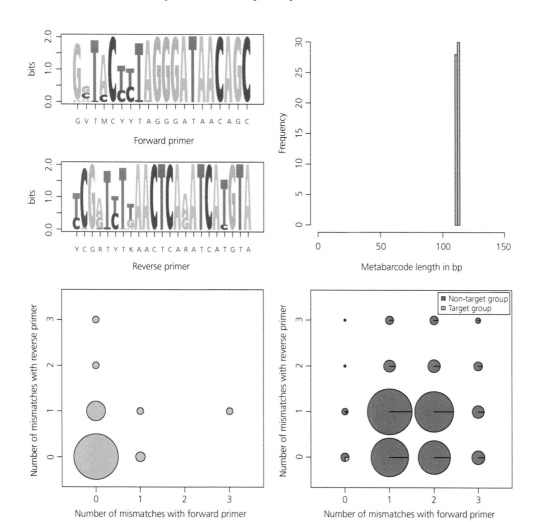

Pera01

Target taxonomic group: Peracarida (Amphipods, Isopods, and related groups)
     NCBI taxid: 6820
Forward primer: TTRYNACCTCGATGTTGAATT
     Reference: this book
Reverse primer: GGTYTGAACTCARATCAYGT
     Reference: this book
Recommended annealing temperature: 49°C
Target gene: 16S mitochondrial rDNA
Coverage for the target group: 100.00% (23 species amplified *in silico* out of 23)
Min. length: 56 bp   **Mean length:** 58 bp   **Max. length:** 60 bp
Taxonomic resolution in the target group:

Species	Genus	Family	Order
83.2% (184)	97.6% (85)	95.8% (48)	100.0% (4)

Comments: Also amplifies many other groups, including mainly insects and arachnids. This metabarcode can be used to complement Euka01, Euka02, Euka03, or Euka04 that are not suitable for Peracarida. Not empirically tested.

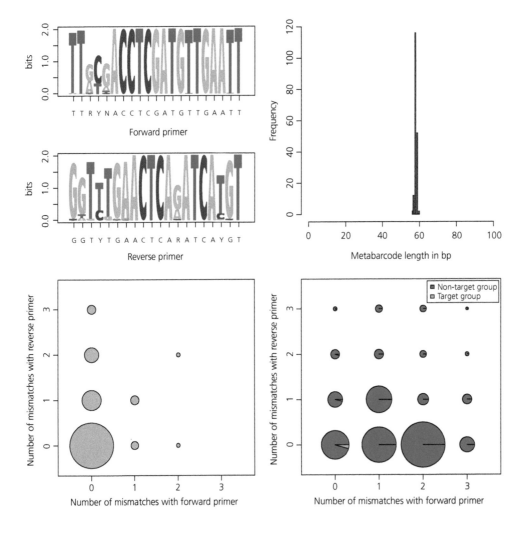

Pera02

Target taxonomic group: Peracarida (Amphipods, Isopods, and related groups)
      NCBI taxid: 6820
Forward primer: CCCTTTGTACACACCGCC
      Reference: this book
Reverse primer: ATGATCCTTCCGCAGGTTCA
      Reference: this book
Recommended annealing temperature: 56°C
Target gene: 18S nuclear ribosomal DNA (V9)
Coverage for the target group: NA
Min. length: 131 bp   **Mean length:** 148 bp   **Max. length:** 231 bp
Taxonomic resolution in the target group:

Species	Genus	Family	Order
74.4% (195)	81.5% (151)	90.6% (85)	100.0% (9)

Comments: Can complement Euka03 or Euka04 that are not working properly for Peracarida. Amplifies many other eukaryotes. Not empirically tested.

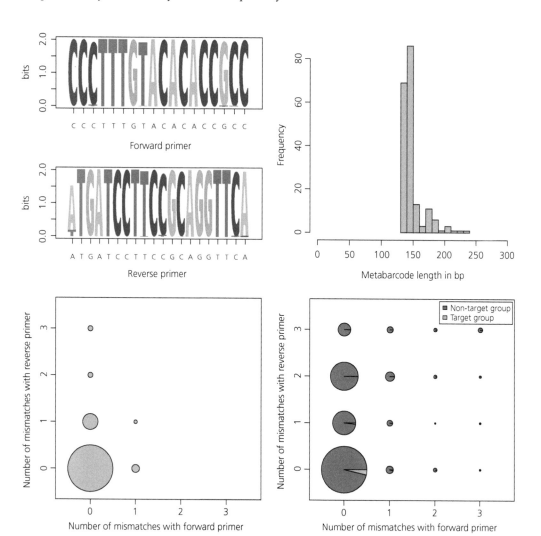

Amph01

Target taxonomic group: Amphipoda
      NCBI taxid: 6821
Forward primer: TTRYNACCTCGATGTTGAATT
      Reference: this book
Reverse primer: GGTYTGAACTCARATCATGTA
      Reference: this book
Recommended annealing temperature: 49°C
Target gene: 16S mitochondrial rDNA
Coverage for the target group: 100% (19 species amplified *in silico* out of 19)
Min. length: 56 bp   **Mean length:** 57 bp   **Max. length:** 58 bp
Taxonomic resolution in the target group:

Species	Genus	Family	Order
94.3% (88)	100.0% (45)	100.0% (26)	100.0% (1)

Comments: Also amplifies many other groups. Not empirically tested.

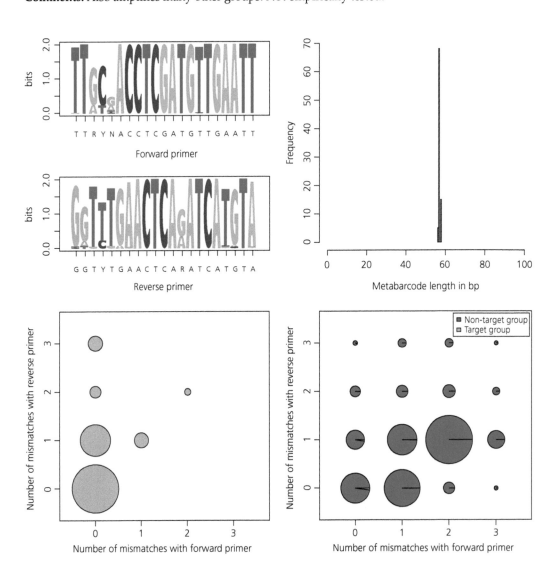

Vert01

Target taxonomic group: Vertebrata (vertebrates)
        NCBI taxid: 7742
Forward primer: TTAGATACCCCACTATGC
        Reference: Riaz *et al.* (2011)
Reverse primer: TAGAACAGGCTCCTCTAG
        Reference: Riaz *et al.* (2011)
Recommended annealing temperature: 49°C
Target gene: 12S mitochondrial rDNA
Coverage for the target group: 97.86% (3,481 species amplified *in silico* out of 3,557)
Min. length: 56 bp   **Mean length:** 97 bp   **Max. length:** 132 bp
Taxonomic resolution in the target group:

Species	Genus	Family	Order
72.4% (11,187)	85.3% (4,179)	94.0% (816)	97.1% (138)

Comments: Highly specific to vertebrates.

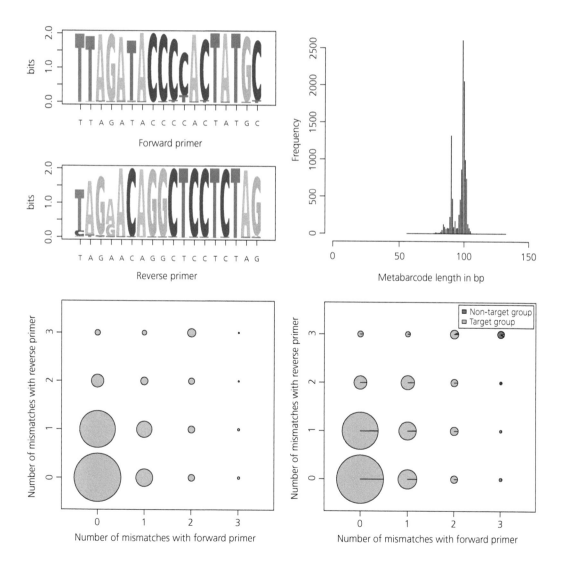

Chon01

Target taxonomic group: Chondrichthyes (cartilaginous fishes)
    NCBI taxid: 7777
Forward primer: ACACCGCCCGTCACTCTC
    Reference: this book
Reverse primer: CATGTTACGACTTGCCTCCTC
    Reference: this book
Recommended annealing temperature: 58°C
Target gene: 12S mitochondrial rDNA
Coverage for the target group: 98.00% (98 species amplified *in silico* out of 100)
Min. length: 41 bp   **Mean length:** 44 bp   **Max. length:** 47 bp
Taxonomic resolution in the target group:

Species	Genus	Family	Order
78.7% (225)	93.8% (97)	100.0% (44)	100.0% (11)

Comments: Good taxonomic resolution for a very short metabarcode. A human blocking oligonucleotide (caccctcCTCAAGTATACTTCAAAGG-C3) can be used if human contamination is a problem. Not empirically tested.

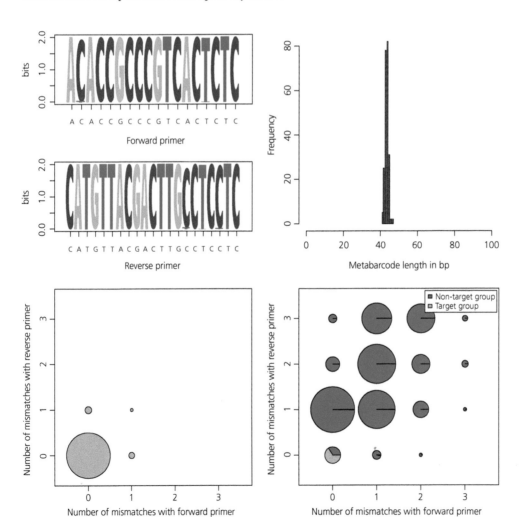

Elas01

Target taxonomic group: Elasmobranchii (rays, sharks, skates)
      NCBI taxid: 7778
Forward primer: GTTGGTAAATCTCGTGCCAGC
      Reference: Miya *et al.* (2015)
Reverse primer: CATAGTGGGGTATCTAATCCTAGTTTG
      Reference: Miya *et al.* (2015)
Recommended annealing temperature: 59°C
Target gene: 12S mitochondrial rDNA
Coverage for the target group: 97.83% (90 species amplified *in silico* out of 92)
Min. length: 170 bp   **Mean length:** 182 bp   **Max. length:** 185 bp
Taxonomic resolution in the target group:

Species	Genus	Family	Order
82.6% (190)	97.7% (88)	100.0% (40)	100.0% (10)

Comments: This primer pair has not been fully optimized (see Elas02 for a slightly better primer pair for elasmobranches). Excellent taxonomic resolution.

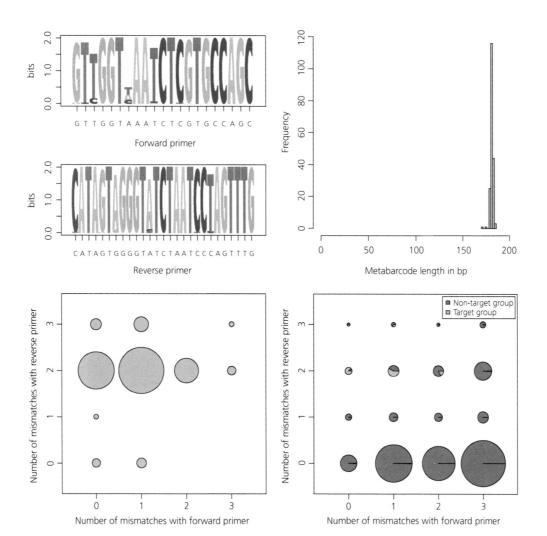

Elas02

Target taxonomic group: Elasmobranchii (rays, sharks, skates)
  NCBI taxid: 7778
Forward primer: GTTGGTHAATCTCGTGCCAGC
  Reference: this book
Reverse primer: CATAGTAGGGTATCTAATCCTAGTTTG
  Reference: this book
Recommended annealing temperature: 59°C
Target gene: 12S mitochondrial rDNA
Coverage for the target group: 97.83% (90 species amplified *in silico* out of 92)
Min. length: 170 bp **Mean length:** 182 bp **Max. length:** 185 bp
Taxonomic resolution in the target group:

Species	Genus	Family	Order
82.6% (190)	97.7% (88)	100.0% (40)	100.0% (10)

Comments: Very close to Elas01, but optimized. Excellent taxonomic resolution. Not empirically tested.

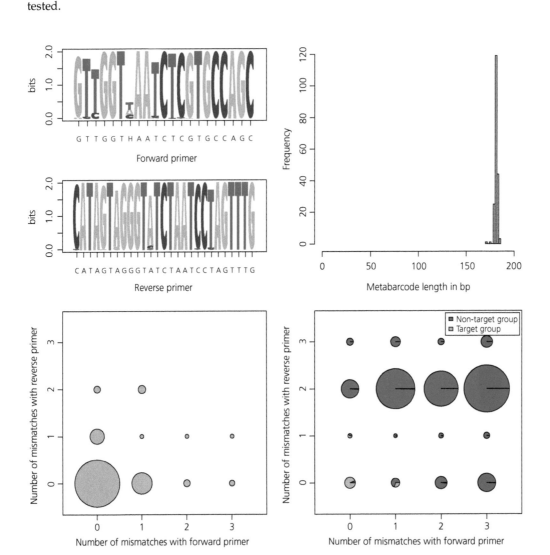

Tele01

Target taxonomic group: Teleostei (teleost fishes)

NCBI taxid: 32443

Forward primer: ACACCGCCCGTCACTCT

Reference: Valentini *et al.* (2016)

Reverse primer: CTTCCGGTACACTTACCATG

Reference: Valentini *et al.* (2016)

Recommended annealing temperature: 55°C

Target gene: 12S mitochondrial rDNA

Coverage for the target group: 98.05% (1,706 species amplified *in silico* out of 1,740)

Min. length: 45 bp   **Mean length:** 64 bp   **Max. length:** 96 bp

Taxonomic resolution in the target group:

Species	Genus	Family	Order
81.5% (3,285)	90.6% (1,610)	98.9% (358)	100.0% (58)

Comments: Excellent taxonomic resolution for a short metabarcode, except for Cyprinidae and Gadidae (see Cypr01 and Gadi01 primer pairs that can be used in complement of Tele01).

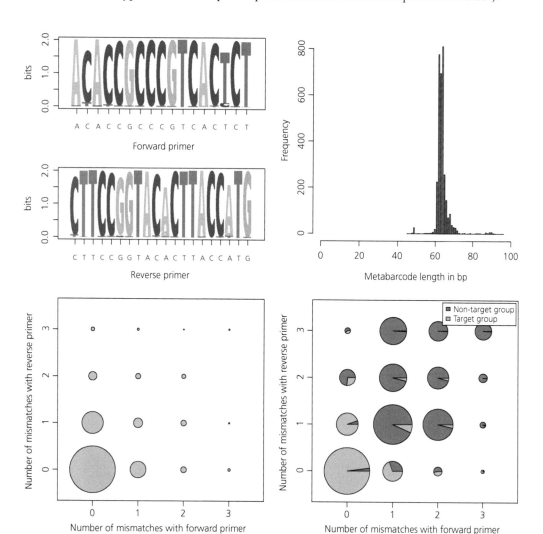

Tele02

Target taxonomic group: Teleostei (teleost fishes)
    NCBI taxid: 32443
Forward primer: AAACTCGTGCCAGCCACC
    Reference: this book
Reverse primer: GGGTATCTAATCCCAGTTTG
    Reference: this book
Recommended annealing temperature: 54°C
Target gene: 12S mitochondrial rDNA
Coverage for the target group: 98.05% (1,706 species amplified *in silico* out of 1,740)
Min. length: 129 bp   **Mean length:** 167 bp   **Max. length:** 209 bp
Taxonomic resolution in the target group:

Species	Genus	Family	Order
87.0% (3,060)	94.1% (1,527)	98.9% (356)	100.0% (59)

Comments: Very close to Tele03, but optimized. Excellent taxonomic resolution. Not empirically tested.

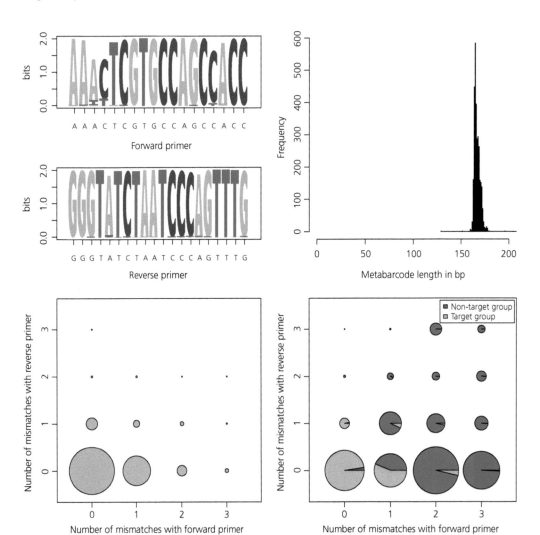

Tele03

Target taxonomic group: Teleostei (teleost fishes)
   NCBI taxid: 32443
Forward primer: GTCGGTAAAACTCGTGCCAGC
   Reference: Miya *et al.* (2015)
Reverse primer: CATAGTGGGGTATCTAATCCCAGTTTG
   Reference: Miya *et al.* (2015)
Recommended annealing temperature: 61°C
Target gene: 12S mitochondrial rDNA
Coverage for the target group: 95.80% (1,667 species amplified *in silico* out of 1,740)
Min. length: 133 bp **Mean length:** 171 bp **Max. length:** 213 bp
Taxonomic resolution in the target group:

Species	Genus	Family	Order
86.8% (2,964)	93.8% (1,474)	98.8% (341)	100.0% (59)

Comments: Very close to Tele02. Excellent taxonomic resolution.

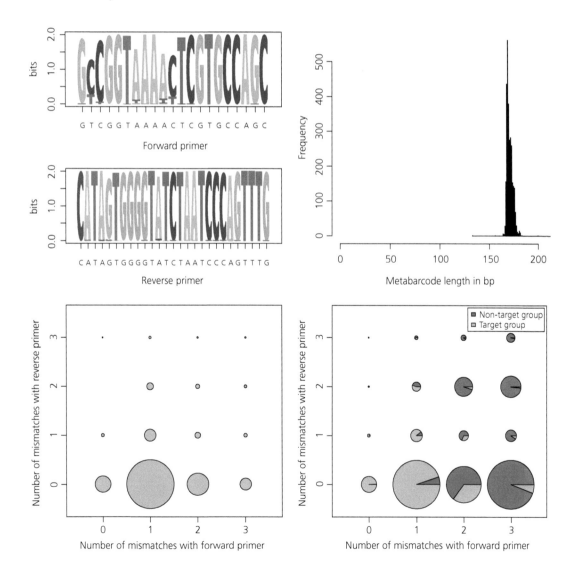

Cypr01

Target taxonomic group: Cyprinidae (carp family)

      NCBI taxid: 7953

Forward primer: AAGACCCTTTGGAGCTTAAGGT

      Reference: this book

Reverse primer: TGGTCGCCCCAACCGAAG

      Reference: this book

Recommended annealing temperature: 58°C

Target gene: 16S mitochondrial rDNA

Coverage for the target group: 93.91% (324 species amplified *in silico* out of 345)

Min. length: 52 bp   **Mean length:** 66 bp   **Max. length:** 80 bp

Taxonomic resolution in the target group:

Species	Genus	Family	Order
69.7% (871)	82.6% (258)	100.0% (1)	100.0% (1)

Comments: Good resolution, better than suggested by the above estimates because of many taxonomic misassignment in the public databases for this family. Not empirically tested.

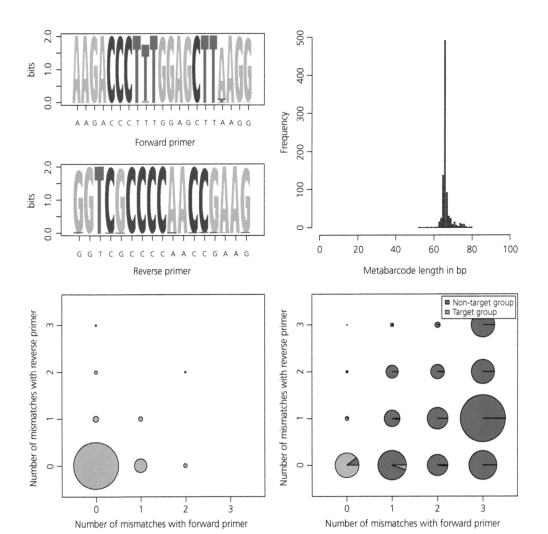

Gadi01

Target taxonomic group: Gadidae (cods)
NCBI taxid: 8045
Forward primer: YCAAATTGATGAAAAACGGCGT
    Reference: this book
Reverse primer: CGRATGCGTATAACTGCTTTG
    Reference: this book
Recommended annealing temperature: 54°C
Target gene: 12S mitochondrial rDNA
Coverage for the target group: 100.00% (10 species amplified *in silico* out of 10)
Min. length: 42 bp   **Mean length:** 44 bp   **Max. length:** 44 bp
Taxonomic resolution in the target group:

Species	Genus	Family	Order
100.0% (15)	100.0% (10)	100.0% (1)	100.0% (1)

Comments: Excellent resolution. Not empirically tested.

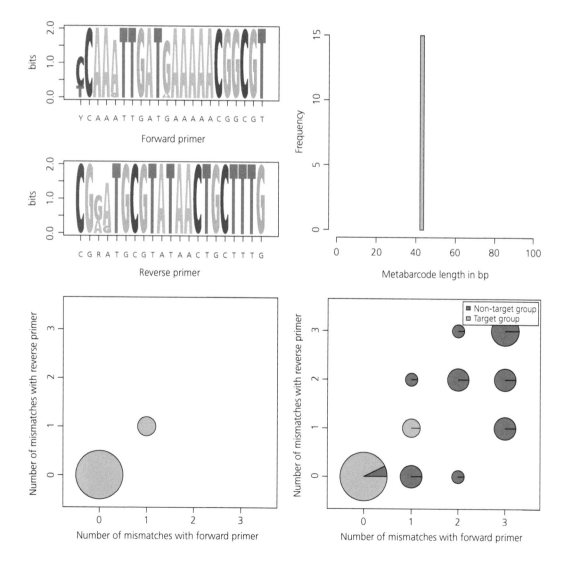

Batr01

Target taxonomic group: Batrachia (Frogs and salamanders)
   NCBI taxid: 41666
Forward primer: ACACCGCCCGTCACCCT
   Reference: Valentini *et al.* (2016)
Reverse primer: GTAYACTTACCATGTTACGACTT
   Reference: Valentini *et al.* (2016)
Recommended annealing temperature: 55°C
Target gene: 12S mitochondrial rDNA
Coverage for the target group: 98.26% (169 species amplified *in silico* out of 172)
Min. length: 16 bp   **Mean length:** 51 bp   **Max. length:** 100 bp
Taxonomic resolution in the target group:

Species	Genus	Family	Order
83.3% (2,802)	93.3% (418)	94.8% (58)	100.0% (2)

Comments: Excellent taxonomic resolution for a short metabarcode.

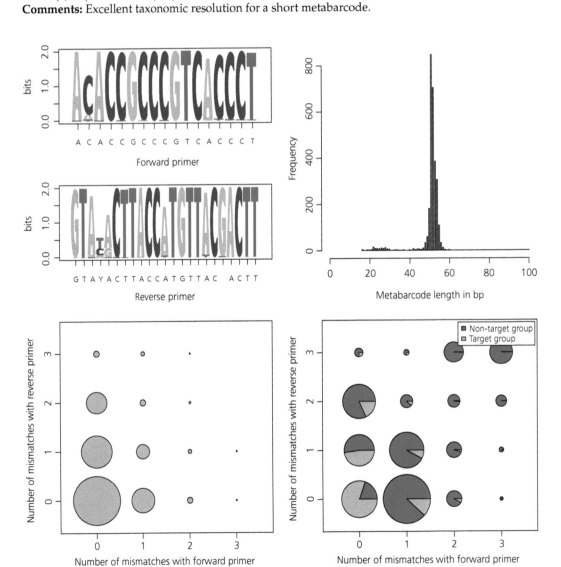

Test01

Target taxonomic group: Testudines (turtles)
            NCBI taxid: 7778
Forward primer: AGACGAGAAGACCCTGTGGAA
            Reference: this book
Reverse primer: TCCGAGGTCRCCCCAACC
            Reference: this book
Recommended annealing temperature: 54°C
Target gene: 16S mitochondrial rDNA
Coverage for the target group: 100.00% (70 species amplified *in silico* out of 70)
Min. length: 69 bp   **Mean length:** 75 bp   **Max. length:** 83 bp
Taxonomic resolution in the target group:

Species	Genus	Family	Order
89.2% (195)	97.5% (79)	100.0% (14)	100.0% (1)

Comments: Excellent taxonomic resolution. Not empirically tested.

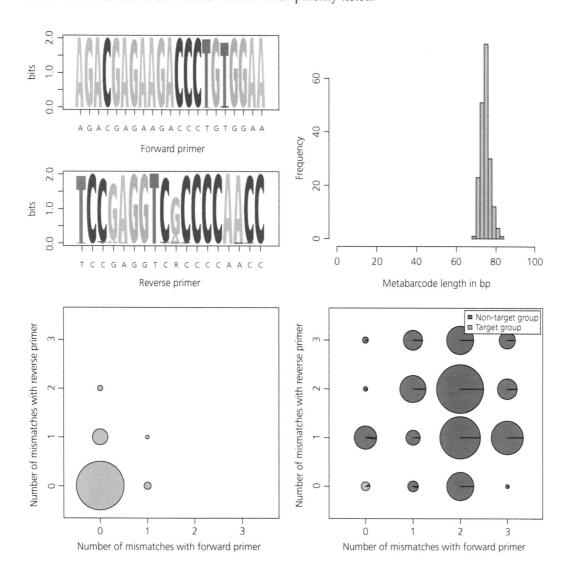

Aves01

Target taxonomic group: Aves (birds)
        NCBI taxid: 8782
Forward primer: GATTAGATACCCCACTATGC
        Reference: Epp *et al.* (2012)
Reverse primer: GTTTTAAGCGTTTGTGCTCG
        Reference: Epp *et al.* (2012)
Recommended annealing temperature: 54°C
Target gene: 12S mitochondrial rDNA
Coverage for the target group: 98.29% (460 species amplified *in silico* out of 468)
Min. length: 39 bp   **Mean length:** 52 bp   **Max. length:** 63 bp
Taxonomic resolution in the target group:

Species	Genus	Family	Order
53.9% (1,608)	64.8% (819)	74.2% (163)	75.0% (36)

Comments: Relatively limited taxonomic resolution, but highly specific to birds.

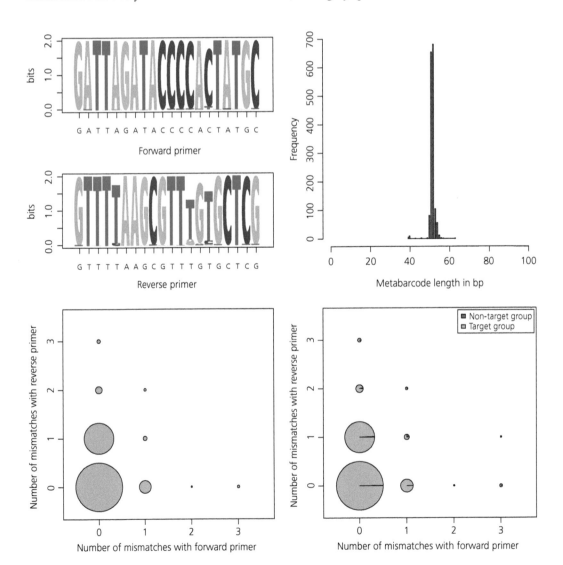

Aves02

Target taxonomic group: Aves (birds)
NCBI taxid: 8782

Forward primer: GAAAATGTAGCCCATTTCTTCC
   Reference: this book
Reverse primer: CATACCGCCGTCGCCAG
   Reference: this book
Recommended annealing temperature: 56°C
Target gene: 12S mitochondrial rDNA
Coverage for the target group: 98.93% (463 species amplified *in silico* out of 468)
Min. length: 66 bp   **Mean length:** 80 bp   **Max. length:** 86 bp
Taxonomic resolution in the target group:

Species	Genus	Family	Order
73.1% (1933)	87.6% (874)	88.9% (171)	94.4% (36)

Comments: Good taxonomic resolution for a relatively short metabarcode. Not as specific to birds as Aves01. Not empirically tested.

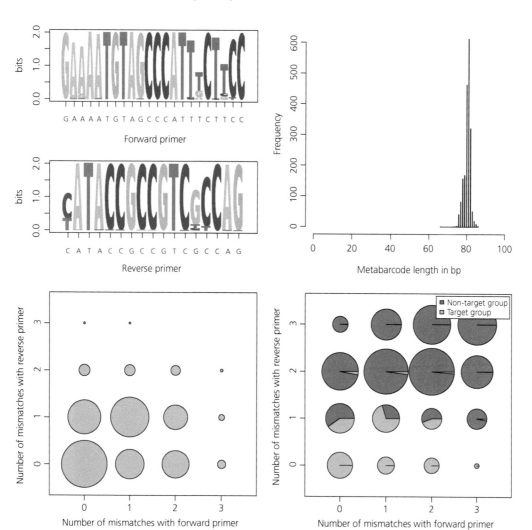

Aves03

Target taxonomic group: Aves (birds)
      NCBI taxid: 8782
Forward primer: TATGATAAAGTGAACATRGAGG
      Reference: this book
Reverse primer: GGTTCRAYTCCTRCTTTTCTA
      Reference: this book
Recommended annealing temperature: 51°C
Target gene: mitochondrial tRNA ile-gln
Coverage for the target group: 100.00% (468 species amplified *in silico* out of 468)
Min. length: 22 bp   **Mean length:** 31 bp   **Max. length:** 46 bp
Taxonomic resolution in the target group:

Species	Genus	Family	Order
68.9% (687)	83.4% (398)	84.1% (132)	71.4% (35)

Comments: Relatively good taxonomic resolution for an extremely short metabarcode. Highly specific to birds (not a single other group amplified *in silico*). Not empirically tested.

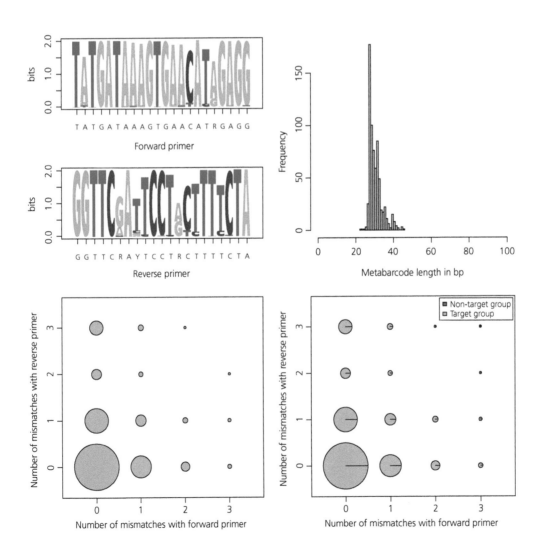

Mamm01

Target taxonomic group: Mammalia (mammals)
        NCBI taxid: 40674
Forward primer: CCGCCCGTCACCCTCCT
        Reference: this book
Reverse primer: GTAYRCTTACCWTGTTACGAC
        Reference: this book
Recommended annealing temperature: 53°C
Target gene: 12S mitochondrial rDNA
Coverage for the target group: 99.73% (734 species amplified *in silico* out of 736)
Min. length: 13 bp   **Mean length:** 58 bp   **Max. length:** 74 bp
Taxonomic resolution in the target group:

Species	Genus	Family	Order
85.7% (1552)	96.4% (693)	98.6% (140)	100.0% (24)

Comments: Excellent taxonomic resolution.

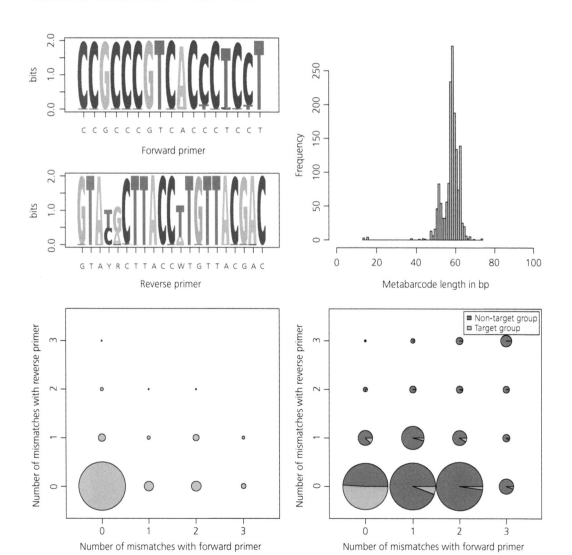

Mamm02

Target taxonomic group: Mammalia (mammals)
        NCBI taxid: 40674
Forward primer: CGAGAAGACCCTRTGGAGCT
        Reference: this book
Reverse primer: CCGAGGTCRCCCCAACC
        Reference: Giguet-Covex *et al.* (2014)
Recommended annealing temperature: 57°C
Target gene: 16S mitochondrial rDNA
Coverage for the target group: 99.86% (735 species amplified *in silico* out of 736)
Min. length: 53 bp   **Mean length:** 74 bp   **Max. length:** 84 bp
Taxonomic resolution in the target group:

Species	Genus	Family	Order
89.7% (1926)	97.4% (763)	100.0% (145)	100.0% (26)

Comments: Excellent taxonomic resolution. Also amplifies some other vertebrates, including mainly amphibians and reptiles. Forward primer slightly modified from Giguet-Covex *et al.* (2014).

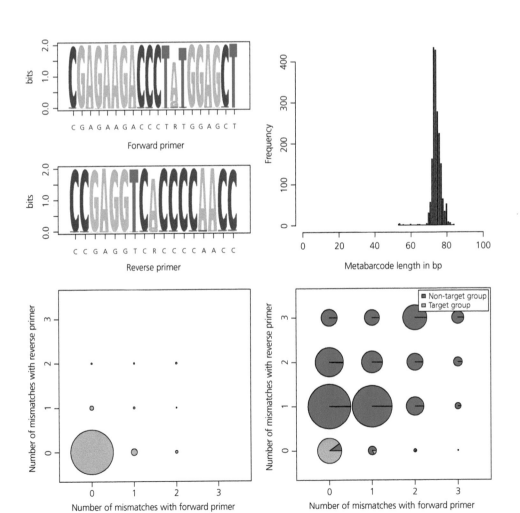

Three-hundred and eighty-four (384) tags of eight nucleotides, with at least three differences among them

These tags can be added on the 5′-end of the primers, allowing the identification of polymerase chain reaction (PCR) products originating from four 96-plates, with the forward and reverse tags used only once for a single PCR.

1: aacaagcc	26: tcatagcg	51: tagccact	76: cgcataga
2: ggaatgag	27: tgaggaca	52: gaggacta	77: cagttctc
3: aattgccg	28: aacaggag	53: agaagagg	78: aacgagtg
4: cgaccata	29: gagtaacc	54: cgatgagt	79: ctgtcaac
5: atgctgac	30: cagctcat	55: gtgtagtc	80: tccaccta
6: tgagacag	31: tgctccaa	56: gagttcct	81: cgcatctt
7: gagcttac	32: tcagtcga	57: acacacag	82: gatacgct
8: ttaccagg	33: ggagaaga	58: aacctagc	83: gtgttgga
9: tgagagct	34: gtgctcaa	59: acacaggt	84: acactagg
10: ctgacctt	35: acaagacc	60: tagagctg	85: cgccaatt
11: atgcttgg	36: caggaaca	61: aacctcag	86: tagcaagg
12: aacaccgt	37: gtgatctc	62: ggatgatc	87: gtcacaga
13: ttaccgct	38: acttggct	63: tatctggc	88: tccagatc
14: ccagtatg	39: cgatacac	64: ctggttga	89: acactcct
15: tgagatgc	40: tcatcctg	65: tccaacac	90: gttaacgg
16: gtgcaact	41: caggctaa	66: gtgtgaag	91: aacgcgat
17: acaaccga	42: tcatcggt	67: aacctgct	92: cgcgtaat
18: tgagccta	43: agttccac	68: ggatgtct	93: ccgatact
19: atggaggt	44: aaccaacg	69: tccaagca	94: agacatcc
20: tcatacgc	45: ctgcgaat	70: agaatgcc	95: gatagagg
21: ctgagtct	46: gtggttag	71: gtgtgtgt	96: ctgttgtg
22: gaggtgaa	47: agaacgtg	72: cacggata	97: aacgcttc
23: ggcatgta	48: tcatgcct	73: caacctca	98: acactgtc
24: gtgccata	49: ggatagca	74: ggattcga	99: tccatacg
25: ttacgcca	50: cagtagac	75: ttaggcac	100: agtgttcc

101: acagaagc
102: agacctgt
103: cgctctta
104: agacgaac
105: tgcaagag
106: aagagtgc
107: catacctc
108: ctgttcgt
109: cgcttatg
110: agacgctt
111: aacgtcga
112: tagcctga
113: aaggttcg
114: atgaggtg
115: gatatcgc
116: gttactcg
117: tccgagat
118: ggctctat
119: tagcgtag
120: tgcactac
121: acaacgct
122: tccgctaa
123: atacagcg
124: aactacgc
125: caacgtac
126: gttaggca
127: catagtcg
128: gtgcctat
129: cggattgt
130: agagaacg
131: ataccacc
132: aatgccag
133: aacagctc
134: cggtacta
135: ataccgga
136: gatcatcg
137: agagagtc
138: aactcctg
139: cgaacaac
140: cggtcatt
141: acagtgag

142: cggttgaa
143: gagctaga
144: tgccagta
145: agagctac
146: aactgagg
147: ggtagttc
148: cgtaagtc
149: atacgtgc
150: catcagca
151: accactca
152: caatggct
153: aactgtcc
154: tcctggta
155: catccagt
156: tgcgaatg
157: gagtacag
158: agagtctg
159: cgtagact
160: tccttcca
161: tgcttgtc
162: catcgatg
163: cgtagcta
164: cttcaacg
165: catcggat
166: aagaaccg
167: gatgaggt
168: cttcaggt
169: tcgaagtg
170: gatgcctt
171: agatcagc
172: tcgactct
173: ttcactgg
174: tgaagtcg
175: cggcataa
176: ggtgcata
177: tcgagaca
178: tagtcgag
179: gatggtca
180: ggactatg
181: tcgagcat
182: gtgcattg

183: attcgcag
184: acaagctg
185: acaggtca
186: gttgagag
187: agatctcg
188: cgtctaca
189: accaacct
190: atgagcac
191: ggttactg
192: agatggag
193: cgtgatac
194: cttcttcc
195: attctcgc
196: tcgattgc
197: catggcaa
198: tccgaaga
199: tctcacc
200: cttgacga
201: tcgcatac
202: catgtacg
203: tgagtgac
204: ggttcgaa
205: tggaacct
206: tagttcgg
207: atagtggc
208: tcgcgatt
209: catgtgga
210: tcgctcta
211: agcaactg
212: cgttacgt
213: ggttgtag
214: agcaaggt
215: cgttagag
216: tggtgatg
217: gagtcagt
218: accatgga
219: ttccgaag
220: tcggatca
221: attgcagc
222: aagccgaa
223: gattccga

224: accgcata
225: ttcctcac
226: accgtaac
227: gtaagcga
228: caggtgtt
229: taaccgac
230: tactctcg
231: agcatcca
232: attggcgt
233: acctaacc
234: aagctcca
235: tcgtatgg
236: ttcgacgt
237: aagctgtg
238: agcattgc
239: aatcacgg
240: ctaactcc
241: tcggactt
242: taacgctc
243: ttcgatcg
244: tggcgtta
245: aaggacac
246: tacagcct
247: gcaatcca
248: acctctgt
249: gaacgact
250: gcaatggt
251: acctgcaa
252: tcgttcac
253: agcgacat
254: aaggcatg
255: ctgctatc
256: caacagtg
257: gcacagaa
258: aaggctgt
259: aaccagga
260: tctaaccg
261: gcacattc
262: gaactcgt
263: caaccaag
264: ttccgctt

265: gaactgca

266: ccaatagc

267: gcactaac

268: ttcgtgct

269: tctaaggc

270: tggtcgta

271: accttgtg

272: agtaacgc

273: ttcgttgc

274: atcaccag

275: caacgaga

276: ggtgtaag

277: gcacttga

278: ctacagac

279: ccacactt

280: gtgtacca

281: tatcctcc

282: ttctaggc

283: gaagatcc

284: agctcaag

285: tgccatct

286: tcgttgct

287: ccacgtta

288: gaagctga

289: ccactgat

290: ccttggaa

291: gaaggaag

292: caagaagg

293: ctgatcca

294: acgaactc

295: caagactc

296: taaggcgt

297: agctgtga

298: aagtcgtc

299: acgatcgt

300: ctactcag

301: aagtgcga

302: agcttcgt

303: caagcact

304: atcagagc

305: gtagcgaa

306: agaggcaa

307: gaagtgtg

308: caagcgta

309: gtagcttc

310: acgcagta

311: ctagacct

312: gcatggtt

313: gcatgtac

314: gacattcc

315: acgcgtat

316: caaggttg

317: gcattacg

318: aggaccta

319: caagtcca

320: ctagcaga

321: atccacca

322: ccatccaa

323: agcagtct

324: tatgcagg

325: tctccaga

326: ctagctag

327: gaatccac

328: acggattg

329: tatgcgca

330: tgtatcgg

331: acggtaga

332: cgagatca

333: caataccg

334: tacgtctc

335: gccagtta

336: taggtagc

337: tctcgtca

338: tgccttga

339: ttgacggt

340: aataccgc

341: tctctcgt

342: ccattgca

343: atcgagac

344: aggcaagt

345: accgtctt

346: ccgaacaa

347: caatccgt

348: gcctatct

349: ttgagagg

350: cagaacgt

351: gtatcgtg

352: acgtcaca

353: ccgacata

354: acgtctac

355: tagaggac

356: atcgctga

357: gaattggc

358: atcggatg

359: acgtgctt

360: tgtcgtac

361: tccttagc

362: ccgtaaga

363: ttgatgcg

364: aggtagtg

365: ccgtagtt

366: cttgtgac

367: gcggtatt

368: ttgcagca

369: cgaaggat

370: aggtgact

371: gcaggata

372: cctaagct

373: gtcaatgc

374: tctgttcg

375: tgcaacga

376: ggagactt

377: cacactct

378: cacagagt

379: actactgg

380: tgtgcttg

381: cctagaga

382: accagtac

383: ctaggtgt

384: acggaact

Checklist when designing a PCR-based DNA metabarcoding experiment

Sampling

- Is the size of the sampling area sufficient to answer the biological question? (Chapter 4)
- Are there enough sampling units to answer the biological question? (Chapter 4)
- Are there enough biological replicates per sampling unit and/or sufficient volume/mass of collected material to ensure the significance of the results? (Chapters 4–6)
- Is the collected material correctly preserved to minimize the risks of microbial community turnover or DNA degradation? (Chapter 4)

DNA extraction and amplification

- Are the experimental conditions and reagents appropriate to minimize contamination risks? (Chapters 5–6)
- Is the DNA extraction protocol appropriate for the studied matrix/organisms? (Chapter 5)
- Is the primer pair suitable to amplify the target community:
  - Does the primer pair have minimal biases and maximal resolution? (Chapter 2)
  - Are there enough references available? Should the reference database be completed? (Chapter 3)
- Have the PCR mixtures and thermocycling conditions been optimized to minimize the production of artifacts and ensure the reliability of the results? (Chapter 6)
- If a proofreading polymerase is used for the amplification, do the primers have between three and five phosphorothioate binds on their 3'-end? (Chapter 6)

- Are there enough technical replicates per sampling unit to demonstrate that the signal obtained differs from experimental noise? (Chapters 4–6)
- Are there enough extraction and PCR negative controls? One to several negative controls per extraction batch and PCR plates are mandatory. (Chapters 5–6)
- Are negative controls distributed randomly in PCR plates to test for possible sporadic contaminations? (Chapter 6)
- Are the tagging sequences different enough to ensure the discrimination of samples during bioinformatics analyses? (Chapter 6)
- If a dual tagging strategy is used (two different tags to differentiate samples), are there enough unused tag combinations to detect eventual tag jumping events? (Chapter 6)
- Are there enough PCR positive controls to adjust the sequence data filtering or allow cross-comparisons across different sequencing runs or experiments? (Chapter 6)
- If several DNA metabarcoding experiments must be carried out within the same study, will the same positive control(s) be included in all experiments? (Chapter 6)

Sequencing

- Is the sequencing technology able to fully cover the amplified region? (Chapter 7)
- Is the sequencing technology able to provide enough sequence reads per sample? (Chapter 8)
- Is the sequencing design appropriate for limiting potential artifacts? (e.g., potential "index jumps" among different libraries; Chapters 6–7)

Bioinformatics analysis

- Is the sequence data of sufficient average quality? (Chapters 7–8)
- Were basic data handling and curation performed? (Chapter 8)
- Were filtration or classification procedures applied to minimize artifactual sequences due to PCR or sequencing errors? (Chapter 8)
- Were experimental positive and negative controls sufficiently considered to further exclude potential artifactual sequences (e.g., external and internal contaminants)? (Chapter 8)
- Were all PCRs screened to remove potential dysfunctional replicates? (Chapter 8)

- Overall, are the chosen bioinformatic tools appropriate for target organisms, for metabarcodes, or for amplification/sequencing protocols? (Chapter 8)

Comparative analyses and inferences

- When comparing sample diversity or other characteristics, are the sequencing depth and coverage comparable and sufficient among samples? (Chapter 8)
- Are the metrics used appropriate to metabarcoding data? (Chapter 8)
- Do the resulting inferences consider the particularity of metabarcoding data? (Chapter 8)

References

Acinas SG, Marcelino LA, Klepac-Ceraj V, Polz MF (2004). Divergence and redundancy of 16S rRNA sequences in genomes with multiple rrn operons. *Journal of Bacteriology*, **186**, 2629–2635.

Adey A, Morrison HG, Asan, et al. (2010). Rapid, low-input, low-bias construction of shotgun fragment libraries by high-density *in vitro* transposition. *Genome Biology*, **11**, R119.

Aerts R, Chapin FS (1999). The mineral nutrition of wild plants revisited: a re-evaluation of processes and patterns. *Advances in Ecological Research*, **30**, 1–67.

Albert CH, Yoccoz NG, Edwards TC Jr, Graham CH, Zimmermann NE, Thuiller W (2010). Sampling in ecology and evolution—bridging the gap between theory and practice. *Ecography*, **33**, 1–10.

Alsos IG, Sjögren P, Edwards ME, et al. (2016). Sedimentary ancient DNA from Lake Skartjørna, Svalbard: assessing the resilience of arctic flora to Holocene climate change. *Holocene*, **26**, 627–642.

Altschul SF, Madden TL, Schäffer AA, et al. (1997). Gapped BLAST and PSI-BLAST: a new generation of protein database search programs. *Nucleic Acids Research*, **25**, 3389–3402.

Amaral-Zettler L, Artigas LF, Baross J, et al. (2010). A global census of marine microbes. In: *Life in the world's oceans—diversity, distribution, and abundance* (ed. McIntyre AD), pp. 223–245. Wiley-Blackwell, Oxford, UK.

Amaral-Zettler LA, McCliment EA, Ducklow HW, Huse SM (2009). A method for studying protistan diversity using massively parallel sequencing of V9 hypervariable regions of small-subunit ribosomal RNA genes. *PLoS One*, **4**, e6372.

Amend AS, Oliver TA, Amaral-Zettler LA, et al. (2013). Macroecological patterns of marine bacteria on a global scale. *Journal of Biogeography*, **40**, 800–811.

Anders S, Huber W (2010). Differential expression analysis for sequence count data. *Genome Biology*, **11**, R106.

Andersen K, Bird KL, Rasmussen M, Haile J, et al. (2012). Meta-barcoding of "dirt" DNA from soil reflects vertebrate biodiversity. *Molecular Ecology*, **21**, 1966–1979.

Anderson KE, Russell JA, Moreau CS, et al. (2012). Highly similar microbial communities are shared among related and trophically similar ant species. *Molecular Ecology*, **21**, 2282–2296.

Anderson-Carpenter LL, McLachlan JS, Jackson ST, Kuch M, Lumibao CY, Poinar HN (2011). Ancient DNA from lake sediments: bridging the gap between paleoecology and genetics. *BMC Evolutionary Biology*, **11**, 30.

Andrews S (2014). *FastQC a quality-control tool for high-throughput sequence data.* https://www.bioinformatics.babraham.ac.uk/projects/fastqc/.

Andújar C, Arribas P, Ruzicka F, Crampton-Platt A, Timmermans MJTN, Vogler AP (2015). Phylogenetic community ecology of soil biodiversity using mitochondrial metagenomics. *Molecular Ecology*, **24**, 3603–3617.

Anslan S, Bahram M, Hiiesalu I, Tedersoo L (2017). PipeCraft: flexible open-source toolkit for bioinformatics analysis of custom high-throughput amplicon sequencing data. *Molecular Ecology Resources*, **17**, in press.

Anthony RG, Smith NS (1974). Comparison of rumen and fecal analysis to describe deer diets. *Journal of Wildlife Management*, **38**, 535–540.

Apothéloz-Perret-Gentil L, Cordonier A, Straub F, Iseli J, Esling P, Pawlowski J (2017). Taxonomy-free molecular diatom index for high-throughput eDNA biomonitoring. *Molecular Ecology Resources*, **17**, in press.

Apprill A, McNally S, Parsons R, Webe L (2015). Minor revision to V4 region SSU rRNA 806R gene primer greatly increases detection of SAR11 bacterioplankton. *Aquatic Microbial Ecology*, **75**, 129–137.

Ardura A, Zaiko A, Martinez JL, Samulioviene A, Semenova A, Garcia-Vazquez E (2015). eDNA and specific primers for early detection of invasive species—A case study on the bivalve *Rangia cuneata*, currently spreading in Europe. *Marine Environmental Research*, **112**, 48–55.

Aronesty E (2011). *Command-line tools for processing biological sequencing data, ea-utils.* https://expressionanalysis.github.io/ea-utils/.

Arribas P, Andújar C, Hopkins K, Shepherd M, Vogler AP (2016). Metabarcoding and mitochondrial metagenomics of endogean arthropods to unveil the mesofauna of the soil. *Methods in Ecology and Evolution*, **7**, 1071–1081.

Ashburner M, Ball CA, Blake JA, et al. (2000). Gene ontology: tool for the unification of biology. The Gene Ontology Consortium. *Nature Genetics*, **25**, 25–29.

Ashelford KE, Chuzhanova NA, Fry JC, Jones AJ, Weightman AJ (2005). At least 1 in 20 16S rRNA sequence records currently held in public repositories is estimated to contain substantial anomalies. *Applied and Environmental Microbiology*, **71**, 7724–7736.

Aylagas E, Borja Á, Irigoien X, Rodríguez-Ezpeleta N (2016). Benchmarking DNA metabarcoding for biodiversity-based monitoring and assessment. *Frontiers in Marine Science*, **3**, 96.

Aßhauer KP, Wemheuer B, Daniel R, Meinicke P (2015). Tax4Fun: predicting functional profiles from metagenomic 16S rRNA data. *Bioinformatics*, **31**, 2882–2884.

Baamrane MAA, Shehzad W, Ouhammou A, et al. (2012). Assessment of the food habits of the Moroccan dorcas gazelle in M'Sabih Talaa, West Central Morocco, using the *trn*L approach. *PLoS One*, **7**, e35643.

Bahram M, Kohout P, Anslan S, Harend H, Abarenkov K, Tedersoo L (2016). Stochastic distribution of small soil eukaryotes resulting from high dispersal and drift in a local environment. *ISME Journal*, **10**, 885–896.

Bai Y, Müller DB, Srinivas G, et al. (2015). Functional overlap of the *Arabidopsis* leaf and root microbiota. *Nature*, **528**, 364–369.

Baird DJ, Hajibabaei M (2012). Biomonitoring 2.0: a new paradigm in ecosystem assessment made possible by next-generation DNA sequencing. *Molecular Ecology*, **21**, 2039–2044.

Baker ME, King RS (2010). A new method for detecting and interpreting biodiversity and ecological community thresholds. *Methods in Ecology and Evolution*, **1**, 25–37.

Baker GC, Smith JJ, Cowan DA (2003). Review and re-analysis of domain-specific 16S primers. *Journal of Microbiological Methods*, **55**, 541–555.

Bakken LR (1985). Separation and purification of bacteria from soil. *Applied and Environmental Microbiology*, **49**, 1482–1487.

Baldigo BP, Sporn LA, George SD, Ball JA (2017). Efficacy of environmental DNA to detect and quantify brook trout populations in headwater streams of the Adirondack Mountains, New York. *Transactions of the American Fisheries Society*, **146**, 99–111.

Baldrian P, Kolařík M, Stursová M, et al. (2012). Active and total microbial communities in forest soil are largely different and highly stratified during decomposition. *ISME Journal*, **6**, 248–258.

Baldwin DS, Colloff MJ, Rees GN, et al. (2013). Impacts of inundation and drought on eukaryote biodiversity in semi-arid floodplain soils. *Molecular Ecology*, **22**, 1746–1758.

Bálint M, Schmidt P-A, Sharma R, Thines M, Schmitt I (2014). An Illumina metabarcoding pipeline for fungi. *Ecology and Evolution*, **4**, 2642–2653.

Bálint M, Tiffin P, Hallström B, et al. (2013). Host genotype shapes the foliar fungal microbiome of balsam poplar (*Populus balsamifera*). *PLoS One*, **8**, e53987.

Baraloto C, Hardy OJ, Paine CET, et al. (2012). Using functional traits and phylogenetic trees to examine the assembly of tropical tree communities. *Journal of Ecology*, **100**, 690–701.

Barberán A, McGuire KL, Wolf JA, et al. (2015). Relating belowground microbial composition to the taxonomic, phylogenetic, and functional trait distributions of trees in a tropical forest. *Ecology Letters*, **18**, 1397–1405.

Barnes MA, Turner CR (2016). The ecology of environmental DNA and implications for conservation genetics. *Conservation Genetics*, **17**, 1–17.

Barnes MA, Turner CR, Jerde CL, Renshaw MA, Chadderton WL, Lodge DM (2014). Environmental conditions influence eDNA persistence in aquatic systems. *Environmental Science & Technology*, **48**, 1819–1827.

Barnosky AD, Matzke N, Tomiya S, et al. (2011). Has the Earth's sixth mass extinction already arrived? *Nature*, **471**, 51–57.

Bartlett MD, Harris JA, James IT, Ritz K (2006). Inefficiency of mustard extraction technique for assessing size and structure of earthworm communities in UK pasture. *Soil Biology & Biochemistry*, **38**, 2990–2992.

Bellemain E, Carlsen T, Brochmann C, Coissac E, Taberlet P, Kauserud H (2010). ITS as an environmental DNA barcode for fungi: an *in silico* approach reveals potential PCR biases. *BMC Microbiology*, **10**, 189.

Bellemain E, Davey ML, Kauserud H, et al. (2013). Fungal palaeodiversity revealed using high-throughput metabarcoding of ancient DNA from arctic permafrost. *Environmental Microbiology*, **15**, 1176–1189.

Berger SA, Krompass D, Stamatakis A (2011). Performance, accuracy, and web server for evolutionary placement of short sequence reads under maximum likelihood. *Systematic Biology*, **60**, 291–302.

Bergmann GT, Craine JM, Robeson MS 2nd, Fierer N (2015). Seasonal shifts in diet and gut microbiota of the American bison (*Bison bison*). *PLoS One*, **10**, e0142409.

Besnard G, Christin P-A, Malé P-JG, et al. (2014). From museums to genomics: old herbarium specimens shed light on a C3 to C4 transition. *Journal of Experimental Botany*, **65**, 6711–6721.

Best K, Oakes T, Heather JM, Shawe-Taylor J, Chain B (2015). Computational analysis of stochastic heterogeneity in PCR amplification efficiency revealed by single molecule barcoding. *Scientific Reports*, **5**, 14629.

Bienert R, de Danieli S, Miquel C, et al. (2012). Tracking earthworm communities from soil DNA. *Molecular Ecology*, **21**, 2017–2030.

Bienhold C, Boetius A, Ramette A (2012). The energy-diversity relationship of complex bacterial communities in Arctic deep-sea sediments. *ISME Journal*, **6**, 724–732.

Bienhold C, Zinger L, Boetius A, Ramette A (2016). Diversity and biogeography of bathyal and abyssal seafloor bacteria. *PLoS One*, **11**, e0148016.

Biggs J, Ewald N, Valentini A, et al. (2015). Using eDNA to develop a national citizen science-based monitoring programme for the great crested newt (*Triturus cristatus*). *Biological Conservation*, **183**, 19–28.

Bik EM, Costello EK, Switzer AD, et al. (2016). Marine mammals harbor unique microbiotas shaped by and yet distinct from the sea. *Nature Communications*, **7**, 10516.

Bilofsky HS, Burks C, Fickett JW, et al. (1986). The GenBank genetic sequence databank. *Nucleic Acids Research*, **14**, 1–4.

Binladen J, Gilbert MTP, Bollback JP, et al. (2007). The use of coded PCR primers enables high-throughput sequencing of multiple homolog amplification products by 454 parallel sequencing. *PLoS One*, **2**, e197.

Birer C, Tysklind N, Zinger L, Duplais C (2017). Comparative analysis of DNA extraction methods to study the body surface microbiota of insects: a case study with ant cuticular bacteria. *Molecular Ecology Resources*, **17**, in press.

Bista I, Carvalho GR, Walsh K, et al. (2017). Annual time-series analysis of aqueous eDNA reveals ecologically relevant dynamics of lake ecosystem biodiversity. *Nature Communications*, **8**, 14087.

Bithell SL, Tran-Nguyen LTT, Hearnden MN, Hartley DM (2014). DNA analysis of soil extracts can be used to investigate fine root depth distribution of trees. *AoB Plants*, **7**, plu091.

Bjugstad AJ, Crawford HS, Neal DL (1970). Determining forage consumption by direct observation of domestic animals. In: *Range and wildlife habitat evaluation*, pp. 101–104. US Department of Agriculture, Washington, DC.

Blaalid R, Carlsen T, Kumar S, et al. (2012). Changes in the root-associated fungal communities along a primary succession gradient analysed by 454 pyrosequencing. *Molecular Ecology*, **21**, 1897–1908.

Blackman RC, Constable D, Hahn C, et al. (2017). Detection of a new non-native freshwater species by DNA metabarcoding of environmental samples—first record of *Gammarus fossarum* in the UK. *Aquatic Invasions*, **17**, in press.

Blanchart E, Albrecht A, Alegre J, et al. (1999). Effects of earthworms on soil structure and physical properties. In: *Earthworm management in tropical agroecosystems* (eds. Lavelle P, Brussaard L, Hendrix P), pp. 149–171. CAB International, Wallingford, UK.

Blankenship LE, Yayanos AA (2005). Universal primers and PCR of gut contents to study marine invertebrate diets. *Molecular Ecology*, **14**, 891–899.

Blazewicz SJ, Barnard RL, Daly RA, Firestone MK (2013). Evaluating rRNA as an indicator of microbial activity in environmental communities: limitations and uses. *ISME Journal*, **7**, 2061–2068.

Boere AC, Rijpstra WIC, De Lange GJ, Damste JSS, Coolen MJL (2011). Preservation potential of ancient plankton DNA in Pleistocene marine sediments. *Geobiology*, **9**, 377–393.

Boessenkool S, Epp LS, Haile J, et al. (2012). Blocking human contaminant DNA during PCR allows amplification of rare mammal species from sedimentary ancient DNA. *Molecular Ecology*, **21**, 1806–1815.

Boessenkool S, McGlynn G, Epp LS, et al. (2014). Use of ancient sedimentary DNA as a novel conservation tool for high-altitude tropical biodiversity. *Conservation Biology*, **28**, 446–455.

Bohmann K, Evans A, Gilbert MTP, et al. (2014). Environmental DNA for wildlife biology and biodiversity monitoring. *Trends in Ecology & Evolution*, **29**, 358–367.

Bokulich NA, Subramanian S, Faith JJ, et al. (2013). Quality-filtering vastly improves diversity estimates from Illumina amplicon sequencing. *Nature Methods*, **10**, 57–59.

Bonen L (1998). Mitochondrial genomes. In: *Advances in genome biology*, pp. 415–461. JAI Press Inc, Greenwich, CT.

Bonen L, Brown GG (1993). Genetic plasticity and its consequences: perspectives on gene organization and expression in plant mitochondria. *Canadian Journal of Botany*, **71**, 645–660.

Boore JL (1999). Animal mitochondrial genomes. *Nucleic Acids Research*, **27**, 1767–1780.

Borcard D, Gillet F, Legendre P (2011). *Numerical ecology with R*. Springer Science & Business Media, Berlin, Germany.

Boreham PFL, Ohiagu CE (1978). The use of serology in evaluating invertebrate prey-predator relationships: a review. *Bulletin of Entomological Research*, **68**, 171.

Bourlat SJ, Borja A, Gilbert J, et al. (2013). Genomics in marine monitoring: new opportunities for assessing marine health status. *Marine Pollution Bulletin*, **74**, 19–31.

Bowers RM, Clum A, Tice H, et al. (2015). Impact of library preparation protocols and template quantity on the metagenomic reconstruction of a mock microbial community. *BMC Genomics*, **16**, 856.

Bowles E, Schulte PM, Tollit DJ, Deagle BE, Trites AW (2011). Proportion of prey consumed can be determined from faecal DNA using real-time PCR. *Molecular Ecology Resources*, **11**, 530–540.

Boyer F, Mercier C, Bonin A, Le Bras Y, Taberlet P, Coissac E (2016). obitools: a unix-inspired software package for DNA metabarcoding. *Molecular Ecology Resources*, **16**, 176–182.

Briggs AW, Stenzel U, Johnson PLF, et al. (2007). Patterns of damage in genomic DNA sequences from a Neandertal. *Proceedings of the National Academy of Sciences of the United States of America*, **104**, 14616–14621.

Briones MJI (2014). Soil fauna and soil functions: a jigsaw puzzle. *Frontiers of Environmental Science & Engineering in China*, **2**, 7.

Broderick NA, Lemaitre B (2012). Gut-associated microbes of *Drosophila melanogaster*. *Gut Microbes*, **3**, 307–321.

Brooks AW, Kohl KD, Brucker RM, van Opstal EJ, Bordenstein SR (2016). Phylosymbiosis: relationships and functional effects of microbial communities across host evolutionary history. *PLoS Biology*, **14**, e2000225.

Brown EA, Chain FJJ, Crease TJ, MacIsaac HJ, Cristescu ME (2015). Divergence thresholds and divergent biodiversity estimates: can metabarcoding reliably describe zooplankton communities? *Ecology and Evolution*, **5**, 2234–2251.

Brown MV, Lauro FM, DeMaere MZ, et al. (2012). Global biogeography of SAR11 marine bacteria. *Molecular Systems Biology*, **8**, 595.

Brum JR, Ignacio-Espinoza JC, Roux S, et al. (2015). Patterns and ecological drivers of ocean viral communities. *Science*, **348**, 1261498.

Buchan A, LeCleir GR, Gulvik CA, González JM (2014). Master recyclers: features and functions of bacteria associated with phytoplankton blooms. *Nature Reviews. Microbiology*, **12**, 686–698.

Buchfink B, Xie C, Huson DH (2015). Fast and sensitive protein alignment using DIAMOND. *Nature Methods*, **12**, 59–60.

Bulgarelli D, Garrido-Oter R, Münch PC, et al. (2015). Structure and function of the bacterial root microbiota in wild and domesticated barley. *Cell Host & Microbe*, **17**, 392–403.

Burke C, Kjelleberg S, Thomas T (2009). Selective extraction of bacterial DNA from the surfaces of macroalgae. *Applied and Environmental Microbiology*, **75**, 252–256.

Buttigieg PL, Ramette A (2014). A guide to statistical analysis in microbial ecology: a community-focused, living review of multivariate data analyses. *FEMS Microbiology Ecology*, **90**, 543–550.

Buée M, De Boer W, Martin F, van Overbeek L, Jurkevitch E (2009). The rhizosphere zoo: an overview of plant-associated communities of microorganisms, including phages, bacteria, archaea, and fungi, and of some of their structuring factors. *Plant and Soil*, **321**, 189–212.

Callahan BJ, McMurdie PJ, Rosen MJ, Han AW, Johnson AJA, Holmes SP (2016). DADA2: high-resolution sample inference from Illumina amplicon data. *Nature Methods*, **13**, 581–583.

Calvignac-Spencer S, Leendertz FH, Gilbert MTP, Schubert G (2013a). An invertebrate stomach's view on vertebrate ecology: certain invertebrates could be used as "vertebrate samplers" and deliver DNA-based information on many aspects of vertebrate ecology. *BioEssays*, **35**, 1004–1013.

Calvignac-Spencer S, Merkel K, Kutzner N, et al. (2013b). Carrion fly-derived DNA as a tool for comprehensive and cost-effective assessment of mammalian biodiversity. *Molecular Ecology*, **22**, 915–924.

Cao Y, Williams DD, Larsen DP (2002). Comparison of ecological communities: the problem of sample representativeness. *Ecological Monographs*, **72**, 41–56.

Capo E, Debroas D, Arnaud F, Domaizon I (2015). Is planktonic diversity well recorded in sedimentary DNA? Toward the reconstruction of past protistan diversity. *Microbial Ecology*, **70**, 865–875.

Caporaso JG, Kuczynski J, Stombaugh J, et al. (2010). QIIME allows analysis of high-throughput community sequencing data. *Nature Methods*, **7**, 335–336.

Caporaso JG, Lauber CL, Walters WA, et al. (2012). Ultra-high-throughput microbial community analysis on the Illumina HiSeq and MiSeq platforms. *ISME Journal*, **6**, 1621–1624.

Caporaso JG, Lauber CL, Walters WA, et al. (2011). Global patterns of 16S rRNA diversity at a depth of millions of sequences per sample. *Proceedings of the National Academy of Sciences of the United States of America*, **108**, 4516–4522.

Cardinale BJ, Duffy JE, Gonzalez A, et al. (2012). Biodiversity loss and its impact on humanity. *Nature*, **486**, 59–67.

Carew ME, Pettigrove VJ, Metzeling L, Hoffmann AA (2013). Environmental monitoring using next generation sequencing: rapid identification of macroinvertebrate bioindicator species. *Frontiers in Zoology*, **10**, 45.

Carini P, Marsden PJ, Leff JW, Morgan EE, Strickland MS, Fierer N (2016). Relic DNA is abundant in soil and obscures estimates of soil microbial diversity. *Nature Microbiology*, **2**, 16242.

CBoL Plant Working Group (2009). A DNA barcode for land plants. *Proceedings of the National Academy of Sciences of the United States of America*, **106**, 12794–12797.

Celio GJ, Padamsee M, Dentinger BTM, Bauer R, McLaughlin DJ (2006). Assembling the fungal tree of life: constructing the structural and biochemical database. *Mycologia*, **98**, 850–859.

Chagnon P-L, Bradley RL, Maherali H, Klironomos JN (2013). A trait-based framework to understand life history of mycorrhizal fungi. *Trends in Plant Science*, **18**, 484–491.

Chan KY (2001). An overview of some tillage impacts on earthworm population abundance and diversity—implications for functioning in soils. *Soil and Tillage Research*, **57**, 179–191.

Chao A, Chiu C-H, Jost L (2014). Unifying species diversity, phylogenetic diversity, functional diversity, and related similarity and differentiation measures through Hill numbers. *Annual Review of Ecology, Evolution, and Systematics*, **45**, 297–324.

Chao A, Jost L (2012). Coverage-based rarefaction and extrapolation: standardizing samples by completeness rather than size. *Ecology*, **93**, 2533–2547.

Chariton AA, Court LN, Hartley DM, Colloff MJ, Hardy CM (2010). Ecological assessment of estuarine sediments by pyrosequencing eukaryotic ribosomal DNA. *Frontiers in Ecology and the Environment*, **8**, 233–238.

Chariton AA, Stephenson S, Morgan MJ, et al. (2015). Metabarcoding of benthic eukaryote communities predicts the ecological condition of estuaries. *Environmental Pollution*, **203**, 165–174.

Chiu C-H, Chao A (2016). Estimating and comparing microbial diversity in the presence of sequencing errors. *PeerJ*, **4**, e1634.

Chow S, Suzuki S, Matsunaga T, Lavery S, Jeffs A, Takeyama H (2011). Investigation on natural diets of larval marine animals using peptide nucleic acid-directed polymerase chain reaction clamping. *Marine Biotechnology*, **13**, 305–313.

Chun J, Lee J-H, Jung Y, et al. (2007). EzTaxon: a web-based tool for the identification of prokaryotes based on 16S ribosomal RNA gene sequences. *International Journal of Systematic and Evolutionary Microbiology*, **57**, 2259–2261.

Civade R (2016). L'ADN environnemental, méthode moléculaire d'étude de la biodiversité aquatique—Une approche innovante. PhD Thesis. AgroParisTech, Paris, France.

Civade R, Dejean T, Valentini A, et al. (2016). Spatial representativeness of environmental DNA metabarcoding signal for fish biodiversity assessment in a natural freshwater system. *PLoS One*, **11**, e0157366.

Clarke LJ, Beard JM, Swadling KM, Deagle BE (2017). Effect of marker choice and thermal cycling protocol on zooplankton DNA metabarcoding studies. *Ecology and Evolution*, **7**, 873–883.

Clarke LJ, Soubrier J, Weyrich LS, Cooper A (2014). Environmental metabarcodes for insects: *in silico* PCR reveals potential for taxonomic bias. *Molecular Ecology Resources*, **14**, 1160–1170.

Clegg MT, Learn GH, Golenberg EM (1991). Molecular evolution of chloroplast DNA. In: *Evolution at the molecular level* (eds. Selander RK, Clark AG, Whittman TS), pp. 135–149. Sinauer, Sunderland, MA.

Clusa L, Ardura A, Gower F, et al. (2016). An easy phylogenetically informative method to trace the globally invasive *Potamopyrgus* mud snail from river's eDNA. *PLoS One*, **11**, e0162899.

Coghlan ML, White NE, Murray DC, et al. (2013). Metabarcoding avian diets at airports: implications for birdstrike hazard management planning. *Investigative Genetics*, **4**, 27.

Coissac E (2012). oligotag: a program for designing sets of tags for next-generation sequencing of multiplexed samples. In: *Data production and analysis in population genomics: methods and protocols* (eds. Pompanon F, Bonin A), pp. 13–31. Springer Science + Business Media, New York, NY.

Coissac E, Hollingsworth PM, Lavergne S, Taberlet P (2016). From barcodes to genomes: extending the concept of DNA barcoding. *Molecular Ecology*, **25**, 1423–1428.

Coissac E, Riaz T, Puillandre N (2012). Bioinformatic challenges for DNA metabarcoding of plants and animals. *Molecular Ecology*, **21**, 1834–1847.

Coja T, Zehetner K, Bruckner A, Watzinger A, Meyer E (2008). Efficacy and side effects of five sampling methods for soil earthworms (Annelida, Lumbricidae). *Ecotoxicology and Environmental Safety*, **71**, 552–565.

Cole JR, Wang Q, Fish JA, et al. (2014). Ribosomal Database Project: data and tools for high throughput rRNA analysis. *Nucleic Acids Research*, **42**, D633–D642.

Coolen MJL, Orsi WD, Balkema C, et al. (2013). Evolution of the plankton paleome in the Black Sea from the Deglacial to Anthropocene. *Proceedings of the National Academy of Sciences of the United States of America*, **110**, 8609–8614.

Cooper A (1994). DNA from museum specimens. In: *Ancient DNA* (eds. Herrmann B, Hummel S), pp. 149–165. Springer, New York, NY.

Cordier T, Robin C, Capdevielle X, Fabreguettes O, Desprez-Loustau M-L, Vacher C (2012). The composition of phyllosphere fungal assemblages of European beech (*Fagus sylvatica*) varies significantly along an elevation gradient. *New Phytologist*, **196**, 510–519.

Corinaldesi C, Barucca M, Luna GM, Dell'Anno A (2011). Preservation, origin and genetic imprint of extracellular DNA in permanently anoxic deep-sea sediments. *Molecular Ecology*, **20**, 642–654.

Corinaldesi C, Tangherlini M, Luna GM, Dell'anno A (2014). Extracellular DNA can preserve the genetic signatures of present and past viral infection events in deep hypersaline anoxic basins. *Proceedings. Biological Sciences*, **281**, 20133299.

Cortes RMV, Hughes SJ, Pereira VR, Varandas S da GP (2013). Tools for bioindicator assessment in rivers: the importance of spatial scale, land use patterns and biotic integration. *Ecological Indicators*, **34**, 460–477.

Costa TRD, Felisberto-Rodrigues C, Meir A, et al. (2015). Secretion systems in Gram-negative bacteria: structural and mechanistic insights. *Nature Reviews. Microbiology*, **13**, 343–359.

Cowart DA, Pinheiro M, Mouchel O, et al. (2015). Metabarcoding is powerful yet still blind: a comparative analysis of morphological and molecular surveys of seagrass communities. *PLoS One*, **10**, e0117562.

Creer S, Deiner K, Frey S, et al. (2016). The ecologist's field guide to sequence-based identification of biodiversity. *Methods in Ecology and Evolution*, **7**, 1008–1018.

Danovaro R, Molari M, Corinaldesi C, Dell'Anno A (2016). Macroecological drivers of archaea and bacteria in benthic deep-sea ecosystems. *Science Advances*, **2**, e1500961.

Deagle BE, Bax N, Hewitt CL, Patil JG (2003). Development and evaluation of a PCR-based test for detection of *Asterias* (Echinodermata: Asteroidea) larvae in Australian plankton samples from ballast water. *Marine and Freshwater Research*, **54**, 709–719.

Deagle BE, Chiaradia A, McInnes J, Jarman SN (2010). Pyrosequencing faecal DNA to determine diet of little penguins: is what goes in what comes out? *Conservation Genetics*, **11**, 2039–2048.

Deagle BE, Eveson JP, Jarman SN (2006). Quantification of damage in DNA recovered from highly degraded samples—a case study on DNA in faeces. *Frontiers in Zoology*, **3**, 11.

Deagle BE, Jarman SN, Coissac E, Pompanon F, Taberlet P (2014). DNA metabarcoding and the COI marker: not a perfect match. *Biology Letters*, **10**, 20140562.

Deagle BE, Kirkwood R, Jarman SN (2009). Analysis of Australian fur seal diet by pyrosequencing prey DNA in faeces. *Molecular Ecology*, **18**, 2022–2038.

Deagle BE, Thomas AC, Shaffer AK, Trites AW, Jarman SN (2013). Quantifying sequence proportions in a DNA-based diet study using Ion Torrent amplicon sequencing: which counts count? *Molecular Ecology Resources*, **13**, 620–633.

Deagle BE, Tollit DJ, Jarman SN, Hindell MA, Trites AW, Gales NJ (2005). Molecular scatology as a tool to study diet: analysis of prey DNA in scats from captive Steller sea lions. *Molecular Ecology*, **14**, 1831–1842.

De Barba M, Miquel C, Boyer F, Rioux D, Coissac E, Taberlet P (2014). DNA metabarcoding multiplexing for omnivorous diet analysis and validation of data accuracy. *Molecular Ecology Resources*, **14**, 306–323.

de Bruijn FJ (ed.) (2011a). *Handbook of molecular microbial ecology I: metagenomics and complementary approaches*. Wiley-Blackwell, Hoboken, NJ.

de Bruijn FJ (ed.) (2011b). *Handbook of molecular microbial ecology II: metagenomics in different habitats*. Wiley-Blackwell, Hoboken, NJ.

Decelle J, Romac S, Stern RF, et al. (2015). PhytoREF: a reference database of the plastidial 16S rRNA gene of photosynthetic eukaryotes with curated taxonomy. *Molecular Ecology Resources*, **15**, 1435–1445.

Deiner K, Altermatt F (2014). Transport distance of invertebrate environmental DNA in a natural river. *PLoS One*, **9**, e88786.

Deiner K, Fronhofer EA, Mächler E, Walser J-C, Altermatt F (2016). Environmental DNA reveals that rivers are conveyer belts of biodiversity information. *Nature Communications*, **7**, 12544.

Deiner K, Walser J-C, Mächler E, Altermatt F (2015). Choice of capture and extraction methods affect detection of freshwater biodiversity from environmental DNA. *Biological Conservation*, **183**, 53–63.

Deininger PL (1983). Random subcloning of sonicated DNA: application to shotgun DNA sequence analysis. *Analytical Biochemistry*, **129**, 216–223.

Dejean T, Valentini A, Duparc A, et al. (2011). Persistence of environmental DNA in freshwater ecosystems. *PLoS One*, **6**, e23398.

Dejean T, Valentini A, Miquel C, Taberlet P, Bellemain E, Miaud C (2012). Improved detection of an alien invasive species through environmental DNA barcoding: the example of the American bullfrog *Lithobates catesbeianus*. *Journal of Applied Ecology*, **49**, 953–959.

Del Fabbro C, Scalabrin S, Morgante M, Giorgi FM (2013). An extensive evaluation of read trimming effects on Illumina NGS data analysis. *PLoS One*, **8**, e85024.

Delcher AL, Harmon D, Kasif S, White O, Salzberg SL (1999). Improved microbial gene identification with GLIMMER. *Nucleic Acids Research*, **27**, 4636–4641.

Dell'Anno A, Carugati L, Corinaldesi C, Riccioni G, Danovaro R (2015). Unveiling the biodiversity of deep-sea nematodes through metabarcoding: are we ready to bypass the classical taxonomy? *PLoS One*, **10**, e0144928.

Dell'Anno A, Danovaro R (2005). Extracellular DNA plays a key role in deep-sea ecosystem functioning. *Science*, **309**, 2179.

Delmont TO, Robe P, Cecillon S, et al. (2011). Accessing the soil metagenome for studies of microbial diversity. *Applied and Environmental Microbiology*, **77**, 1315–1324.

DeLong EF (1992). Archaea in coastal marine environments. *Proceedings of the National Academy of Sciences of the United States of America*, **89**, 5685–5689.

DeSantis TZ, Hugenholtz P, Larsen N, et al. (2006). Greengenes, a chimera-checked 16S rRNA gene database and workbench compatible with ARB. *Applied and Environmental Microbiology*, **72**, 5069–5072.

Desrosiers C, Leflaive J, Eulin A, Ten-Hage L (2013). Bioindicators in marine waters: benthic diatoms as a tool to assess water quality from eutrophic to oligotrophic coastal ecosystems. *Ecological Indicators*, **32**, 25–34.

De Vargas C, Audic S, Henry N, et al. (2015). Eukaryotic plankton diversity in the sunlit ocean. *Science*, **348**, 1261605.

Díaz S, Cabido M (2001). Vive la différence: plant functional diversity matters to ecosystem processes. *Trends in Ecology & Evolution*, **16**, 646–655.

Díaz S, Kattge J, Cornelissen JHC, et al. (2016). The global spectrum of plant form and function. *Nature*, **529**, 167–171.

Dillehay TD, Ocampo C, Saavedra J, et al. (2015). New archaeological evidence for an early human presence at Monte Verde, Chile. *PLoS One*, **10**, e0141923.

Dillehay TD, Ramírez C, Pino M, Collins MB, Rossen J, Pino-Navarro JD (2008). Monte Verde: seaweed, food, medicine, and the peopling of South America. *Science*, **320**, 784–786.

Dineen SM, Aranda R 4th, Anders DL, Robertson JM (2010). An evaluation of commercial DNA extraction kits for the isolation of bacterial spore DNA from soil. *Journal of Applied Microbiology*, **109**, 1886–1896.

Dinsdale EA, Edwards RA, Hall D, et al. (2008). Functional metagenomic profiling of nine biomes. *Nature*, **452**, 629–632.

Dodsworth S (2015). Genome skimming for next-generation biodiversity analysis. *Trends in Plant Science*, **20**, 525–527.

Dodt M, Roehr JT, Ahmed R, Dieterich C (2012). FLEXBAR—flexible barcode and adapter processing for next-generation sequencing platforms. *Biology*, **1**, 895–905.

Doi H, Inui R, Akamatsu Y, et al. (2017). Environmental DNA analysis for estimating the abundance and biomass of stream fish. *Freshwater Biology*, **62**, 30–39.

Douglas AE (2015). Multiorganismal insects: diversity and function of resident microorganisms. *Annual Review of Entomology*, **60**, 17–34.

Douville M, Gagné F, Blaise C, André C (2007). Occurrence and persistence of *Bacillus thuringiensis* (Bt) and transgenic Bt corn cry1Ab gene from an aquatic environment. *Ecotoxicology and Environmental Safety*, **66**, 195–203.

Dove H, Mayes RW (1996). Plant wax components: a new approach to estimating intake and diet composition in herbivores. *Journal of Nutrition*, **126**, 13–26.

Dunker KJ, Sepulveda AJ, Massengill RL, et al. (2016). Potential of environmental DNA to evaluate northern pike (*Esox lucius*) eradication efforts: an experimental test and case study. *PLoS One*, **11**, e0162277.

Dunshea G (2009). DNA-based diet analysis for any predator. *PLoS One*, **4**, e5252.

Durbin EG, Casas MC, Rynearson TA (2012). Copepod feeding and digestion rates using prey DNA and qPCR. *Journal of Plankton Research*, **34**, 72–82.

Eckert KA, Kunkel TA (1990). High fidelity DNA synthesis by the *Thermus aquaticus* DNA polymerase. *Nucleic Acids Research*, **18**, 3739–3744.

Edgar RC (2013). UPARSE: highly accurate OTU sequences from microbial amplicon reads. *Nature Methods*, **10**, 996–998.

Edgar RC, Haas BJ, Clemente JC, Quince C, Knight R (2011). UCHIME improves sensitivity and speed of chimera detection. *Bioinformatics*, **27**, 2194–2200.

Edwards CA (ed.) (2004). *Earthworm ecology*. CRC Press, Boca Raton, FL.

Edwards CA, Bohlen PJ (1996). *Biology and ecology of earthworms*. Chapman & Hall, London, UK.

Edwards J, Johnson C, Santos-Medellín C, et al. (2015). Structure, variation, and assembly of the root-associated microbiomes of rice. *Proceedings of the National Academy of Sciences of the United States of America*, **112**, E911–E920.

Eichmiller JJ, Miller LM, Sorensen PW (2016). Optimizing techniques to capture and extract environmental DNA for detection and quantification of fish. *Molecular Ecology Resources*, **16**, 56–68.

Eitzinger B, Traugott M (2011). Which prey sustains cold-adapted invertebrate generalist predators in arable land? Examining prey choices by molecular gut-content analysis. *Journal of Applied Ecology*, **48**, 591–599.

Elbrecht V, Leese F (2015). Can DNA-based ecosystem assessments quantify species abundance? Testing primer bias and biomass—sequence relationships with an innovative metabarcoding protocol. *PLoS One*, **10**, e0130324.

Elbrecht V, Leese F (2017). Validation and development of COI metabarcoding primers for freshwater macroinvertebrate bioassessment. *Frontiers of Environmental Science*, **5**, 11.

Ellison SLR, English CA, Burns MJ, Keer JT (2006). Routes to improving the reliability of low level DNA analysis using real-time PCR. *BMC Biotechnology*, **6**, 33.

England LS, Vincent ML, Trevors JT, Holmes SB (2004). Extraction, detection and persistence of extracellular DNA in forest litter microcosms. *Molecular and Cellular Probes*, **18**, 313–319.

Enright AJ, Van Dongen S, Ouzounis CA (2002). An efficient algorithm for large-scale detection of protein families. *Nucleic Acids Research*, **30**, 1575–1584.

Epp LS, Boessenkool S, Bellemain EP, et al. (2012). New environmental metabarcodes for analysing soil DNA: potential for studying past and present ecosystems. *Molecular Ecology*, **21**, 1821–1833.

Epp LS, Gussarova G, Boessenkool S, et al. (2015). Lake sediment multi-taxon DNA from North Greenland records early post-glacial appearance of vascular plants and accurately tracks environmental changes. *Quaternary Science Reviews*, **117**, 152–163.

Escalas A, Bouvier T, Mouchet MA, et al. (2013). A unifying quantitative framework for exploring the multiple facets of microbial biodiversity across diverse scales. *Environmental Microbiology*, **15**, 2642–2657.

Esling P, Lejzerowicz F, Pawlowski J (2015). Accurate multiplexing and filtering for high-throughput amplicon-sequencing. *Nucleic Acids Research*, **43**, 2513–2524.

Ettema CH, Wardle DA (2002). Spatial soil ecology. *Trends in Ecology & Evolution*, **17**, 177–183.

European Council (2000). Directive 2000/60/EC of the European Parliament and of the Council of 23 October 2000 establishing a framework for Community action in the field of water policy.

Evans NT, Olds BP, Renshaw MA, et al. (2016). Quantification of mesocosm fish and amphibian species diversity via environmental DNA metabarcoding. *Molecular Ecology Resources*, **16**, 29–41.

Fahner NA, Shokralla S, Baird DJ, Hajibabaei M (2016). Large-scale monitoring of plants through environmental DNA metabarcoding of soil: recovery, resolution, and annotation of four DNA markers. *PLoS One*, **11**, e0157505.

Fakruddin M, Mannan KSB, Chowdhury A, et al. (2013). Nucleic acid amplification: alternative methods of polymerase chain reaction. *Journal of Pharmacy & Bioallied Sciences*, **5**, 245–252.

Ficetola GF, Coissac E, Zundel S, et al. (2010). An *in silico* approach for the evaluation of DNA barcodes. *BMC Genomics*, **11**, 434.

Ficetola GF, Miaud C, Pompanon F, Taberlet P (2008). Species detection using environmental DNA from water samples. *Biology Letters*, **4**, 423–425.

Ficetola GF, Pansu J, Bonin A, et al. (2015). Replication levels, false presences and the estimation of the presence/absence from eDNA metabarcoding data. *Molecular Ecology Resources*, **15**, 543–556.

Ficetola GF, Taberlet P, Coissac E (2016). How to limit false positives in environmental DNA and metabarcoding? *Molecular Ecology Resources*, **16**, 604–607.

Fierer N, Leff JW, Adams BJ, et al. (2012). Cross-biome metagenomic analyses of soil microbial communities and their functional attributes. *Proceedings of the National Academy of Sciences of the United States of America*, **109**, 21390–21395.

Fieseler L, Quaiser A, Schleper C, Hentschel U (2006). Analysis of the first genome fragment from the marine sponge-associated, novel candidate phylum Poribacteria by environmental genomics. *Environmental Microbiology*, **8**, 612–624.

Finlay BJ (2002). Global dispersal of free-living microbial eukaryote species. *Science*, **296**, 1061–1063.

Finn RD, Coggill P, Eberhardt RY, et al. (2016). The Pfam protein families database: towards a more sustainable future. *Nucleic Acids Research*, **44**, D279–285.

Fliegerova K, Tapio I, Bonin A, et al. (2014). Effect of DNA extraction and sample preservation method on rumen bacterial population. *Anaerobe*, **29**, 80–84.

Foley WJ, McIlwee A, Lawler I, Aragones L, Woolnough AP, Berding N (1998). Ecological applications of near infrared reflectance spectroscopy—a tool for rapid, cost-effective prediction of the composition of plant and animal tissues and aspects of animal performance. *Oecologia*, **116**, 293–305.

Folmer O, Black M, Hoeh W, Lutz R, Vrijenhoek R (1994). DNA primers for amplification of mitochondrial cytochrome c oxidase subunit I from diverse metazoan invertebrates. *Molecular Marine Biology and Biotechnology*, **3**, 294–299.

Fonseca VG, Carvalho GR, Sung W, et al. (2010). Second-generation environmental sequencing unmasks marine metazoan biodiversity. *Nature Communications*, **1**, 1–8.

Fonseca VG, Lallias D (2016). Metabarcoding marine sediments: preparation of amplicon libraries. In: *Marine genomics: methods and protocols*. (ed. Bourlat SJ), pp. 183–196. Springer, New York, NY.

Fonseca VG, Nichols B, Lallias D, et al. (2012). Sample richness and genetic diversity as drivers of chimera formation in nSSU metagenetic analyses. *Nucleic Acids Research*, **40**, e66.

Foote AD, Thomsen PF, Sveegaard S, et al. (2012). Investigating the potential use of environmental DNA (eDNA) for genetic monitoring of marine mammals. *PLoS One*, **7**, e41781.

Fox GE, Pechman KR, Woese CR (1977). Comparative cataloging of 16S ribosomal ribonucleic acid: molecular approach to procaryotic systematics. *International Journal of Systematic and Evolutionary Microbiology*, **27**, 44–57.

Fröhlich-Nowoisky J, Burrows SM, Xie Z, et al. (2012). Biogeography in the air: fungal diversity over land and oceans. *Biogeosciences*, **9**, 1125–1136.

Frossard A, Hammes F, Gessner MO (2016). Flow cytometric assessment of bacterial abundance in soils, sediments and sludge. *Frontiers in Microbiology*, **7**, 903.

Frostegård A, Courtois S, Ramisse V, et al. (1999). Quantification of bias related to the extraction of DNA directly from soils. *Applied and Environmental Microbiology*, **65**, 5409–5420.

Fuertes Aguilar J, Rosselló JA, Nieto Feliner G (1999). Nuclear ribosomal DNA (nrDNA) concerted evolution in natural and artificial hybrids of *Armeria* (Plumbaginaceae). *Molecular Ecology*, **8**, 1341–1346.

Fuhrman JA (2009). Microbial community structure and its functional implications. *Nature*, **459**, 193–199.

Fuhrman JA, Cram JA, Needham DM (2015). Marine microbial community dynamics and their ecological interpretation. *Nature Reviews. Microbiology*, **13**, 133–146.

Fuhrman JA, McCallum K, Davis AA (1992). Novel major archaebacterial group from marine plankton. *Nature*, **356**, 148–149.

Fujiwara A, Matsuhashi S, Doi H, Yamamoto S, Minamoto T (2016). Use of environmental DNA to survey the distribution of an invasive submerged plant in ponds. *Freshwater Science*, **35**, 748–754.

Galimberti A, Spinelli S, Bruno A, et al. (2016). Evaluating the efficacy of restoration plantings through DNA barcoding of frugivorous bird diets. *Conservation Biology*, **30**, 763–773.

Gansauge M-T, Gerber T, Glocke I, et al. (2017). Single-stranded DNA library preparation from highly degraded DNA using T4 DNA ligase. *Nucleic Acids Research*, **45**, e79.

Gansauge M-T, Meyer M (2013). Single-stranded DNA library preparation for the sequencing of ancient or damaged DNA. *Nature Protocols*, **8**, 737–748.

Gariepy TD, Lindsay R, Ogden N, Gregory TR (2012). Identifying the last supper: utility of the DNA barcode library for bloodmeal identification in ticks. *Molecular Ecology Resources*, **12**, 646–652.

Geisen S, Laros I, Vizcaíno A, Bonkowski M, de Groot GA (2015). Not all are free-living: high-throughput DNA metabarcoding reveals a diverse community of protists parasitizing soil metazoa. *Molecular Ecology*, **24**, 4556–4569.

Geml J, Pastor N, Fernandez L, et al. (2014). Large-scale fungal diversity assessment in the Andean Yungas forests reveals strong community turnover among forest types along an altitudinal gradient. *Molecular Ecology*, **23**, 2452–2472.

Gibbons SM, Caporaso JG, Pirrung M, Field D, Knight R, Gilbert JA (2013). Evidence for a persistent microbial seed bank throughout the global ocean. *Proceedings of the National Academy of Sciences of the United States of America*, **110**, 4651–4655.

Giguet-Covex C, Arnaud F, Poulenard J, et al. (2011). Changes in erosion patterns during the Holocene in a currently treeless subalpine catchment inferred from lake sediment geochemistry (Lake Anterne, 2063 m asl, NW French Alps): the role of climate and human activities. *Holocene*, **21**, 651–665.

Giguet-Covex C, Pansu J, Arnaud F, et al. (2014). Long livestock farming history and human landscape shaping revealed by lake sediment DNA. *Nature Communications*, **5**, 3211.

Gilbert JA, Jansson JK, Knight R (2014). The Earth microbiome project: successes and aspirations. *BMC Biology*, **12**, 69.

Gilbert MTP, Jenkins DL, Götherstrom A, et al. (2008). DNA from pre-Clovis human coprolites in Oregon, North America. *Science*, **320**, 786–789.

Gilbert JA, Meyer F, Antonopoulos D, et al. (2010). Meeting report: the terabase metagenomics workshop and the vision of an Earth microbiome project. *Standards in Genomic Sciences*, **3**, 243–248.

Giovannoni SJ, Britschgi TB, Moyer CL, Field KG (1990). Genetic diversity in Sargasso Sea bacterioplankton. *Nature*, **345**, 60–63.

Glenn TC (2011). Field guide to next-generation DNA sequencers. *Molecular Ecology Resources*, **11**, 759–769.

Goldberg CS, Pilliod DS, Arkle RS, Waits LP (2011). Molecular detection of vertebrates in stream water: a demonstration using Rocky Mountain tailed frogs and Idaho giant salamanders. *PLoS One*, **6**, e22746.

Goldberg CS, Sepulveda A, Ray A, Baumgardt J, Waits LP (2013). Environmental DNA as a new method for early detection of New Zealand mudsnails (*Potamopyrgus antipodarum*). *Freshwater Science*, **32**, 792–800.

Goldfarb KC, Karaoz U, Hanson CA, et al. (2011). Differential growth responses of soil bacterial taxa to carbon substrates of varying chemical recalcitrance. *Frontiers in Microbiology*, **2**, 94.

Good IJ (1953). The population frequencies of species and the estimation of population parameters. *Biometrika*, **40**, 237–264.

Gotelli NJ, Colwell RK (2011). Estimating species richness. In: *Biological diversity—frontiers in measurement and assessment* (eds. Magurran AE, McGill BJ), pp. 39–54. Oxford University Press, New York, NY.

Gravel D, Albouy C, Thuiller W (2016). The meaning of functional trait composition of food webs for ecosystem functioning. *Philosophical Transactions of the Royal Society of London. Series B, Biological Sciences*, **371**, 20150268.

Grealy A, Douglass K, Haile J, Bruwer C, Gough C, Bunce M (2016). Tropical ancient DNA from bulk archaeological fish bone reveals the subsistence practices of a historic coastal community in southwest Madagascar. *Journal of Archaeological Science*, **75**, 82–88.

Grealy AC, McDowell MC, Scofield P, et al. (2015). A critical evaluation of how ancient DNA bulk bone metabarcoding complements traditional morphological analysis of fossil assemblages. *Quaternary Science Reviews*, **128**, 37–47.

Green SJ, Minz D (2005). Suicide polymerase endonuclease restriction, a novel technique for enhancing PCR amplification of minor DNA templates. *Applied and Environmental Microbiology*, **71**, 4721–4727.

Green HC, Shanks OC, Sivaganesan M, Haugland RA, Field KG (2011). Differential decay of human faecal Bacteroides in marine and freshwater. *Environmental Microbiology*, **13**, 3235–3249.

Griffiths RI, Whiteley AS, O'Donnell AG, Bailey MJ (2000). Rapid method for coextraction of DNA and RNA from natural environments for analysis of ribosomal DNA-

and rRNA-based microbial community composition. *Applied and Environmental Microbiology*, **66**, 5488–5491.

Groussin M, Mazel F, Sanders JG, et al. (2017). Unraveling the processes shaping mammalian gut microbiomes over evolutionary time. *Nature Communications*, **8**, 14319.

Guardiola M, Uriz MJ, Taberlet P, Coissac E, Wangensteen OS, Turon X (2015). Deep-sea, deep-sequencing: metabarcoding extracellular DNA from sediments of marine canyons. *PLoS One*, **10**, e0139633.

Guardiola M, Wangensteen OS, Taberlet P, Coissac E, Uriz MJ, Turon X (2016). Spatio-temporal monitoring of deep-sea communities using metabarcoding of sediment DNA and RNA. *PeerJ*, **4**, e2807.

Guest MA, Nichols PD, Frusher SD, Hirst AJ (2008). Evidence of abalone (*Haliotis rubra*) diet from combined fatty acid and stable isotope analyses. *Marine Biology*, **153**, 579–588.

Guillou L, Bachar D, Audic S, et al. (2013). The Protist Ribosomal Reference database (PR2): a catalog of unicellular eukaryote Small Sub-Unit rRNA sequences with curated taxonomy. *Nucleic Acids Research*, **41**, D597–D604.

Gunnigle E, Frossard A, Ramond J-B, Guerrero L, Seely M, Cowan DA (2017). Diel-scale temporal dynamics recorded for bacterial groups in Namib Desert soil. *Scientific Reports*, **7**, 40189.

Gutiérrez-Cacciabue D, Cid AG, Rajal VB (2016). How long can culturable bacteria and total DNA persist in environmental waters? The role of sunlight and solid particles. *Science of the Total Environment*, **539**, 494–502.

Haas BJ, Gevers D, Earl AM, et al. (2011). Chimeric 16S rRNA sequence formation and detection in Sanger and 454-pyrosequenced PCR amplicons. *Genome Research*, **21**, 494–504.

Hacquard S (2016). Disentangling the factors shaping microbiota composition across the plant holobiont. *New Phytologist*, **209**, 454–457.

Haegeman B, Hamelin J, Moriarty J, Neal P, Dushoff J, Weitz JS (2013). Robust estimation of microbial diversity in theory and in practice. *ISME Journal*, **7**, 1092–1101.

Haile J, Froese DG, MacPhee RDE, et al. (2009). Ancient DNA reveals late survival of mammoth and horse in interior Alaska. *Proceedings of the National Academy of Sciences of the United States of America*, **106**, 22352–22357.

Haile J, Holdaway R, Oliver K, et al. (2007). Ancient DNA chronology within sediment deposits: are paleobiological reconstructions possible and is DNA leaching a factor? *Molecular Biology and Evolution*, **24**, 982–989.

Hajibabaei M (2012). The golden age of DNA metasystematics. *Trends in Genetics*, **28**, 535–537.

Hajibabaei M, Shokralla S, Zhou X, Singer GAC, Baird DJ (2011). Environmental barcoding: a next-generation sequencing approach for biomonitoring applications using river benthos. *PLoS One*, **6**, e17497.

Hamady M, Walker JJ, Harris JK, Gold NJ, Knight R (2008). Error-correcting barcoded primers for pyrosequencing hundreds of samples in multiplex. *Nature Methods*, **5**, 235–237.

Hamm GH, Cameron GN (1986). The EMBL data library. *Nucleic Acids Research*, **14**, 5–9.

Hammer TJ, Dickerson JC, Fierer N (2015). Evidence-based recommendations on storing and handling specimens for analyses of insect microbiota. *PeerJ*, **3**, e1190.

Handelsman J, Rondon MR, Brady SF, Clardy J, Goodman RM (1998). Molecular biological access to the chemistry of unknown soil microbes: a new frontier for natural products. *Chemistry & Biology*, **5**, R245–R249.

Hänfling B, Lawson Handley L, Read DS, et al. (2016). Environmental DNA metabarcoding of lake fish communities reflects long-term data from established survey methods. *Molecular Ecology*, **25**, 3101–3119.

Hanson CA, Allison SD, Bradford MA, Wallenstein MD, Treseder KK (2008). Fungal taxa target different carbon sources in forest soil. *Ecosystems*, **11**, 1157–1167.

Hardy CM, Krull ES, Hartley DM, Oliver RL (2010). Carbon source accounting for fish using combined DNA and stable isotope analyses in a regulated lowland river weir pool. *Molecular Ecology*, **19**, 197–212.

Hargreaves SK, Roberto AA, Hofmockel KS (2013). Reaction- and sample-specific inhibition affect standardization of qPCR assays of soil bacterial communities. *Soil Biology & Biochemistry*, **59**, 89–97.

Harper GL, Sheppard SK, Harwood JD, et al. (2006). Evaluation of temperature gradient gel electrophoresis for the analysis of prey DNA within the guts of invertebrate predators. *Bulletin of Entomological Research*, **96**, 295–304.

Harwood VJ, Staley C, Badgley BD, Borges K, Korajkic A (2014). Microbial source tracking markers for detection of fecal contamination in environmental waters: relationships between pathogens and human health outcomes. *FEMS Microbiology Reviews*, **38**, 1–40.

Hata H, Umezawa Y (2011). Food habits of the farmer damselfish *Stegastes nigricans* inferred by stomach content, stable isotope, and fatty acid composition analyses. *Ecological Research*, **26**, 809–818.

Hatzenbuhler C, Kelly JR, Martinson J, Okum S, Pilgrim E (2017). Sensitivity and accuracy of high-throughput metabarcoding methods for early detection of invasive fish species. *Scientific Reports*, **7**, 46393.

He X, Chen H, Shi W, Cui Y, Zhang X-X (2015). Persistence of mitochondrial DNA markers as fecal indicators in water environments. *Science of the Total Environment*, **533**, 383–390.

He S, Ivanova N, Kirton E, et al. (2013). Comparative metagenomic and metatranscriptomic analysis of hindgut paunch microbiota in wood- and dung-feeding higher termites. *PLoS One*, **8**, e61126.

Head SR, Komori HK, LaMere SA, et al. (2014). Library construction for next-generation sequencing: overviews and challenges. *BioTechniques*, **56**, 61–77.

Hebert PDN, Cywinska A, Ball SL, deWaard JR (2003a). Biological identifications through DNA barcodes. *Proceedings. Biological Sciences*, **270**, 313–321.

Hebert PDN, Ratnasingham S, deWaard JR (2003b). Barcoding animal life: cytochrome c oxidase subunit 1 divergences among closely related species. *Proceedings. Biological Sciences*, **270**, S96–S99.

Heid CA, Stevens J, Livak KJ, Williams PM (1996). Real time quantitative PCR. *Genome Research*, **6**, 986–994.

Heidelberg KB, Gilbert JA, Joint I (2010). Marine genomics: at the interface of marine microbial ecology and biodiscovery. *Microbial Biotechnology*, **3**, 531–543.

Higuchi R, Fockler C, Dollinger G, Watson R (1993). Kinetic PCR analysis: real-time monitoring of DNA amplification reactions. *Biotechnology*, **11**, 1026–1030.

Hiiesalu I, Öpik M, Metsis M, et al. (2012). Plant species richness belowground: higher richness and new patterns revealed by next-generation sequencing. *Molecular Ecology*, **21**, 2004–2016.

Hofreiter M, Jaenicke V, Serre D, von Haeseler A, Pääbo S (2001). DNA sequences from multiple amplifications reveal artifacts induced by cytosine deamination in ancient DNA. *Nucleic Acids Research*, **29**, 4793–4799.

Hofreiter M, Poinar HN, Spaulding WG, et al. (2000). A molecular analysis of ground sloth diet through the last glaciation. *Molecular Ecology*, **9**, 1975–1984.

Holdaway RJ, Wood JR, Dickie IA, et al. (2017). Using DNA metabarcoding to assess New Zealand's terrestrial biodiversity. *New Zealand Journal of Ecology*, **41**, in press.

Holechek JL, Vavra M, Pieper RD (1982). Botanical composition determination of range herbivore diets: a review. *Journal of Range Management*, **35**, 309–315.

Holland PM, Abramson RD, Watson R, Gelfand DH (1991). Detection of specific polymerase chain reaction product by utilizing the 5′-3′ exonuclease activity of *Thermus aquaticus* DNA polymerase. *Proceedings of the National Academy of Sciences of the United States of America*, **88**, 7276–7280.

Horton TR, Bruns TD (2001). The molecular revolution in ectomycorrhizal ecology: peeking into the black-box. *Molecular Ecology*, **10**, 1855–1871.

Höss M, Kohn M, Pääbo S, Knauer F, Schröder W (1992). Excrement analysis by PCR. *Nature*, **359**, 199.

Howe A, Chain PSG (2015). Challenges and opportunities in understanding microbial communities with metagenome assembly (accompanied by IPython notebook tutorial). *Frontiers in Microbiology*, **6**, 678.

Hugenholtz P, Goebel BM, Pace NR (1998). Impact of culture-independent studies on the emerging phylogenetic view of bacterial diversity. *Journal of Bacteriology*, **180**, 4765–4774.

Human Microbiome Project Consortium (2012). A framework for human microbiome research. *Nature*, **486**, 215–221.

Huse SM, Huber JA, Morrison HG, Sogin ML, Welch DM (2007). Accuracy and quality of massively-parallel DNA pyrosequencing. *Genome Biology*, **8**, R143.

Huse SM, Welch DM, Morrison HG, Sogin ML (2010). Ironing out the wrinkles in the rare biosphere through improved OTU clustering. *Environmental Microbiology*, **12**, 1889–1898.

Hwang C, Ling F, Andersen GL, LeChevallier MW, Liu W-T (2012). Evaluation of methods for the extraction of DNA from drinking water distribution system biofilms. *Microbes and Environments*, **27**, 9–18.

Ibanez S, Manneville O, Miquel C, et al. (2013). Plant functional traits reveal the relative contribution of habitat and food preferences to the diet of grasshoppers. *Oecologia*, **173**, 1459–1470.

Ishige T, Miya M, Ushio M, et al. (2017). Tropical-forest mammals as detected by environmental DNA at natural saltlicks in Borneo. *Biological Conservation*, **210**, 281–285.

Jain M, Fiddes IT, Miga KH, Olsen HE, Paten B, Akeson M (2015). Improved data analysis for the MinION nanopore sequencer. *Nature Methods*, **12**, 351–356.

Jain M, Olsen HE, Paten B, Akeson M (2016). The Oxford Nanopore MinION: delivery of nanopore sequencing to the genomics community. *Genome Biology*, **17**, 239.

Jane SF, Wilcox TM, McKelvey KS, et al. (2015). Distance, flow and PCR inhibition: eDNA dynamics in two headwater streams. *Molecular Ecology Resources*, **15**, 216–227.

Janssen P, Bec S, Fuhr M, Taberlet P, Brun J-J, Bouget C (2017). Present conditions may mediate the legacy effect of past land-use changes on species richness and composition of above- and below-ground assemblages. *Journal of Ecology*, **105**, in press.

Jarman SN, Gales NJ, Tierney M, Gill PC, Elliott NG (2002). A DNA-based method for identification of krill species and its application to analysing the diet of marine vertebrate predators. *Molecular Ecology*, **11**, 2679–2690.

Jerde CL, Mahon AR, Chadderton WL, Lodge DM (2011). "Sight-unseen" detection of rare aquatic species using environmental DNA. *Conservation Letters*, **4**, 150–157.

Ji YQ, Ashton L, Pedley SM, et al. (2013). Reliable, verifiable and efficient monitoring of biodiversity via metabarcoding. *Ecology Letters*, **16**, 1245–1257.

Jiang H, Lei R, Ding S-W, Zhu S (2014). Skewer: a fast and accurate adapter trimmer for next-generation sequencing paired-end reads. *BMC Bioinformatics*, **15**, 182.

Johnson M, Zaretskaya I, Raytselis Y, Merezhuk Y, McGinnis S, Madden TL (2008). NCBI BLAST: a better web interface. *Nucleic Acids Research*, **36**, W5–W9.

Jones RT, Sanchez LG, Fierer N (2013). A cross-taxon analysis of insect-associated bacterial diversity. *PLoS One*, **8**, e61218.

Jónsson H, Ginolhac A, Schubert M, Johnson PLF, Orlando L (2013). mapDamage2.0: fast approximate Bayesian estimates of ancient DNA damage parameters. *Bioinformatics*, **29**, 1682–1684.

Jørgensen T, Haile J, Möller P, et al. (2012). A comparative study of ancient sedimentary DNA, pollen and macrofossils from permafrost sediments of northern Siberia reveals long-term vegetational stability. *Molecular Ecology*, **21**, 1989–2003.

Jumpponen A, Jones KL (2009). Massively parallel 454 sequencing indicates hyperdiverse fungal communities in temperate *Quercus macrocarpa* phyllosphere. *New Phytologist*, **184**, 438–448.

Jurado-Rivera JA, Vogler AP, Reid CAM, Petitpierre E, Gomez-Zurita J (2009). DNA barcoding insect-host plant associations. *Proceedings. Biological Sciences*, **276**, 639–648.

Kamenova S, Mayer R, Coissac E, Plantegenest M, Traugott M (2017). Comparing three types of dietary samples for prey DNA decay in an insect generalist predator. *bioRxiv*, 098806.

Kanehisa M (2000). *Post-genome informatics*. Oxford University Press, Oxford, UK.

Kaneko H, Lawler IR (2006). Can near infrared spectroscopy be used to improve assessment of marine mammal diets via fecal analysis? *Marine Mammal Science*, **22**, 261–275.

Karsenti E, Acinas SG, Bork P, et al. (2011). A holistic approach to marine eco-systems biology. *PLoS Biology*, **9**, e1001177.

Kartzinel TR, Chen PA, Coverdale TC, et al. (2015). DNA metabarcoding illuminates dietary niche partitioning by African large herbivores. *Proceedings of the National Academy of Sciences of the United States of America*, **112**, 8019–8024.

Katoch R, Thakur N (2012). Insect gut nucleases: a challenge for RNA interference mediated insect control strategies. *International Journal of Biotechnology and Biochemistry*, **1**, 198–203.

Kattge J, Díaz S, Lavorel S, et al. (2011). TRY—a global database of plant traits. *Global Change Biology*, **17**, 2905–2935.

Keck F, Bouchez A, Franc A, Rimet F (2016). Linking phylogenetic similarity and pollution sensitivity to develop ecological assessment methods: a test with river diatoms. *Journal of Applied Ecology*, **53**, 856–864.

Kelly RP, Port JA, Yamahara KM, Crowder LB (2014a). Using environmental DNA to census marine fishes in a large mesocosm. *PLoS One*, **9**, e86175.

Kelly RP, Port JA, Yamahara KM, et al. (2014b). Harnessing DNA to improve environmental management. *Science*, **344**, 1455–1456.

Kembel SW, O'Connor TK, Arnold HK, Hubbell SP, Wright SJ, Green JL (2014). Relationships between phyllosphere bacterial communities and plant functional traits in a neotropical forest. *Proceedings of the National Academy of Sciences of the United States of America*, **111**, 13715–13720.

Kembel SW, Wu M, Eisen JA, Green JL (2012). Incorporating 16S gene copy number information improves estimates of microbial diversity and abundance. *PLoS Computational Biology*, **8**, e1002743.

Kent RJ (2009). Molecular methods for arthropod blood-meal identification and applications to ecological and vector-borne disease studies. *Molecular Ecology Resources*, **9**, 4–18.

Kim H, Jebrail MJ, Sinha A, et al. (2013a). A microfluidic DNA library preparation platform for next-generation sequencing. *PLoS One*, **8**, e68988.

Kim J, Lim J, Lee C (2013b). Quantitative real-time PCR approaches for microbial community studies in waste-water treatment systems: applications and considerations. *Biotechnology Advances*, **31**, 1358–1373.

King RA, Vaughan IP, Bell JR, Bohan DA, Symondson WOC (2010). Prey choice by carabid beetles feeding on an earthworm community analysed using species- and lineage-specific PCR primers. *Molecular Ecology*, **19**, 1721–1732.

Kisand V, Valente A, Lahm A, Tanet G, Lettieri T (2012). Phylogenetic and functional metagenomic profiling for assessing microbial biodiversity in environmental monitoring. *PLoS One*, **7**, e43630.

Kitson JJN, Warren BH, Florens FBV, Baider C, Strasberg D, Emerson BC (2013). Molecular characterization of trophic ecology within an island radiation of insect herbivores (Curculionidae: Entiminae: *Cratopus*). *Molecular Ecology*, **22**, 5441–5455.

Kocher TD, Thomas WK, Meyer A, Edwards SV, Pääbo S, Villablanca FX, Wilson AC (1989). Dynamics of mitochondrial DNA evolution in animals: amplification and sequencing with conserved primers. *Proceedings of the National Academy of Sciences of the United States of America*, **86**, 6196–6200.

Kohl KD (2012). Diversity and function of the avian gut microbiota. *Journal of Comparative Physiology. B, Biochemical, Systemic, and Environmental Physiology*, **182**, 591–602.

Kohl KD, Brun A, Magallanes M, et al. (2017). Gut microbial ecology of lizards: insights into diversity in the wild, effects of captivity, variation across gut regions and transmission. *Molecular Ecology*, **26**, 1175–1189.

Kõljalg U, Larsson K-H, Abarenkov K, et al. (2005). UNITE: a database providing web-based methods for the molecular identification of ectomycorrhizal fungi. *New Phytologist*, **166**, 1063–1068.

Kõljalg U, Nilsson RH, Abarenkov K, et al. (2013). Towards a unified paradigm for sequence-based identification of fungi. *Molecular Ecology*, **22**, 5271–5277.

Kolman CJ, Tuross N (2000). Ancient DNA analysis of human populations. *American Journal of Physical Anthropology,* **111**, 5–23.

Kopylova E, Navas-Molina JA, Mercier C, et al. (2016). Open-source sequence clustering methods improve the state of the art. *mSystems,* **1**, e00003–15.

Kopylova E, Noé L, Touzet H (2012). SortMeRNA: fast and accurate filtering of ribosomal RNAs in metatranscriptomic data. *Bioinformatics,* **28**, 3211–3217.

Kowalczyk R, Taberlet P, Coissac E, et al. (2011). Influence of management practices on large herbivore diet—case of European bison in Białowieża Primeval Forest (Poland). *Forest Ecology and Management,* **261**, 821–828.

Kraaijeveld K, de Weger LA, Ventayol García M, et al. (2015). Efficient and sensitive identification and quantification of airborne pollen using next-generation DNA sequencing. *Molecular Ecology Resources,* **15**, 8–16.

Kunin V, Engelbrektson A, Ochman H, Hugenholtz P (2010). Wrinkles in the rare biosphere: pyrosequencing errors can lead to artificial inflation of diversity estimates. *Environmental Microbiology,* **12**, 118–123.

Kvist S (2013). Barcoding in the dark? A critical view of the sufficiency of zoological DNA barcoding databases and a plea for broader integration of taxonomic knowledge. *Molecular Phylogenetics and Evolution,* **69**, 39–45.

Kwok S, Higuchi R (1989). Avoiding false positives with PCR. *Nature,* **339**, 237–238.

Kwok S, Kellogg DE, McKinney N, et al. (1990). Effects of primer-template mismatches on the polymerase chain reaction: human immunodeficiency virus type 1 model studies. *Nucleic Acids Research,* **18**, 999–1005.

Lacoursière-Roussel A, Côté G, Leclerc V, Bernatchez L (2016a). Quantifying relative fish abundance with eDNA: a promising tool for fisheries management. *Journal of Applied Ecology,* **53**, 1148–1157.

Lacoursière-Roussel A, Rosabal M, Bernatchez L (2016b). Estimating fish abundance and biomass from eDNA concentrations: variability among capture methods and environmental conditions. *Molecular Ecology Resources,* **16**, 1401–1414.

Lahoz-Monfort JJ, Guillera-Arroita G, Tingley R (2016). Statistical approaches to account for false-positive errors in environmental DNA samples. *Molecular Ecology Resources,* **16**, 673–685.

Langille MGI, Zaneveld J, Caporaso JG, et al. (2013). Predictive functional profiling of microbial communities using 16S rRNA marker gene sequences. *Nature Biotechnology,* **31**, 814–821.

Lapage SP, Sneath PHA, Lessel EF, Skerman VBD, Seeliger HPR, Clark WA (eds.) (1992). *International code of nomenclature of bacteria: bacteriological code, 1990 revision.* ASM Press, Washington, DC.

Laramie MB, Pilliod DS, Goldberg CS (2015). Characterizing the distribution of an endangered salmonid using environmental DNA analysis. *Biological Conservation,* **183**, 29–37.

Larsen PE, Field D, Gilbert JA (2012). Predicting bacterial community assemblages using an artificial neural network approach. *Nature Methods,* **9**, 621–625.

Lavelle P, Bignell D, Lepage M, et al. (1997). Soil function in a changing world: the role of invertebrate ecosystem engineers. *European Journal of Soil Biology,* **33**, 159–193.

Lawrence AP, Bowers MA (2002). A test of the "hot" mustard extraction method of sampling earthworms. *Soil Biology & Biochemistry,* **34**, 549–552.

Lawson Handley L (2015). How will the "molecular revolution" contribute to biological recording? *Biological Journal of the Linnean Society,* **115**, 750–766.

Lee OO, Yang J, Bougouffa S, et al. (2012). Spatial and species variations in bacterial communities associated with corals from the Red Sea as revealed by pyrosequencing. *Applied and Environmental Microbiology,* **78**, 7173–7184.

Lejzerowicz F, Esling P, Pillet L, Wilding TA, Black KD, Pawlowski J (2015). High-throughput sequencing and morphology perform equally well for benthic monitoring of marine ecosystems. *Scientific Reports,* **5**, 13932.

Lennon JT, Jones SE (2011). Microbial seed banks: the ecological and evolutionary implications of dormancy. *Nature Reviews. Microbiology,* **9**, 119–130.

Lennon JT, Placella SA, Muscarella ME (2017). Relic DNA contributes minimally to estimates of microbial diversity. *bioRxiv,* 131284.

Leonard JA, Shanks O, Hofreiter M, et al. (2007). Animal DNA in PCR reagents plagues ancient DNA research. *Journal of Archaeological Science,* **34**, 1361–1366.

Leray M, Knowlton N (2015). DNA barcoding and metabarcoding of standardized samples reveal patterns of marine benthic diversity. *Proceedings of the National Academy of Sciences of the United States of America,* **112**, 2076–2081.

Leray M, Yang JY, Meyer CP, et al. (2013). A new versatile primer set targeting a short fragment of the mitochondrial COI region for metabarcoding metazoan diversity: application for characterizing coral reef fish gut contents. *Frontiers in Zoology,* **10**, 34.

Levin LA (2006). Recent progress in understanding larval dispersal: new directions and digressions. *Integrative and Comparative Biology,* **46**, 282–297.

Levy-Booth DJ, Campbell RG, Gulden RH, et al. (2007). Cycling of extracellular DNA in the soil environment. *Soil Biology & Biochemistry,* **39**, 2977–2991.

Ley RE, Hamady M, Lozupone C, et al. (2008a). Evolution of mammals and their gut microbes. *Science,* **320**, 1647–1651.

Ley RE, Lozupone CA, Hamady M, Knight R, Gordon JI (2008b). Worlds within worlds: evolution of the vertebrate gut microbiota. *Nature Reviews. Microbiology*, **6**, 776–788.

Lim NKM, Tay YC, Srivathsan A, et al. (2016). Next-generation freshwater bioassessment: eDNA metabarcoding with a conserved metazoan primer reveals species-rich and reservoir-specific communities. *Royal Society Open Science*, **3**, 160635.

Linard B, Crampton-Platt A, Gillett CPDT, Timmermans MJTN, Vogler AP (2015). Metagenome skimming of insect specimen pools: potential for comparative genomics. *Genome Biology and Evolution*, **7**, 1474–1489.

Liu S, Wang X, Xie L, et al. (2016). Mitochondrial capture enriches mito-DNA 100 fold, enabling PCR-free mitogenomics biodiversity analysis. *Molecular Ecology Resources*, **16**, 470–479.

Liu B, Yuan J, Yiu S-M, et al. (2012). COPE: an accurate k-mer-based pair-end reads connection tool to facilitate genome assembly. *Bioinformatics*, **28**, 2870–2874.

Lobo EA, Heinrich CG, Schuch M, Wetzel CE, Ector L (2016). Diatoms as bioindicators in rivers. In: *River algae* (ed. Necchi O Jr), pp. 245–271. Springer International Publishing, Switzerland.

Logan JMJ, Edwards KJ, Saunders NA (2009). *Real-time PCR: current technology and applications*. Horizon Scientific Press, Norwich, UK.

Logares R, Sunagawa S, Salazar G, et al. (2014). Metagenomic 16S rDNA Illumina tags are a powerful alternative to amplicon sequencing to explore diversity and structure of microbial communities. *Environmental Microbiology*, **16**, 2659–2671.

Longmire JL, Maltbie M, Baker RJ (1997). *Use of "lysis buffer" in DNA isolation and its implication for museum collections*. Museum of Texas Tech University, Lubbock, TX.

Lopes CM, De Barba M, Boyer F, et al. (2015). DNA metabarcoding diet analysis for species with parapatric vs sympatric distribution: a case study on subterranean rodents. *Heredity*, **114**, 525–536.

Lowe S, Browne M, Boudjelas S, De Poorter M (2000). *100 of the world's worst invasive alien species: a selection from the global invasive species database*. IUCN, Gland, Switzerland.

Lozupone CA, Stombaugh JI, Gordon JI, Jansson JK, Knight R (2012). Diversity, stability and resilience of the human gut microbiota. *Nature*, **489**, 220–230.

Ludwig W, Strunk O, Westram R, et al. (2004). ARB: a software environment for sequence data. *Nucleic Acids Research*, **32**, 1363–1371.

Lukashin AV, Borodovsky M (1998). GeneMark.hmm: new solutions for gene finding. *Nucleic Acids Research*, **26**, 1107–1115.

Luna GM, Manini E, Danovaro R (2002). Large fraction of dead and inactive bacteria in coastal marine sediments: comparison of protocols for determination and ecological significance. *Applied and Environmental Microbiology*, **68**, 3509–3513.

Mächler E, Deiner K, Steinmann P, Altermatt F (2014). Utility of environmental DNA for monitoring rare and indicator macroinvertebrate species. *Freshwater Science*, **33**, 1174–1183.

MacKenzie DI, Nichols JD, Royle JA, Pollock KH, Bailey LL, Hines JE (2006). *Occupancy estimation and modeling: inferring patterns and dynamics of species occurrence*. Academic Press, Burlington, MA.

Magoč T, Salzberg SL (2011). FLASH: fast length adjustment of short reads to improve genome assemblies. *Bioinformatics*, **27**, 2957–2963.

Magurran AE, McGill BJ (eds.) (2011). *Biological diversity—frontiers in measurement and assessment*. Oxford University Press, New York, NY.

Mahon AR, Jerde CL, Chadderton WL, Lodge DM (2011). Using environmental DNA to elucidate the Asian carp (genus *Hypophthalmichthys*) invasion front in the Chicago Area Waterway System. *Integrative and Comparative Biology*, **51**, E85–E85.

Mahon AR, Jerde CL, Galaska M, et al. (2013). Validation of eDNA surveillance sensitivity for detection of Asian carps in controlled and field experiments. *PLoS One*, **8**, e58316.

Mahé F, Mayor J, Bunge J, et al. (2015). Comparing high-throughput platforms for sequencing the V4 region of SSU-rDNA in environmental microbial eukaryotic diversity surveys. *Journal of Eukaryotic Microbiology*, **62**, 338–345.

Mahé F, Rognes T, Quince C, De Vargas C, Dunthorn M (2014). Swarm: robust and fast clustering method for amplicon-based studies. *PeerJ*, **2**, e593.

Malmström H, Storå J, Dalén L, Holmlund G, Götherström A (2005). Extensive human DNA contamination in extracts from ancient dog bones and teeth. *Molecular Biology and Evolution*, **22**, 2040–2047.

Manel S, Albert CH, Yoccoz NG (2012). Sampling in Landscape Genomics. In: *Data production and analysis in population genomics*. (eds. Pompanon F, Bonin A), pp. 3–12. Humana Press, New York, NY.

Margulies M, Egholm M, Altman WE, et al. (2005). Genome sequencing in microfabricated high-density picolitre reactors. *Nature*, **437**, 376–380.

Marston MF, Taylor S, Sme N, Parsons RJ, Noyes TJE, Martiny JBH (2013). Marine cyanophages exhibit local and regional biogeography. *Environmental Microbiology*, **15**, 1452–1463.

Martellini A, Payment P, Villemur R (2005). Use of eukaryotic mitochondrial DNA to differentiate human, bovine, porcine and ovine sources in fecally contaminated surface water. *Water Research*, **39**, 541–548.

Martin-Laurent F, Philippot L, Hallet S, et al. (2001). DNA extraction from soils: old bias for new microbial diversity analysis methods. *Applied and Environmental Microbiology*, **67**, 2354–2359.

Martinson VG, Douglas AE, Jaenike J (2017). Community structure of the gut microbiota in sympatric species of wild *Drosophila*. *Ecology Letters*, **20**, 629–639.

Martiny JBH, Jones SE, Lennon JT, Martiny AC (2015). Microbiomes in light of traits: a phylogenetic perspective. *Science*, **350**, aac9323.

Masella AP, Bartram AK, Truszkowski JM, Brown DG, Neufeld JD (2012). PANDAseq: paired-end assembler for illumina sequences. *BMC Bioinformatics*, **13**, 31.

Matsen FA, Kodner RB, Armbrust EV (2010). pplacer: linear time maximum-likelihood and Bayesian phylogenetic placement of sequences onto a fixed reference tree. *BMC Bioinformatics*, **11**, 538.

Matsuhashi S, Doi H, Fujiwara A, Watanabe S, Minamoto T (2016). Evaluation of the environmental DNA method for estimating distribution and biomass of submerged aquatic plants. *PLoS One*, **11**, e0156217.

Mattick JS (2004). RNA regulation: a new genetics? *Nature Reviews. Genetics*, **5**, 316–323.

McCarthy A, Chiang E, Schmidt ML, Denef VJ (2015). RNA preservation agents and nucleic acid extraction method bias perceived bacterial community composition. *PLoS One*, **10**, e0121659.

McDonald D, Clemente JC, Kuczynski J, et al. (2012). The Biological Observation Matrix (BIOM) format or: how I learned to stop worrying and love the ome-ome. *GigaScience*, **1**, 7.

McGuire KL, Bent E, Borneman J, Majumder A, Allison SD, Tresederi KK (2010). Functional diversity in resource use by fungi. *Ecology*, **91**, 2324–2332.

McHardy AC, Martín HG, Tsirigos A, Hugenholtz P, Rigoutsos I (2007). Accurate phylogenetic classification of variable-length DNA fragments. *Nature Methods*, **4**, 63–72.

McInnes JC, Emmerson L, Southwell C, Faux C, Jarman SN (2016). Simultaneous DNA-based diet analysis of breeding, non-breeding and chick Adélie penguins. *Royal Society Open Science*, **3**, 150443.

McInnis ML, Vavra M, Krueger WC (1983). A comparison of four methods used to determine the diets of large herbivores. *Journal of Range Management*, **36**, 302–306.

McMurdie PJ, Holmes S (2014). Waste not, want not: why rarefying microbiome data is inadmissible. *PLoS Computational Biology*, **10**, e1003531.

Mercier C, Boyer F, Bonin A, Coissac E (2013). *SUMATRA and SUMACLUST: fast and exact comparison and clustering of sequences*. https://git.metabarcoding.org/obitools/sumatra/wikis/home. https://git.metabarcoding.org/obitools/sumaclust/wikis/home.

Metzker ML (2010). Sequencing technologies—the next generation. *Nature Reviews. Genetics*, **11**, 31–46.

Meyer CP (2003). Molecular systematics of cowries (Gastropoda: Cypraeidae) and diversification patterns in the tropics. *Biological Journal of the Linnean Society*, **79**, 401–459.

Meyer F, Paarmann D, D'Souza M, et al. (2008). The metagenomics RAST server—a public resource for the automatic phylogenetic and functional analysis of metagenomes. *BMC Bioinformatics*, **9**, 386.

Mitchell A, Bucchini F, Cochrane G, et al. (2016). EBI metagenomics in 2016—an expanding and evolving resource for the analysis and archiving of metagenomic data. *Nucleic Acids Research*, **44**, D595–603.

Miya M, Sato Y, Fukunaga T, et al. (2015). MiFish, a set of universal PCR primers for metabarcoding environmental DNA from fishes: detection of more than 230 subtropical marine species. *Royal Society Open Science*, **2**, 150088.

Moens T, Vincx M (1997). Observations on the feeding ecology of estuarine nematodes. *Journal of the Marine Biological Association of the United Kingdom*, **77**, 211–227.

Mokany K, Ferrier S (2011). Predicting impacts of climate change on biodiversity: a role for semi-mechanistic community-level modelling. *Diversity and Distributions*, **17**, 374–380.

Monier A, Comte J, Babin M, Forest A, Matsuoka A, Lovejoy C (2015). Oceanographic structure drives the assembly processes of microbial eukaryotic communities. *ISME Journal*, **9**, 990–1002.

Moran NA, Sloan DB (2015). The hologenome concept: helpful or hollow? *PLoS Biology*, **13**, e1002311.

Moreby SJ (1988). An aid to the identification of arthropod fragments in the faeces of gamebird chicks (Galliformes). *Ibis*, **130**, 519–526.

Morgan MJ, Chariton AA, Hartley DM, Court LN, Hardy CM (2013). Improved inference of taxonomic richness from environmental DNA. *PLoS One*, **8**, e71974.

Mullis KB, Faloona FA (1987). Specific synthesis of DNA *in vitro* via a polymerase-catalyzed chain reaction. *Methods in enzymology*, **155**, 335–350.

Naeem S, Duffy JE, Zavaleta E (2012). The functions of biological diversity in an age of extinction. *Science*, **336**, 1401–1406.

Namiki T, Hachiya T, Tanaka H, Sakakibara Y (2012). MetaVelvet: an extension of Velvet assembler to *de novo* metagenome assembly from short sequence reads. *Nucleic Acids Research*, **40**, e155.

Nathan LR, Jerde CL, Budny ML, Mahon AR (2015). The use of environmental DNA in invasive species surveillance of the Great Lakes commercial bait trade. *Conservation Biology*, **29**, 430–439.

Navarro SP, Jurado-Rivera JA, Gómez-Zurita J, Lyal CHC, Vogler AP (2010). DNA profiling of host–herbivore interactions in tropical forests. *Ecological Entomology*, **35**, 18–32.

Neufeld JD, Vohra J, Dumont MG, et al. (2007). DNA stable-isotope probing. *Nature Protocols*, **2**, 860–866.

Nguyen NH, Song Z, Bates ST, et al. (2016). FUNGuild: an open annotation tool for parsing fungal community datasets by ecological guild. *Fungal Ecology*, **20**, 241–248.

Ni J, Yan Q, Yu Y (2013). How much metagenomic sequencing is enough to achieve a given goal? *Scientific Reports*, **3**, 1968.

Nichols RV, Cromsigt JPGM, Spong G (2015). Using eDNA to experimentally test ungulate browsing preferences. *SpringerPlus*, **4**, 489.

Nichols RV, Åkesson M, Kjellander P (2016). Diet assessment based on rumen contents: a comparison between DNA metabarcoding and macroscopy. *PLoS One*, **11**, e0157977.

Nielsen KM, Johnsen PJ, Bensasson D, Daffonchio D (2007). Release and persistence of extracellular DNA in the environment. *Environmental Biosafety Research*, **6**, 37–53.

Nilsson RH, Ryberg M, Kristiansson E, Abarenkov K, Larsson K-H, Kõljalg U (2006). Taxonomic reliability of DNA sequences in public sequence databases: a fungal perspective. *PLoS One*, **1**, e59.

Nilsson RH, Tedersoo L, Lindahl BD, et al. (2011). Towards standardization of the description and publication of next-generation sequencing datasets of fungal communities. *New Phytologist*, **191**, 314–318.

Nocker A, Sossa-Fernandez P, Burr MD, Camper AK (2007). Use of propidium monoazide for live/dead distinction in microbial ecology. *Applied and Environmental Microbiology*, **73**, 5111–5117.

Nsubuga AM, Robbins MM, Roeder AD, Morin PA, Boesch C, Vigilant L (2004). Factors affecting the amount of genomic DNA extracted from ape faeces and the identification of an improved sample storage method. *Molecular Ecology*, **13**, 2089–2094.

O'Brien SL, Gibbons SM, Owens SM, et al. (2016). Spatial scale drives patterns in soil bacterial diversity. *Environmental Microbiology*, **18**, 2039–2051.

O'Donnell JL, Kelly RP, Lowell NC, Port JA (2016). Indexed PCR primers induce template-specific bias in large-scale DNA sequencing studies. *PLoS One*, **11**, e0148698.

Ogram A, Sayler GS, Barkay T (1987). The extraction and purification of microbial DNA from sediments. *Journal of Microbiological Methods*, **7**, 57–66.

Oksanen J, Kindt R, Legendre P, et al. (2007). The vegan package. *Community Ecology Package*, **10**, 631–637.

Oliver AK, Brown SP, Callaham MA Jr, Jumpponen A (2015). Polymerase matters: non-proofreading enzymes inflate fungal community richness estimates by up to 15%. *Fungal Ecology*, **15**, 86–89.

Olsen GJ, Lane DJ, Giovannoni SJ, Pace NR, Stahl DA (1986). Microbial ecology and evolution: a ribosomal RNA approach. *Annual Review of Microbiology*, **40**, 337–365.

Olsen GJ, Overbeek R, Larsen N, et al. (1992). The Ribosomal Database Project. *Nucleic Acids Research*, **20 Suppl**, 2199–2200.

Öpik M, Vanatoa A, Vanatoa E, et al. (2010). The online database MaarjAM reveals global and ecosystemic distribution patterns in arbuscular mycorrhizal fungi (Glomeromycota). *New Phytologist*, **188**, 223–241.

Oulhen N, Schulz BJ, Carrier TJ (2016). English translation of Heinrich Anton de Bary's 1878 speech, "Die Erscheinung der Symbiose" ("De la symbiose"). *Symbiosis*, **69**, 131–139.

Pääbo S, Poinar H, Serre D, et al. (2004). Genetic analyses from ancient DNA. *Annual Review of Genetics*, **38**, 645–679.

Pace NR (1997). A molecular view of microbial diversity and the biosphere. *Science*, **276**, 734–740.

Pace NR (2009). Mapping the tree of life: progress and prospects. *Microbiology and Molecular Biology Reviews*, **73**, 565–576.

Pace NR, Stahl DA, Lane DJ, Olsen GJ (1986). The analysis of natural microbial populations by ribosomal RNA sequences. In: *Advances in microbial ecology* (ed. Marshall KC), pp. 1–55. Springer, New York, NY.

Palumbi S, Martin A, Romano S, McMillan WO, Stice L, Grabowski G (2002). *The simple fool's guide to PCR*. Department of Zoology and Kewalo Marine Laboratory, University of Hawaii, Honolulu, HI 96822.

Pansu J, De Danieli S, Puissant J, et al. (2015a). Landscape-scale distribution patterns of earthworms inferred from soil DNA. *Soil Biology & Biochemistry*, **83**, 100–105.

Pansu J, Giguet-Covex C, Ficetola GF, et al. (2015b). Reconstructing long-term human impacts on plant communities: an ecological approach based on lake sediment DNA. *Molecular Ecology*, **24**, 1485–1498.

Papadopoulou A, Taberlet P, Zinger L (2015). Metagenome skimming for phylogenetic community ecology: a new era in biodiversity research. *Molecular Ecology*, **24**, 3515–3517.

Parada AE, Needham DM, Fuhrman JA (2016). Every base matters: assessing small subunit rRNA primers for marine microbiomes with mock communities, time series and global field samples. *Environmental Microbiology*, **18**, 1403–1414.

Parducci L, Matetovici I, Fontana SL, et al. (2013). Molecular- and pollen-based vegetation analysis in lake sediments from central Scandinavia. *Molecular Ecology*, **22**, 3511–3524.

Parducci L, Valiranta M, Salonen JS, et al. (2015). Proxy comparison in ancient peat sediments: pollen, macro-fossil and plant DNA. *Philosophical Transactions of the Royal Society B-Biological Sciences*, **370**, 20130382.

Patil KR, Roune L, McHardy AC (2012). The PhyloPythiaS web server for taxonomic assignment of metagenome sequences. *PLoS One*, **7**, e38581.

Pawlowski J, Esling P, Lejzerowicz F, Cedhagen T, Wilding TA (2014). Environmental monitoring through protist next-generation sequencing metabarcoding: assessing the impact of fish farming on benthic foraminifera communities. *Molecular Ecology Resources*, **14**, 1129–1140.

Pedersen MW, Ginolhac A, Orlando L, et al. (2013). A comparative study of ancient environmental DNA to pollen and macrofossils from lake sediments reveals taxonomic overlap and additional plant taxa. *Quaternary Science Reviews*, **75**, 161–168.

Pedersen MW, Overballe-Petersen S, Ermini L, et al. (2015). Ancient and modern environmental DNA. *Philosophical Transactions of the Royal Society of London. Series B, Biological Sciences*, **370**, 20130383.

Pedersen MW, Ruter A, Schweger C, et al. (2016). Postglacial viability and colonization in North America's ice-free corridor. *Nature*, **537**, 45–49.

Peiffer JA, Spor A, Koren O, et al. (2013). Diversity and heritability of the maize rhizosphere microbiome under field conditions. *Proceedings of the National Academy of Sciences of the United States of America*, **110**, 6548–6553.

Peng Y, Leung HCM, Yiu SM, Chin FYL (2011). Meta-IDBA: a de novo assembler for metagenomic data. *Bioinformatics*, **27**, i94–101.

Pérès G, Vandenbulcke F, Guernion M, et al. (2011). Earthworm indicators as tools for soil monitoring, characterization and risk assessment. An example from the national bioindicator programme (France). *Pedobiologia*, **54**, S77–S87.

Philippot L, Andersson SGE, Battin TJ, et al. (2010). The ecological coherence of high bacterial taxonomic ranks. *Nature Reviews. Microbiology*, **8**, 523–529.

Philippot L, Bru D, Saby NPA, et al. (2009). Spatial patterns of bacterial taxa in nature reflect ecological traits of deep branches of the 16S rRNA bacterial tree. *Environmental Microbiology*, **11**, 3096–3104.

Philippot L, Ritz K, Pandard P, Hallin S, Martin-Laurent F (2012). Standardisation of methods in soil microbiology: progress and challenges. *FEMS Microbiology Ecology*, **82**, 1–10.

Piaggio AJ, Engeman RM, Hopken MW, et al. (2014). Detecting an elusive invasive species: a diagnostic PCR to detect Burmese python in Florida waters and an assessment of persistence of environmental DNA. *Molecular Ecology Resources*, **14**, 374–380.

Pietramellara G, Ascher J, Borgogni F, Ceccherini MT, Guerri G, Nannipieri P (2009). Extracellular DNA in soil and sediment: fate and ecological relevance. *Biology and Fertility of Soils*, **45**, 219–235.

Pilliod DS, Goldberg CS, Arkle RS, Waits LP (2013). Estimating occupancy and abundance of stream amphibians using environmental DNA from filtered water samples. *Canadian Journal of Fisheries and Aquatic Sciences*, **70**, 1123–1130.

Pilliod DS, Goldberg CS, Arkle RS, Waits LP (2014). Factors influencing detection of eDNA from a stream-dwelling amphibian. *Molecular Ecology Resources*, **14**, 109–116.

Poinar HN, Cooper A (2000). Ancient DNA: do it right or not at all. *Science*, **5482**, 1139.

Polz MF, Cavanaugh CM (1998). Bias in template-to-product ratios in multitemplate PCR. *Applied and Environmental Microbiology*, **64**, 3724–3730.

Pompanon F, Coissac E, Taberlet P (2011). Metabarcoding a new way to analyze biodiversity. *Biofutur*, 30–32.

Pompanon F, Deagle BE, Symondson WOC, Brown DS, Jarman SN, Taberlet P (2012). Who is eating what: diet assessment using next generation sequencing. *Molecular Ecology*, **21**, 1931–1950.

Ponsard S, Arditi R (2000). What can stable isotopes (δ ^{15}N and δ ^{13}C) tell about the food web of soil macro-invertebrates? *Ecology*, **81**, 852–864.

Porazinska DL, Giblin-Davis RM, Esquivel A, Powers TO, Sung W, Thomas WK (2010) Ecometagenetics confirms high tropical rainforest nematode diversity. *Molecular Ecology*, **19**, 5521–5530.

Portillo MC, Leff JW, Lauber CL, Fierer N (2013). Cell size distributions of soil bacterial and archaeal taxa. *Applied and Environmental Microbiology*, **79**, 7610–7617.

Pringle EG, Moreau CS (2017). Community analysis of microbial sharing and specialization in a Costa Rican ant-plant-hemipteran symbiosis. *Proceedings. Biological Sciences*, **284**, 20162770.

Prosser JI (2010). Replicate or lie. *Environmental Microbiology*, **12**, 1806–1810.

Prosser SWJ, Hebert PDN (2017). Rapid identification of the botanical and entomological sources of honey using DNA metabarcoding. *Food Chemistry*, **214**, 183–191.

Pruesse E, Quast C, Knittel K, et al. (2007). SILVA: a comprehensive online resource for quality checked and aligned ribosomal RNA sequence data compatible with ARB. *Nucleic Acids Research*, **35**, 7188–7196.

Puillandre N, Lambert A, Brouillet S, Achaz G (2012). ABGD, Automatic Barcode Gap Discovery for primary species delimitation. *Molecular Ecology*, **21**, 1864–1877.

Putman RJ (1984). Facts from faeces. *Mammal Review*, **14**, 79–97.

Quast C, Pruesse E, Yilmaz P, et al. (2013). The SILVA ribosomal RNA gene database project: improved data processing and web-based tools. *Nucleic Acids Research*, **41**, D590–D596.

Quince C, Lanzen A, Davenport RJ, Turnbaugh PJ (2011). Removing noise from pyrosequenced amplicons. *BMC Bioinformatics*, **12**, 38.

R Development Core Team (2016). R: a language and environment for statistical computing. R Foundation for Statistical Computing. Vienna, Austria. https://www.R-project.org/.

Radajewski S, Ineson P, Parekh NR, Murrell JC (2000). Stable-isotope probing as a tool in microbial ecology. *Nature*, **403**, 646–649.

Ramgren AC, Newhall HS, James KE (2015). DNA barcoding and metabarcoding with the Oxford Nanopore MinION. *Genome/National Research Council Canada = Genome/Conseil National de Recherches Canada*, **58**, 268–268.

Ramirez KS, Leff JW, Barberan A, et al. (2014). Biogeographic patterns in below-ground diversity in New York City's Central Park are similar to those observed globally. *Proceedings. Biological Sciences*, **281**, 20141988.

Ranjard L, Lejon DPH, Mougel C, Schehrer L, Merdinoglu D, Chaussod R (2003). Sampling strategy in molecular microbial ecology: influence of soil sample size on DNA fingerprinting analysis of fungal and bacterial communities. *Environmental Microbiology*, **5**, 1111–1120.

Ransome E, Geller JB, Timmers M, et al. (2017). The importance of standardization for biodiversity comparisons: a case study using autonomous reef monitoring structures (ARMS) and metabarcoding to measure cryptic diversity on Mo'orea coral reefs, French Polynesia. *PLoS One*, **12**, e0175066.

Rao CR (1986). Rao's axiomatization of diversity measures. In: *Encyclopedia of statistical sciences* (eds. Kotz S, Johnson NL), pp. 614–617. Wiley, New York, NY.

Ratnasingham S, Hebert PDN (2007). BOLD: the barcode of life data system (http://www.barcodinglife.org). *Molecular Ecology Notes*, **7**, 355–364.

Ray JL, Althammer J, Skaar KS, et al. (2016). Metabarcoding and metabolome analyses of copepod grazing reveal feeding preference and linkage to metabolite classes in dynamic microbial plankton communities. *Molecular Ecology*, **25**, 5585–5602.

Rayé G, Miquel C, Coissac E, Redjadj C, Loison A, Taberlet P (2011). New insights on diet variability revealed by DNA barcoding and high-throughput pyrosequencing: chamois diet in autumn as a case study. *Ecological Research*, **26**, 265–276.

Reeder J, Knight R (2009). The "rare biosphere": a reality check. *Nature Methods*, **6**, 636–637.

Riaz T, Shehzad W, Viari A, Pompanon F, Taberlet P, Coissac E (2011). ecoPrimers: inference of new DNA barcode markers from whole genome sequence analysis. *Nucleic Acids Research*, **39**, e145.

Richards TA, Jones MDM, Leonard G, Bass D (2012). Marine fungi: their ecology and molecular diversity. *Annual Review of Marine Science*, **4**, 495–522.

Richards TA, Leonard G, Mahé F, et al. (2015). Molecular diversity and distribution of marine fungi across 130 European environmental samples. *Proceedings. Biological Sciences*, **282**.

Richardson RT, Lin C-H, Sponsler DB, Quijia JO, Goodell K, Johnson RM (2015). Application of ITS2 metabarcoding to determine the provenance of pollen collected by honey bees in an agroecosystem. *Applications in Plant Sciences*, **3**, 1400066.

Richter-Heitmann T, Eickhorst T, Knauth S, Friedrich MW, Schmidt H (2016). Evaluation of strategies to separate root-associated microbial communities: a crucial choice in rhizobiome research. *Frontiers in Microbiology*, **7**, 773.

Ririe KM, Rasmussen RP, Wittwer CT (1997). Product differentiation by analysis of DNA melting curves during the polymerase chain reaction. *Analytical Biochemistry*, **245**, 154–160.

Robinson CJ, Bohannan BJM, Young VB (2010). From structure to function: the ecology of host-associated microbial communities. *Microbiology and Molecular Biology Reviews*, **74**, 453–476.

Robinson MD, Smyth GK (2008). Small-sample estimation of negative binomial dispersion, with applications to SAGE data. *Biostatistics*, **9**, 321–332.

Robson HLA, Noble TH, Saunders RJ, Robson SKA, Burrows DW, Jerry DR (2016). Fine-tuning for the tropics: application of eDNA technology for invasive fish detection in tropical freshwater ecosystems. *Molecular Ecology Resources*, **16**, 922–932.

Rodgers TW, Mock KE (2015). Drinking water as a source of environmental DNA for the detection of terrestrial wildlife species. *Conservation Genetics Resources*, **7**, 693–696.

Römbke J, Jänsch S, Didden W (2005). The use of earthworms in ecological soil classification and assessment concepts. *Ecotoxicology and Environmental Safety*, **62**, 249–265.

Rosenberg DM, Resh V (1993). Introduction to freshwater biomonitoring and benthic macroinvertebrates. In: *Freshwater biomonitoring and benthic macroinvertebrates* (eds. Rosenberg DM, Resh V). Chapman & Hall, New York, NY.

Rosselló-Móra R (2012). Towards a taxonomy of Bacteria and Archaea based on interactive and cumulative data repositories. *Environmental Microbiology*, **14**, 318–334.

Rosvall M, Bergstrom CT (2008). Maps of random walks on complex networks reveal community structure. *Pro-

ceedings of the National Academy of Sciences of the United States of America, **105**, 1118–1123.

Roy J, Albert CH, Ibanez S, et al. (2013). Microbes on the cliff: alpine cushion plants structure bacterial and fungal communities. *Frontiers in Microbiology*, **4**, 64.

Ruff SE, Biddle JF, Teske AP, Knittel K, Boetius A, Ramette A (2015). Global dispersion and local diversification of the methane seep microbiome. *Proceedings of the National Academy of Sciences of the United States of America*, **112**, 4015–4020.

Rusch DB, Halpern AL, Sutton G, et al. (2007). The Sorcerer II Global Ocean Sampling expedition: northwest Atlantic through eastern tropical Pacific. *PLoS Biology*, **5**, e77.

Saiki RK, Gelfand DH, Stoffel S, et al. (1988). Primer-directed enzymatic amplification of DNA with a thermostable DNA polymerase. *Science*, **239**, 487–491.

Saiki RK, Scharf S, Faloona F, et al. (1985). Enzymatic amplification of beta-globin genomic sequences and restriction site analysis for diagnosis of sickle cell anemia. *Science*, **230**, 1350–1354.

Saint-Béat B, Baird D, Asmus H, et al. (2015). Trophic networks: how do theories link ecosystem structure and functioning to stability properties? A review. *Ecological Indicators*, **52**, 458–471.

Saitoh S, Aoyama H, Fujii S, et al. (2016). A quantitative protocol for DNA metabarcoding of springtails (Collembola). *Genome*, **59**, 705–723.

Salter SJ, Cox MJ, Turek EM, et al. (2014). Reagent and laboratory contamination can critically impact sequence-based microbiome analyses. *BMC Biology*, **12**, 87.

Sanchez G, Holliday VT, Gaines EP, et al. (2014). Human (Clovis)–gomphothere (*Cuvieronius* sp.) association ~13,390 calibrated yBP in Sonora, Mexico. *Proceedings of the National Academy of Sciences of the United States of America*, **111**, 10972–10977.

Sanger F, Nicklen S, Coulson AR (1977). DNA sequencing with chain-terminating inhibitors. *Proceedings of the National Academy of Sciences of the United States of America*, **74**, 5463–5467.

Santamaria M, Fosso B, Consiglio A, et al. (2012). Reference databases for taxonomic assignment in metagenomics. *Briefings in Bioinformatics*, **13**, 682–695.

Sapp J (1994). *Evolution by association: a history of symbiosis*. Oxford University Press, New York, NY.

Schadt EE, Turner S, Kasarskis A (2010). A window into third-generation sequencing. *Human Molecular Genetics*, **19**, R227–R240.

Schloss PD (2010). The effects of alignment quality, distance calculation method, sequence filtering, and region on the analysis of 16S rRNA gene-based studies. *PLoS Computational Biology*, **6**, e1000844.

Schloss PD, Gevers D, Westcott SL (2011). Reducing the effects of PCR amplification and sequencing artifacts on 16S rRNA-based studies. *PLoS One*, **6**, e27310.

Schloss PD, Westcott SL, Ryabin T, et al. (2009). Introducing mothur: open-source, platform-independent, community-supported software for describing and comparing microbial communities. *Applied and Environmental Microbiology*, **75**, 7537–7541.

Schmidt PA, Bálint M, Greshake B, Bandow C, Römbke J, Schmitt I (2013). Illumina metabarcoding of a soil fungal community. *Soil Biology & Biochemistry*, **65**, 128–132.

Schneider TD, Stephens RM (1990). Sequence logos: a new way to display consensus sequences. *Nucleic Acids Research*, **18**, 6097–6100.

Schneider J, Valentini A, Dejean T, et al. (2016). Detection of invasive mosquito vectors using environmental DNA (eDNA) from water samples. *PLoS One*, **11**, e0162493.

Schnell IB, Bohmann K, Gilbert MTP (2015a). Tag jumps illuminated—reducing sequence-to-sample misidentifications in metabarcoding studies. *Molecular Ecology Resources*, **15**, 1289–1303.

Schnell IB, Sollmann R, Calvignac-Spencer S, et al. (2015b). iDNA from terrestrial haematophagous leeches as a wildlife surveying and monitoring tool—prospects, pitfalls and avenues to be developed. *Frontiers in Zoology*, **12**, 24.

Schnell IB, Thomsen PF, Wilkinson N, et al. (2012). Screening mammal biodiversity using DNA from leeches. *Current Biology*, **22**, R262–R263.

Schoch CL, Seifert KA, Huhndorf S, et al. (2012). Nuclear ribosomal internal transcribed spacer (ITS) region as a universal DNA barcode marker for Fungi. *Proceedings of the National Academy of Sciences of the United States of America*, **109**, 6241–6246.

Schubert G, Stockhausen M, Hoffmann C, et al. (2015). Targeted detection of mammalian species using carrion fly-derived DNA. *Molecular Ecology Resources*, **15**, 285–294.

Scriver M, Marinich A, Wilson C, Freeland J (2015). Development of species-specific environmental DNA (eDNA) markers for invasive aquatic plants. *Aquatic Botany*, **122**, 27–31.

Searle D, Sible E, Cooper A, Putonti C (2016). 18S rDNA dataset profiling microeukaryotic populations within Chicago area nearshore waters. *Data in Brief*, **6**, 526–529.

Seersholm FV, Pedersen MW, Søe MJ, et al. (2016). DNA evidence of bowhead whale exploitation by Greenlandic Paleo-Inuit 4,000 years ago. *Nature Communications*, **7**, 13389.

Selinger DW, Saxena RM, Cheung KJ, Church GM, Rosenow C (2003). Global RNA half-life analysis in *Escherichia coli* reveals positional patterns of transcript degradation. *Genome Research*, **13**, 216–223.

Sessitsch A, Hardoim P, Döring J, et al. (2012). Functional characteristics of an endophyte community colonizing rice roots as revealed by metagenomic analysis. *Molecular Plant-Microbe Interactions*, **25**, 28–36.

Sharpton TJ (2014). An introduction to the analysis of shotgun metagenomic data. *Frontiers in Plant Science*, **5**, 209.

Shaw JLA, Clarke LJ, Wedderburn SD, Barnes TC, Weyrich LS, Cooper A (2016). Comparison of environmental DNA metabarcoding and conventional fish survey methods in a river system. *Biological Conservation*, **197**, 131–138.

Shehzad W, McCarthy TM, Pompanon F, et al. (2012a). Prey preference of snow leopard (*Panthera uncia*) in South Gobi, Mongolia. *PLoS One*, **7**, e32104.

Shehzad W, Nawaz MA, Pompanon F, et al. (2015). Forest without prey: livestock sustain a leopard *Panthera pardus* population in Pakistan. *Oryx*, **49**, 248–253.

Shehzad W, Riaz T, Nawaz MA, et al. (2012b). Carnivore diet analysis based on next-generation sequencing: application to the leopard cat (*Prionailurus bengalensis*) in Pakistan. *Molecular Ecology*, **21**, 1951–1965.

Shendure J, Ji H (2008). Next-generation DNA sequencing. *Nature Biotechnology*, **26**, 1035–1045.

Shokralla S, Gibson J, King I, et al. (2016). Environmental DNA barcode sequence capture: targeted, PCR-free sequence capture for biodiversity analysis from bulk environmental samples. *bioRxiv*, 087437.

Shokralla S, Spall JL, Gibson JF, Hajibabaei M (2012). Next-generation sequencing technologies for environmental DNA research. *Molecular Ecology*, **21**, 1794–1805.

Sigsgaard EE, Nielsen IB, Bach SS, et al. (2016). Population characteristics of a large whale shark aggregation inferred from seawater environmental DNA. *Nature Ecology & Evolution*, **1**, 0004.

Simon C, Daniel R (2011). Metagenomic analyses: past and future trends. *Applied and Environmental Microbiology*, **77**, 1153–1161.

Sintes E, De Corte D, Ouillon N, Herndl GJ (2015). Macroecological patterns of archaeal ammonia oxidizers in the Atlantic Ocean. *Molecular Ecology*, **24**, 4931–4942.

Slon V, Hopfe C, Weiß CL, et al. (2017). Neandertal and Denisovan DNA from Pleistocene sediments. *Science*, **356**, 605–608.

Smets W, Leff JW, Bradford MA, McCulley RL, Lebeer S, Fierer N (2016). A method for simultaneous measurement of soil bacterial abundances and community composition via 16S rRNA gene sequencing. *Soil Biology and Biochemistry*, **96**, 145–151.

Smith O, Momber G, Bates R, et al. (2015). Sedimentary DNA from a submerged site reveals wheat in the British Isles 8000 years ago. *Science*, **347**, 998–1001.

Smith CJ, Osborn AM (2009). Advantages and limitations of quantitative PCR (Q-PCR)-based approaches in microbial ecology. *FEMS Microbiology Ecology*, **67**, 6–20.

Soejima T, Iida K-I, Qin T, et al. (2007). Photoactivated ethidium monoazide directly cleaves bacterial DNA and is applied to PCR for discrimination of live and dead bacteria. *Microbiology and Immunology*, **51**, 763–775.

Sogin ML, Morrison HG, Huber JA, et al. (2006). Microbial diversity in the deep sea and the underexplored "rare biosphere." *Proceedings of the National Academy of Sciences of the United States of America*, **103**, 12115–12120.

Soininen EM, Valentini A, Coissac E, et al. (2009). Analysing diet of small herbivores: the efficiency of DNA barcoding coupled with high-throughput pyrosequencing for deciphering the composition of complex plant mixtures. *Frontiers in Zoology*, **6**, 16.

Soininen EM, Zinger L, Gielly L, et al. (2013). Shedding new light on the diet of Norwegian lemmings: DNA metabarcoding of stomach content. *Polar Biology*, **36**, 1069–1076.

Sommeria-Klein G, Zinger L, Taberlet P, Coissac E, Chave J (2016). Inferring neutral biodiversity parameters using environmental DNA data sets. *Scientific Reports*, **6**, 35644.

Sønstebø JH, Gielly L, Brysting A, et al. (2010). Using next-generation sequencing for molecular reconstruction of past Arctic vegetation and climate. *Molecular Ecology Resources*, **10**, 1009–1018.

Srivathsan A, Ang A, Vogler AP, Meier R (2016). Fecal metagenomics for the simultaneous assessment of diet, parasites, and population genetics of an understudied primate. *Frontiers in Zoology*, **13**, 17.

Srivathsan A, Sha JCM, Vogler AP, Meier R (2015). Comparing the effectiveness of metagenomics and metabarcoding for diet analysis of a leaf-feeding monkey (*Pygathrix nemaeus*). *Molecular Ecology Resources*, **15**, 250–261.

Steinbock LJ, Radenovic A (2015). The emergence of nanopores in next-generation sequencing. *Nanotechnology*, **26**, 074003.

Stevenson RJ, Pan Y, van Dam H (2010). Assessing environmental conditions in rivers and streams with diatoms. In: *The diatoms: applications for the environmental and earth sciences, 2nd Edition* (eds. Smol JP, Stoermer EF), pp. 57–85. Cambridge University Press, Cambridge, UK.

Straub SCK, Parks M, Weitemier K, Fishbein M, Cronn RC, Liston A (2012). Navigating the tip of the genomic iceberg: next-generation sequencing for plant systematics. *American Journal of Botany*, **99**, 349–364.

Strickler KM, Fremier AK, Goldberg CS (2015). Quantifying effects of UV-B, temperature, and pH on eDNA degradation in aquatic microcosms. *Biological Conservation*, **183**, 85–92.

Suen G, Scott JJ, Aylward FO, et al. (2010). An insect herbivore microbiome with high plant biomass-degrading capacity. *PLoS Genetics*, **6**, e1001129.

Sullam KE, Essinger SD, Lozupone CA, et al. (2012). Environmental and ecological factors that shape the gut bacterial communities of fish: a meta-analysis. *Molecular Ecology*, **21**, 3363–3378.

Sun Y, Cai Y, Liu L, et al. (2009). ESPRIT: estimating species richness using large collections of 16S rRNA pyrosequences. *Nucleic Acids Research*, **37**, e76.

Sunagawa S, Coelho LP, Chaffron S, et al. (2015). Structure and function of the global ocean microbiome. *Science*, **348**, 1261359.

Symondson WOC (2002). Molecular identification of prey in predator diets. *Molecular Ecology*, **11**, 627–641.

Taberlet P, Coissac E, Hajibabaei M, Rieseberg LH (2012a). Environmental DNA. *Molecular Ecology*, **21**, 1789–1793.

Taberlet P, Coissac E, Pompanon F, Brochmann C, Willerslev E (2012b). Towards next-generation biodiversity assessment using DNA metabarcoding. *Molecular Ecology*, **21**, 2045–2050.

Taberlet P, Coissac E, Pompanon F, et al. (2007). Power and limitations of the chloroplast *trn*L(UAA) intron for plant DNA barcoding. *Nucleic Acids Research*, **35**, e14.

Taberlet P, Gielly L, Pautou G, Bouvet J (1991). Universal primers for amplification of three non-coding regions of chloroplast DNA. *Plant Molecular Biology*, **17**, 1105–1109.

Taberlet P, Prud'homme S, Campione E, et al. (2012c). Soil sampling and isolation of extracellular DNA from large amount of starting material suitable for metabarcoding studies. *Molecular Ecology*, **21**, 1816–1820.

Tada Y, Grossart H-P (2014). Community shifts of actively growing lake bacteria after N-acetyl-glucosamine addition: improving the BrdU-FACS method. *ISME Journal*, **8**, 441–454.

Takahara T, Minamoto T, Doi H (2013). Using environmental DNA to estimate the distribution of an invasive fish species in ponds. *PLoS One*, **8**, e56584.

Takahara T, Minamoto T, Yamanaka H, Doi H, Kawabata Z (2012). Estimation of fish biomass using environmental DNA. *PLoS One*, **7**, e35868.

Tan B, Ng C, Nshimyimana JP, Loh LL, Gin KY-H, Thompson JR (2015). Next-generation sequencing (NGS) for assessment of microbial water quality: current progress, challenges, and future opportunities. *Frontiers in Microbiology*, **6**, 1027.

Tedersoo L, Bahram M, Cajthaml T, et al. (2016). Tree diversity and species identity effects on soil fungi, protists and animals are context dependent. *ISME Journal*, **10**, 346–362.

Tedersoo L, Bahram M, Põlme S, et al. (2014). Global diversity and geography of soil fungi. *Science*, **346**, 1078–1090.

Tedersoo L, Ramirez KS, Nilsson RH, Kaljuvee A, Kõljalg U, Abarenkov K (2015). Standardizing metadata and

taxonomic identification in metabarcoding studies. *GigaScience*, **4**, 34.

Terrat S, Christen R, Dequiedt S, et al. (2012). Molecular biomass and MetaTaxogenomic assessment of soil microbial communities as influenced by soil DNA extraction procedure. *Microbial Biotechnology*, **5**, 135–141.

The UniProt Consortium (2017). UniProt: the universal protein knowledgebase. *Nucleic Acids Research*, **45**, D158–D169.

Thomas AC, Deagle BE, Eveson JP, Harsch CH, Trites AW (2016). Quantitative DNA metabarcoding: improved estimates of species proportional biomass using correction factors derived from control material. *Molecular Ecology Resources*, **16**, 714–726.

Thomas T, Gilbert J, Meyer F (2012). Metagenomics—a guide from sampling to data analysis. *Microbial Informatics and Experimentation*, **2**, 3.

Thomas AC, Jarman SN, Haman KH, Trites AW, Deagle BE (2014). Improving accuracy of DNA diet estimates using food tissue control materials and an evaluation of proxies for digestion bias. *Molecular Ecology*, **23**, 3706–3718.

Thomas CM, Nielsen KM (2005). Mechanisms of, and barriers to, horizontal gene transfer between bacteria. *Nature Reviews. Microbiology*, **3**, 711–721.

Thomsen PF, Kielgast J, Iversen LL, Moller PR, Rasmussen M, Willerslev E (2012a). Detection of a diverse marine fish fauna using environmental DNA from seawater samples. *PLoS One*, **7**, e41732.

Thomsen PF, Kielgast J, Iversen LL, et al. (2012b). Monitoring endangered freshwater biodiversity using environmental DNA. *Molecular Ecology*, **21**, 2565–2573.

Thomsen PF, Møller PR, Sigsgaard EE, Knudsen SW, Jørgensen OA, Willerslev E (2016). Environmental DNA from seawater samples correlate with trawl catches of subarctic, deepwater fishes. *PLoS One*, **11**, e0165252.

Thomsen PF, Willerslev E (2015). Environmental DNA as an emerging tool in conservation for monitoring past and present biodiversity. *Biological Conservation*, **183**, 4–18.

Tindall KR, Kunkel TA (1988). Fidelity of DNA synthesis by the *Thermus aquaticus* DNA polymerase. *Biochemistry*, **27**, 6008–6013.

Tindall BJ, Rosselló-Móra R, Busse H-J, Ludwig W, Kämpfer P (2010). Notes on the characterization of prokaryote strains for taxonomic purposes. *International Journal of Systematic and Evolutionary Microbiology*, **60**, 249–266.

Tisthammer KH, Cobian GM, Amend AS (2016). Global biogeography of marine fungi is shaped by the environment. *Fungal Ecology*, **19**, 39–46.

Torsvik V, Goksøyr J, Daae FL (1990). High diversity in DNA of soil bacteria. *Applied and Environmental Microbiology*, **56**, 782–787.

Torti A, Lever MA, Jørgensen BB (2015). Origin, dynamics, and implications of extracellular DNA pools in marine sediments. *Marine Genomics*, **24**, 185–196.

Tringe SG, von Mering C, Kobayashi A, et al. (2005). Comparative metagenomics of microbial communities. *Science*, **308**, 554–557.

Tréguier A, Paillisson J-M, Dejean T, Valentini A, Schlaepfer MA, Roussel J-M (2014). Environmental DNA surveillance for invertebrate species: advantages and technical limitations to detect invasive crayfish *Procambarus clarkii* in freshwater ponds. *Journal of Applied Ecology*, **51**, 871–879.

Tsuji S, Yamanaka H, Minamoto T (2017). Effects of water pH and proteinase K treatment on the yield of environmental DNA from water samples. *Limnology*, **18**, 1–7.

Turner CR, Barnes MA, Xu CCY, Jones SE, Jerde CL, Lodge DM (2014). Particle size distribution and optimal capture of aqueous macrobial eDNA. *Methods in Ecology and Evolution*, **5**, 676–684.

Turner CR, Uy KL, Everhart RC (2015). Fish environmental DNA is more concentrated in aquatic sediments than surface water. *Biological Conservation*, **183**, 93–102.

Urbach E, Vergin KL, Giovannoni SJ (1999). Immunochemical detection and isolation of DNA from metabolically active bacteria. *Applied and Environmental Microbiology*, **65**, 1207–1213.

Urban MC, Bocedi G, Hendry AP, et al. (2016). Improving the forecast for biodiversity under climate change. *Science*, **353**.

Ushio M, Murakami H, Masuda R, et al. (2017). Quantitative monitoring of multispecies fish environmental DNA using high-throughput sequencing. *bioRxiv*, 113472.

Ushio M, Yamasaki E, Takasu H, et al. (2015). Microbial communities on flower surfaces act as signatures of pollinator visitation. *Scientific Reports*, **5**, 8695.

Usseglio-Polatera P, Richoux P, Bournaud M, Tachet H (2001). A functional classification of benthic macroinvertebrates based on biological and ecological traits: application to river condition assessment and stream management. *Archiv Für Hydrobiologie*, **Suppl. 139**, 53–83.

Valasek MA, Repa JJ (2005). The power of real-time PCR. *Advances in Physiology Education*, **29**, 151–159.

Valentini A, Miquel C, Nawaz MA, et al. (2009). New perspectives in diet analysis based on DNA barcoding and parallel pyrosequencing: the *trn*L approach. *Molecular Ecology Resources*, **9**, 51–60.

Valentini A, Miquel C, Taberlet P (2010). DNA barcoding for honey biodiversity. *Diversity*, **2**, 610–617.

Valentini A, Taberlet P, Miaud C, et al. (2016). Next-generation monitoring of aquatic biodiversity using environmental DNA metabarcoding. *Molecular Ecology*, **25**, 929–942.

Valière N, Taberlet P (2000). Urine collected in the field as a source of DNA for species and individual identification. *Molecular Ecology*, **9**, 2150–2152.

Vandenkoornhuyse P, Dufresne A, Quaiser A, et al. (2010). Integration of molecular functions at the ecosystemic level: breakthroughs and future goals of environmental genomics and post-genomics. *Ecology Letters*, **13**, 776–791.

van Dijk EL, Auger H, Jaszczyszyn Y, Thermes C (2014). Ten years of next-generation sequencing technology. *Trends in Genetics*, **30**, 418–426.

Van Dongen SM (2001). Graph clustering by flow simulation. PhD Thesis. University of Utrecht, Utrecht, the Netherlands.

Vanreusel A, De Groote A, Gollner S, Bright M (2010). Ecology and biogeography of free-living nematodes associated with chemosynthetic environments in the deep sea: a review. *PLoS One*, **5**, e12449.

Venter JC, Remington K, Heidelberg JF, et al. (2004). Environmental genome shotgun sequencing of the Sargasso Sea. *Science*, **304**, 66–74.

Vermeulen ET, Lott MJ, Eldridge MDB, Power ML (2016). Evaluation of next generation sequencing for the analysis of *Eimeria* communities in wildlife. *Journal of Microbiological Methods*, **124**, 1–9.

Vesterinen EJ, Ruokolainen L, Wahlberg N, et al. (2016). What you need is what you eat? Prey selection by the bat *Myotis daubentonii*. *Molecular Ecology*, **25**, 1581–1594.

Vestheim H, Deagle BE, Jarman SN (2011). Application of blocking oligonucleotides to improve signal-to-noise ratio in a PCR. In: *PCR protocols*. (ed. Park DJ), pp. 265–274. Humana Press, New York, NY.

Vestheim H, Edvardsen B, Kaartvedt S (2005). Assessing feeding of a carnivorous copepod using species-specific PCR. *Marine Biology*, **147**, 381–385.

Vestheim H, Jarman SN (2008). Blocking primers to enhance PCR amplification of rare sequences in mixed samples—a case study on prey DNA in Antarctic krill stomachs. *Frontiers in Zoology*, **5**, 12.

Větrovský T, Baldrian P (2013). The variability of the 16S rRNA gene in bacterial genomes and its consequences for bacterial community analyses. *PLoS One*, **8**, e57923.

Vilgalys R (2003). Taxonomic misidentification in public DNA databases. *New Phytologist*, **160**, 4–5.

Visco JA, Apothéloz-Perret-Gentil L, Cordonier A, Esling P, Pillet L, Pawlowski J (2015). Environmental monitoring: inferring the diatom index from next-generation sequencing data. *Environmental Science & Technology*, **49**, 7597–7605.

Wagner AO, Praeg N, Reitschuler C, Illmer P (2015). Effect of DNA extraction procedure, repeated extraction and ethidium monoazide (EMA)/propidium monoazide (PMA) treatment on overall DNA yield and impact on microbial fingerprints for bacteria, fungi and archaea in a reference soil. *Applied Soil Ecology*, **93**, 56–64.

Walters WA, Caporaso JG, Lauber CL, Berg-Lyons D, Fierer N, Knight R (2011). PrimerProspector: *de novo* design and taxonomic analysis of barcoded polymerase chain reaction primers. *Bioinformatics*, **27**, 1159–1161.

Wang Q, Garrity GM, Tiedje JM, Cole JR (2007). Naive Bayesian classifier for rapid assignment of rRNA sequences into the new bacterial taxonomy. *Applied and Environmental Microbiology*, **73**, 5261–5267.

Wang Y, Hayatsu M, Fujii T (2012). Extraction of bacterial RNA from soil: challenges and solutions. *Microbes and Environments*, **27**, 111–121.

Wang GC, Wang Y (1997). Frequency of formation of chimeric molecules as a consequence of PCR coamplification of 16S rRNA genes from mixed bacterial genomes. *Applied and Environmental Microbiology*, **63**, 4645–4650.

Ward DM, Brassell SC, Eglinton G (1985). Archaebacterial lipids in hot-spring microbial mats. *Nature*, **318**, 656–659.

Ward DM, Weller R, Bateson MM (1990). 16S ribosomal-RNA sequences reveal numerous uncultured microorganisms in a natural community. *Nature*, **345**, 63–65.

Wayne LG, Brenner DJ, Colwell RR, et al. (1987). Report of the *ad hoc* committee on reconciliation of approaches to bacterial systematics. *International Journal of Systematic and Evolutionary Microbiology*, **37**, 463–464.

Webster NS, Taylor MW (2012). Marine sponges and their microbial symbionts: love and other relationships. *Environmental Microbiology*, **14**, 335–346.

Wegleitner BJ, Jerde CL, Tucker A, Lindsay Chadderton W, Mahon AR (2015). Long duration, room temperature preservation of filtered eDNA samples. *Conservation Genetics Resources*, **7**, 789–791.

Weiß CL, Dannemann M, Prüfer K, Burbano HA (2015). Contesting the presence of wheat in the British Isles 8,000 years ago by assessing ancient DNA authenticity from low-coverage data. *eLife*, **4**, e10005.

White TJ, Arnheim N, Erlich HA (1989). The polymerase chain reaction. *Trends in Genetics*, **5**, 185–189.

White TJ, Bruns T, Lee S, Taylor JW (1990). Amplification and direct sequencing of fungal ribosomal RNA genes for phylogenetics. In: *PCR protocols: a guide to methods and applications* (eds. Innis MA, Gelfand DH, Sninski JJ, White TJ), pp. 315–322. Academic Press, San Diego, CA.

Wilcox TM, McKelvey KS, Young MK, et al. (2013). Robust detection of rare species using environmental DNA: the importance of primer specificity. *PLoS One*, **8**, e59520.

Wilcox TM, McKelvey KS, Young MK, et al. (2016). Understanding environmental DNA detection probabilities: a case study using a stream-dwelling char *Salvelinus fontinalis*. *Biological Conservation*, **194**, 209–216.

Will KW, Mishler BD, Wheeler QD (2005). The perils of DNA barcoding and the need for integrative taxonomy. *Systematic Biology*, **54**, 844–851.

Willerslev E, Cappellini E, Boomsma W, et al. (2007). Ancient biomolecules from deep ice cores reveal a forested Southern Greenland. *Science*, **317**, 111–114.

Willerslev E, Cooper A (2005). Ancient DNA. *Proceedings. Biological Sciences*, **272**, 3–16.

Willerslev E, Davison J, Moora M, et al. (2014). Fifty-five thousand years of arctic vegetation and megafaunal diet. *Nature*, **506**, 47–51.

Willerslev E, Hansen AJ, Binladen J, et al. (2003). Diverse plant and animal genetic records from Holocene and Pleistocene sediments. *Science*, **300**, 791–795.

Wilson IG (1997). Inhibition and facilitation of nucleic acid amplification. *Applied and Environmental Microbiology*, **63**, 3741–3751.

Wittwer CT, Herrmann MG, Moss AA, Rasmussen RP (1997). Continuous fluorescence monitoring of rapid cycle DNA amplification. *BioTechniques*, **22**, 130–131, 134–138.

Woese CR (1987). Bacterial evolution. *Microbiological Reviews*, **51**, 221–271.

Wong ML, Medrano JF (2005). Real-time PCR for mRNA quantitation. *BioTechniques*, **39**, 75–85.

Woodward G, Ebenman B, Emmerson M, et al. (2005). Body size in ecological networks. *Trends in Ecology & Evolution*, **20**, 402–409.

Wright ES, Yilmaz LS, Noguera DR (2012). DECIPHER, a search-based approach to chimera identification for 16S rRNA sequences. *Applied and Environmental Microbiology*, **78**, 717–725.

Wu TH, Ayres E, Bardgett RD, Wall DH, Garey JR (2011). Molecular study of worldwide distribution and diversity of soil animals. *Proceedings of the National Academy of Sciences of the United States of America*, **108**, 17720–17725.

Wu J-H, Hong P-Y, Liu W-T (2009). Quantitative effects of position and type of single mismatch on single base primer extension. *Journal of Microbiological Methods*, **77**, 267–275.

Wu L, Wen C, Qin Y, et al. (2015). Phasing amplicon sequencing on Illumina Miseq for robust environmental microbial community analysis. *BMC Microbiology*, **15**, 125.

Xiao L, Feng Q, Liang S, et al. (2015). A catalog of the mouse gut metagenome. *Nature Biotechnology*, **33**, 1103–1108.

Xiong M, Wang D, Bu H, et al. (2017). Molecular dietary analysis of two sympatric felids in the mountains of Southwest China biodiversity hotspot and conservation implications. *Scientific Reports*, **7**, 41909.

Yamamoto S, Masuda R, Sato Y, et al. (2017). Environmental DNA metabarcoding reveals local fish communities in a species-rich coastal sea. *Scientific Reports*, **7**, 40368.

Yang CX, Wang XY, Miller JA, et al. (2014). Using metabarcoding to ask if easily collected soil and leaf-litter samples can be used as a general biodiversity indicator. *Ecological Indicators*, **46**, 379–389.

Yilmaz S, Allgaier M, Hugenholtz P (2010). Multiple displacement amplification compromises quantitative analysis of metagenomes. *Nature Methods*, **7**, 943–944.

Yilmaz P, Gilbert JA, Knight R, et al. (2011a). The genomic standards consortium: bringing standards to life for microbial ecology. *ISME Journal*, **5**, 1565–1567.

Yilmaz P, Kottmann R, Field D, et al. (2011b). Minimum information about a marker gene sequence (MIMARKS) and minimum information about any (x) sequence (MIxS) specifications. *Nature Biotechnology*, **29**, 415–420.

Yilmaz P, Parfrey LW, Yarza P, et al. (2014). The SILVA and "All-species Living Tree Project (LTP)" taxonomic frameworks. *Nucleic Acids Research*, **42**, D643–D648.

Yoccoz NG, Bråthen KA, Gielly L, et al. (2012). DNA from soil mirrors plant taxonomic and growth form diversity. *Molecular Ecology*, **21**, 3647–3655.

Yooseph S, Sutton G, Rusch DB, et al. (2007). The Sorcerer II Global Ocean Sampling expedition: expanding the universe of protein families. *PLoS Biology*, **5**, e16.

Yu DW, Ji YQ, Emerson BC, et al. (2012). Biodiversity soup: metabarcoding of arthropods for rapid biodiversity assessment and biomonitoring. *Methods in Ecology and Evolution*, **3**, 613–623.

Yu Y, Lee C, Kim J, Hwang S (2005). Group-specific primer and probe sets to detect methanogenic communities using quantitative real-time polymerase chain reaction. *Biotechnology and Bioengineering*, **89**, 670–679.

Yun J-H, Roh SW, Whon TW, et al. (2014). Insect gut bacterial diversity determined by environmental habitat, diet, developmental stage, and phylogeny of host. *Applied and Environmental Microbiology*, **80**, 5254–5264.

Zaiko A, Samuiloviene A, Ardura A, Garcia-Vazquez E (2015). Metabarcoding approach for nonindigenous species surveillance in marine coastal waters. *Marine Pollution Bulletin*, **100**, 53–59.

Zhang T, Fang HH (2005). 16S rDNA clone library screening of environmental sample using melting curve analysis. *Journal of the Chinese Institute of Engineers*, **28**, 1153–1155.

Zhang T, Fang HHP (2006). Applications of real-time polymerase chain reaction for quantification of microorganisms in environmental samples. *Applied Microbiology and Biotechnology*, **70**, 281–289.

Zhang J, Kobert K, Flouri T, Stamatakis A (2014). PEAR: a fast and accurate Illumina Paired-End reAd mergeR. *Bioinformatics*, **30**, 614–620.

Zheng H, Powell JE, Steele MI, Dietrich C, Moran NA (2017). Honeybee gut microbiota promotes host weight gain via bacterial metabolism and hormonal signaling. *Proceedings of the National Academy of Sciences of the United States of America*, **114**, 4775–4780.

Zhou X, Li Y, Liu S, et al. (2013). Ultra-deep sequencing enables high-fidelity recovery of biodiversity for bulk arthropod samples without PCR amplification. *GigaScience*, **2**, 4.

Zhou J, Wu L, Deng Y, et al. (2011). Reproducibility and quantitation of amplicon sequencing-based detection. *ISME Journal*, **5**, 1303–1313.

Zhu W, Lomsadze A, Borodovsky M (2010). *Ab initio* gene identification in metagenomic sequences. *Nucleic Acids Research*, **38**, e132.

Zilber-Rosenberg I, Rosenberg E (2008). Role of microorganisms in the evolution of animals and plants: the hologenome theory of evolution. *FEMS Microbiology Reviews*, **32**, 723–735.

Zimmermann J, Glöckner G, Jahn R, Enke N, Gemeinholzer B (2015). Metabarcoding vs. morphological identification to assess diatom diversity in environmental studies. *Molecular Ecology Resources*, **15**, 526–542.

Zimmermann J, Jahn R, Gemeinholzer B (2011). Barcoding diatoms: evaluation of the V4 subregion on the 18S rRNA gene, including new primers and protocols. *Organisms, Diversity & Evolution*, **11**, 173–192.

Zinger L, Amaral-Zettler LA, Fuhrman JA, et al. (2011). Global patterns of bacterial beta-diversity in seafloor and seawater ecosystems. *PLoS One*, **6**, e24570.

Zinger L, Boetius A, Ramette A (2014). Bacterial taxa-area and distance-decay relationships in marine environments. *Molecular Ecology*, **23**, 954–964.

Zinger L, Chave J, Coissac E, et al. (2016). Extracellular DNA extraction is a fast, cheap and reliable alternative for multi-taxa surveys based on soil DNA. *Soil Biology & Biochemistry*, **96**, 16–19.

Zinger L, Coissac E, Choler P, Geremia RA (2009a). Assessment of microbial communities by graph partitioning in a study of soil fungi in two alpine meadows. *Applied and Environmental Microbiology*, **75**, 5863–5870.

Zinger L, Gobet A, Pommier T (2012). Two decades of describing the unseen majority of aquatic microbial diversity. *Molecular Ecology*, **21**, 1878–1896.

Zinger L, Shahnavaz B, Baptist F, Geremia RA, Choler P (2009b). Microbial diversity in alpine tundra soils correlates with snow cover dynamics. *ISME Journal*, **3**, 850–859.

Zinger L, Taberlet P, Schimann H, et al. (2017). Soil community assembly varies across body sizes in a tropical forest. *bioRxiv*, 154278.

Index

References to tables and figures are in *italics*